A SPECIALTY CONFERENCE ON:

RECEPTOR MODELS APPLIED TO COMTEMPORARY POLLUTION PROBLEMS

EDITED BY

THE AIR POLLUTION CONTROL ASSOCIATION

SPONSORED BY

APCA's TP-5 Receptor/Source Apportionment Committee

APCA's TP-6 Ambient Measurements Committee

APCA's TP-8 Data Analysis Committee

HOSTED BY

NORTHEAST ATLANTIC INTERNATIONAL SECTION

of

THE AIR POLLUTION CONTROL ASSOCIATION

October 17-20, 1982

Radisson-Ferncroft Hotel

Danvers, Massachusetts

SP-48

APCA Specialty Conference Proceedings

PUBLICATION POLICY

This Proceedings contains technical papers published mostly as they were presented at a recent Specialty Conference of the Air Pollution Control Association. Although the papers have been subjected to peer review essentially equivalent to, if not more critical than, the adopted procedures of the Association, the opinions expressed therein are not to be interpreted as having the endorsement or support of the Air Pollution Control Association.

Manufactured by Publishers Choice Book Mfg. Co.
Mars, Pennsylvania 16046

CONFERENCE COMMITTEE

GENERAL CHAIRMAN
Robert A. Boisselle
Massachusetts DEQE

TECHNICAL PROGRAM CHAIRMAN
Stuart L. Dattner
Texas Air Control Board

LOCAL TECHNICAL PROGRAM COORDINATOR
Richard A. Knapp
Environmental Research & Technology, Inc.

FACILITIES
Charles E. Kitson
Charles T. Main, Inc.

HOUSING AND REGISTRATION
Waterfront Travel Service
Boston, Massachusetts

FINANCE
Edmond G. Benoit
Massachusetts State Agency

EXHIBITS
Richard A. Knapp
Environmental Research & Technology, Inc.

PRINTING
Edward R. Frederick
James D. Morton
Donna E. DeTurck
Air Pollution Control Association

NORTHEAST ATLANTIC INTERNATIONAL SECTION

Chairman: Dr. Richard A. Knapp
Vice Chairman: Lane W. Kirkpatrick
Secretary: Philip A. Morrow
Treasurer: Edmond G. Benoit

A Specialty Conference On:

RECEPTOR MODELS APPLIED TO CONTEMPORARY POLLUTION PROBLEMS

October 17-20, 1982

CONFERENCE PROGRAM

Sunday, October 17, 1982 - Registration

Monday, October 18, 1982 - Registration
Welcome and Introductory Remarks
Richard R. Knapp - Section Chairman
Robert A. Boisselle - General Chairman
Stuart L. Dattner - Technical Program Chairman

SESSION I - Use of Receptor Model Concepts in Government and Industry
Session Chairman - John G. Watson, Desert Research Institute, Reno, NV

OVERVIEW OF RECEPTOR MODEL PRINCIPLES
John G. Watson, Desert Research Institute, Reno, NV

THE ROLE OF RECEPTOR MODELS FOR REVISED PARTICULATE MATTER STANDARD
Thompson G. Pace, U. S. EPA, Research Triangle Park, NC

RECEPTOR MODELING CONCEPTS APPLICABLE TO THE UTILITY INDUSTRY
Peter K. Mueller/Jacques P. Guertin, EPRI, Palto Alto, CA

APPLICATION OF THE RECEPTOR MODEL APPROACH TO LEAD SOURCE APPORTIONMENT IN THREE AIRSHEDS WITH LEAD SMELTERS
James Houck, NEA, Beaverton, OR

NEEDS AND USES OF RECEPTOR MODELS IN THE STEEL INDUSTRY
Richard Lee, U. S. Steel Corporation, Monroeville, PA

AN EVALUATION OF RECEPTOR MODELS FOR USES IN STATE AND LOCAL AGENCIES
John H. Matteson, Massachusetts Department of Environmental Quality Engineering, Springfield, MA

SESSION II - Receptor Model Validation Techniques
Session Chairman - Lloyd Currie, National Bureau of Standards, Gaithersburg, MD

MODEL VALIDATION USING SIMULATED DATA SETS
Lloyd Currie, National Bureau of Standards, Gaithersburg, MD

EVALUATION OF MULTIVARIATE AND CHEMICAL MASS BALANCE APPROACHES TO AEROSOL SOURCE APPORTIONMENT, USING SYNTHETIC DATA AND AN EXPANDED PACS DATA SET
Richard T. DeCesar, Oregon Graduate Center, Beaverton, OR

STATUS OF SOURCE APPORTIONMENT METHODS: QUAIL ROOST II
Robert K. Stevens/Thompson G. Pace, U. S. EPA, Research Triangle Park, NC

INTERCOMPARISON OF SOURCE APPORTIONMENT RESULTS FOR HOUSTON AEROSOL
T. G. Dzubay/Robert K. Stevens, U. S. EPA, Research Triangle Park, NC

STABILITY ANALYSIS OF RECEPTOR MODELS THAT USE LEAST SQUARES FITTING
Ronald C. Henry, Environmental Research & Technology, Westlake Village, CA

POSTER SESSION - George D. Thurston - Late Breaking Topics

Tuesday, October 19, 1982 - Registration

SESSION III - Acid Precipitation and Long Range Transport
Session Chairman - Joan M. Daisey, New York University Medical Center, Tuxedo, NY

ELEMENTAL TRACERS OF CANADIAN SMELTER AEROSOL TRANSPORTED INTO THE NORTHEASTERN UNITED STATES
Kenneth A. Rahn, University of Rhode Island, Narragansett, RI

AEROSOL COMPOSITION FOR RECEPTOR MODELING IN CHINA
John W. Winchester, Florida State University, Tallahassee, FL

AN ASSESSMENT OF REGIONAL POLLUTION TRANSPORT USING ELEMENTAL TRACERS AND WIND TRAJECTORY ANALYSIS
George D. Thurston, Harvard School of Public Health, Boston, MA

ACID PRECIPITATION SOURCE IDENTIFICATION IN SOUTHERN CALIFORNIA
Howard M. Liljestrand, University of Texas, Austin, TX

AMBIENT PARTICULATE CONTRIBUTIONS TO ACID PRECIPITATION IN TEXAS
Michael A. Stafford,Texas Air Control Board, Austin, TX

A RECEPTOR MODEL FOR DIAGNOSING SOURCE CONTRIBUTIONS TO AMBIENT SULFATE CONCENTRATIONS AND SULFUR WET DEPOSITION ON A REGUINAL SCALE
Perry Samson, University of Michigan, Ann Arbor, MI

SESSION IV - Receptor Model Measurements
Session Chairman - Russell Crutcher, Boeing Aerospace Co., Seattle, WA

AMBIENT AEROSOL APPORTIONMENT BY STATISTICAL TREATMENT OF SEM-AIA RESULTS
David L. Johnson, Dept. of Chemistry, State University of New York, Syracuse, NY

AN EVALUATION OF VARIOUS INSTRUMENTAL TECHNIQUES FOR SOURCE APPORTIONMENT OF ATMOSPHERIC AEROSOLS
Ronald G. Draftz, IIT Research Institute, Chicago, IL

HYBRID MODEL FOR SOURCE APPORTIONMENT OF THE HOUSTON AEROSOL
Briant L. Davis, South Dakota School of Mines & Technology, Rapid City, SD

QUANTITATIVE X-RAY POWDER DIFFRACTION ANALYSIS OF AIR PARTICULATE SAMPLES
J. M. Jaklevic, Lawrence Berkeley Laboratory, Berkeley, CA

THE EL PASO AIRSHED: SOURCE APPORTIONMENT USING COMPLEMENTARY ANALYSIS AND RECEPTOR MODELS
Phillip B. Janocko, Energy Technology Consultants, Murrysville, PA

PARTICLE CLASS BALANCE APPROACH TO RECEPTOR SITE MODELING OF AIRBORNE PARTICULATE MATTER
Brian L. McIntyre, Arizona State University, Tempe, AZ

THE REGION V TSP NON-ATTAINMENT STUDY FOR REVISING STATE IMPLEMENTATION PLANS
Jean L. Graf, IIT Research Institute, Chicago, IL

Wednesday, October 20, 1982 - Registration

SESSION V - Urban Scale Studies
Session Chairman - John Core, NEA Lab., Beaverton, OR

AN ASSESSMENT OF SOURCES OF AIRBORNE PARTICULATE MATTER IN HAMILTON, ONTARIO
D. M. Kane, UTS, Ontario Research Foundation, Mississauga, Ontario

FORMULATION AND APPLICATION OF A HYBRID SOURCE-RECEPTOR MODEL
Robert Y. Yarmartino, Environmental Research & Technology, Concord, MA

TIME SERIES ANALYSIS FOR IDENTIFICATION OF PARTICLE EMISSION SOURCES
Chang-Sik Hwang, University of Illinois, Urbana, IL

RECEPTOR-ORIENTED SOURCE-RECEPTOR ANALYSIS
Douglas W. Cooper, Harvard School of Public Health, Boston, MA

USE OF MULTIPLE REGRESSION SOURCE APPORTIONMENT MODELS TO IDENTIFY PROBABLE SOURCES OF PARTICULATE N-NITROSO COMPOUNDS IN NEW YORK CITY
R. J. Hershman, New York University Medical Center, New York, NY

PANEL DISCUSSION: Future Developments in Receptor Modeling
Thompson G. Pace, Ronald R. Landreth, Inland Steel Co., East Chicago, IN; Ted Kneip, NYU Medical Center, Texedo, NY

ADJOURN

TABLE OF CONTENTS

Receptor Models Applied to Contemporary Pollution Problems: An Introduction	Stuart L. Dattner Philip K. Hopke	1
Overview of Receptor Model Principles	John G. Watson	6
The Role of Receptor Methods for Revised Particulate Matter Standards	Thompson G. Pace	18
Receptor Modeling Concepts Applicable to the Utility Industry	Jacques P. Guertin	29
Receptor Model Source Apportionment of Lead in Three Airsheds with Lead Smelters	John A. Cooper	34
Status of Source Apportionment Methods: Quail Roost II	Robert K. Stevens	46
Intercomparison of Results of Several Receptor Models for Apportioning Houston Aerosol	Thomas G. Dzubay	60
Quantitative X-Ray Powder Diffraction Analysis of Air Particulate Samples	Albert C. Thompson	72
Source Characterization and Apportionment of the Houston Aerosol by Means of X-Ray Diffraction Techniques	Briant L. Davis	84
Review of the Quail Roost II Receptor Model Simulation Exercise	R. W. Gerlach	96
Application and Verification Studies of Target Transformation Factor Analysis as an Aerosol Receptor Model	Philip K. Hopke	110
Evaluation of Multivariate and Chemical Mass Balance Approaches to Aerosol Source Apportionment, Using Synthetic Data and an Expanded PACS Data Set	Richard T. DeCesar	127
Stability Analysis of Receptor Models that Use Least Squares Fitting	Ronald C. Henry	141
Response to: Stability Analysis of Receptor Models that Use Least Squares Fitting	Hugh J. Williamson	158
Elemental Tracers of Canadian Smelter Aerosol Transported into the Northeastern United States	Kenneth A. Rahn	163

An Assessment of the Relationship Between Regional Pollution Transport and Trace Elements Using Wind Trajectory Analysis	George D. Thurston	175
A Receptor-Oriented Model for Diagnosing Source Contributions to Ambient Sulfate Concentrations and Sulfur Wet Deposition on a Regional-Scale	Perry J. Samson	189
Ambient Particulate Matter Contributions to Acidic Precipitation	Michael A. Stafford	200
Acidic Precipitation Source Identification by Chemical Mass Balance Methods Employing Fractionation Factors	Howard M. Liljestrand	212
Statistical Considerations in the Employment of SAX Results for Receptor Models	David L. Johnson	224
A Particle Class Balance Receptor Model for Aerosol Apportionment in Syracuse, NY	David L. Johnson	238
The El Paso Airshed; Source Apportionment Using Complementary Analyses and Receptor Models	Philip B. Janocko	249
Light Microscopy as an Analytical Approach to Receptor Modeling	E. R. Crutcher	266
Formulation and Application of a Hybrid Source Receptor Model	Robert J. Yamartino	285
Receptor-Oriented Source-Receptor Analysis	Douglas W. Cooper	296
Time Series Analysis for Identification of Particle Emission Sources	Chang Sik Hwang	317
Application of Receptor Modeling to Volatile Organic Compounds	Ivar Tombach	338
Receptor Source Apportionment Models for Two Polycyclic Aromatic Hydrocarbons	Joan M. Daisey	348
Receptor Models Applied to Contemporary Air Pollution Problems A Panel Discussion	Theo J. Kneip Jacques Guertin Thompson G. Pace John G. Watson	358
Registrants		362
Index		366

Receptor Models Applied to Contemporary Pollution Problems: An Introduction

Stuart L. Dattner
Texas Air Control Board
Austin, TX

and

Philip K. Hopke
University of Illinois
Urbana, IL

On October 18-20, 1982, APCA's Committee TP-5 sponsored a specialty meeting on "Receptor Models Applied to Contemporary Pollution Problems" hosted by the Northeast Atlantic International Section. This highly successful meeting marks the decade that has past since the paper by Miller, Friedlander, and Hidy (1) first formally proposed the mass balance model that has lead to these further developments.

This meeting is the third major public symposium on the field of receptor modeling. These meetings have introduced many of the important developments in the emerging receptor modeling field. The first of these meetings held in 1979, "Aerosols: Anthropogenic and Natural, Source and Transport", was sponsored by the New York Academy of Sciences and the Air Pollution Control Association and was organized by Theodore J. Kneip and Paul J. Lioy (2). The second was the Symposium on Atmospheric Aerosols: Source/Air Quality Relationships held at the national American Chemical Society meeting in 1980 and organized by Edward S. Macias and Philip K. Hopke (3).

In addition to these large meetings, there have been two EPA-sponsored workshops to begin to organize receptor modeling, identify important research directions, and to begin the verification and intercomparison studies needed to demonstrate the validity of these methods. These workshops known as Quail Roost I and Quail Roost II because they were held at the Quail's Roost Conference Center of the University of North Carolina have helped to focus the efforts of the receptor model research community. The results of the first workshop were reported by Watson (4). Some of the major results of the Quail Roost II Workshop are reported in these Proceedings and will be published in a more complete form in the future.

This meeting is the culmination of the efforts of the relatively new TP-5 Committee on Source/Receptor Models. This committee has formally existed for only two and a half years, but has sponsored a highly successful two session program at the 74th National APCA meeting, a single session program at the 75th National APCA meeting and the specialty conference whose Proceedings are presented in this volume. The rapid growth of the interest and participation in this field demonstrates the advancements that have been made recently in the ability to utilize air quality data at a receptor site to quantitatively infer the types and amounts of sources of pollution impacting that site and the recognition that these methods will be useful to the management of air quality. It was the purpose of this specialty meeting to both review the current state of this rapidly developing field as well as to look to the future for new or expanded applications and new developments.

Receptor-oriented models have been developed to utilize chemical and physical properties of airborne pollutants to identify those sources, natural and anthropogenic, that contribute to the observed concentrations at the receptor site. Although most applications have been to the apportionment of suspended particulate matter mass, these methods are certainly applicable to other air quality problems.

These proceedings have been organized around several different groups of papers. The first group including those by Watson, Pace, Mueller and Guertin and Houck and Cooper are designed to provide an overview of receptor modeling and various perspectives on their use for solving air quality management problems. Watson's paper clearly describes the conceptual framework and logical development of receptor models. The report by Thompson Pace outlined the role that receptor models can play in the development of State Implementation Plans (SIPS) for attaining air quality within regulatory limits set by ambient air quality standards. Mueller and Guertin of the Electric Power Research Institute (EPRI) present EPRI's plans for investigating the potential use of receptor models in the electric power industry. Houck and Cooper described the application of currently available sampling, analysis and chemical mass balance models to problems of identifying airborne lead sources in three air sheds where large lead processing facilities are located. This report represents state-of-the-art receptor models as currently practiced in field studies.

The next group of papers presents the results of the first major validation and interlaboratory comparison study in receptor modeling. These are the results of the Quail Roost II workshop and the objectives of the study and workshop are discussed by Stephens and Pace. The two major parts of this study involved the examination of analytical results on actual samples taken in Houston, TX in September 1980 and the analysis of simulated particle composition data sets created for this workshop by the National Bureau of Standards. Dzubay and Stephens present the overall results of the Houston sample analyses that include both mathematical modeling of the composition data and physical examination of the samples by microscopic and x-ray diffraction (XRD) techniques. The improvements in x-ray diffraction technology that permit examination of dichotomous sampler samples are described by Thompson and Jaklevic. Alternative ways of using the XRD results are discussed by Davis.

The development of the simulated data sets and an evaluation of the results of the various mathematical methods applied to resolve the data are presented by Currie and Gerlach. Hopke then reviews two approaches to the analysis of these data with a particular model, Target Transformation Factor Analysis. DeCesar and Cooper used several mathematical models on these N.B.S. data sets and on an expanded set of data from the Portland Aerosol Characterization Study (PACS).

The final papers in this group discuss the possible limitations on the ability of mathematical receptor models. Ronald Henry presents results on studies of the limitations of mathematical models to resolve sources emitting particles with similar compositions. His work suggests that the ability to separate contributions from source materials with similar chemical compositions can be tested objectively and that a framework can be established for extracting the maximum information than can be reliably obtained. He also describes some problems he believes exist in some other approaches to resolving similar composition sources. Williamson responds to these criticisms. One of the critical problems in expanding the utilization of the current mathematical receptor models will be the rigorous evaluation of the methods and the determination of their fundamental limitations. The paper by Henry and the response by Williamson illustrate that these are disagreements as to the capabilities of the ridge regression approach to solve the problem of multicollinearity in the source profile matrix and further studies of it and the aproach suggested by Henry are clearly required to resolve the question. As editors, we felt that this Proceedings could serve as an opportunity to present these differing viewpoints.

Most of the prior applications of receptor models have been to urban scale problems. The next series of papers provides discussion of the possible extention of these methods to longer distance transport and the problems of acidic deposition. Kenneth Rahn and coworkers present results on his work to identify the impacts of smelter operations in Ontario, Canada, to air quality in New England. George Thurston also presents results indicating that receptor models may be able to separate the impacts of long range transported particles from local source emissions. Perry Sampson provides a description of an approach to receptor modeling of acid precipitation starting from a dispersion modeling and back-trajectory approach. Michael Stafford discusses his application of multivariate receptor models to precipitation chemistry data from various locations in Texas. Howard Liljestrand has examined the problem of apportioning sources of acidity in precipitation by developing a physical-chemical model of the unsteady mass-continuity equation. In this work he tries to identify the important atmospheric processes as well as the effects of precipitation scavenging on the fractionation factors describing the change in material composition between the emission and collection locations. These papers indicate that there are substantial possibilities for the use of source/receptor models in examining long range transport and acid deposition although further developmental work is needed.

As shown in the Quail Roost Houston study, another set of primary tools, in addition to chemical composition data and mathematical models, include available physical characterization techniques including microscopic examination of individual particles and determination of mineral phases with x-ray diffraction methods. David Johnson is principal author on two papers describing work on the development and application of computer-controlled scanning electron microscopy. The ability to examine many particles makes this method appear to have great potential for source apportionment studies. In the first report, Johnson gives the mathematical framework for utilizing the data obtained from such measurements, and in the other with B. McIntyre, presents a study of Syracuse, NY. Philip Janocko and Gary Cassucio in conjunction with other coworkers present results on a study using similar techniques to examine airborne samples for the impacts of particulate matter sources in El Paso, TX. Included in this report are an intercomparison of results of both electron and optical microscopy on several selected samples indicating good agreement between microscopists and between microscopy and the x-ray fluorescence analysis of the bulk filter sample. Russell Crutcher presents a detailed introduction to particle identification and characterization using light microscopy.

The final group of papers presents methods under development. Robert Yarmartino and Douglas Cooper are independently working on ways to merge dispersion and receptor models. By incorporating the additional information available about the dispersion characteristics, it may be possible to better identify and understand source contributions to the level of pollutant measured at a particular site. Chang-Sik Hwang discusses initial results in interpreting airborne particulate elemental concentration patterns using time-series methods. Ivar Tombach has examined the use of receptor models to problems of airborne organic compounds. Joan Daisey describes results of partitioning the amounts of two airborne polycyclic hydrocarbons to sources in New York City. It is clear from these results that receptor models can readily be applied to problems other than TSP source apportionment.

The meeting concluded with a panel discussion. The panel members were Theodore Kneip, John Watson, Thompson Pace, and Jacques Guertin. The group reviewed the scope of the meeting and highlighted several of the presentations. They discussed future developmental needs for the increased use of receptor models. There was emphasis on the need to provide a better understanding of the strengths and weaknesses of these methods to the air quality management community and to look more to the integration of these methods into an overall framework involving dispersion and receptor modeling in order to develop cost-effective air quality management plans. A summary of the panel discussions developed by Kneip concludes these Proceedings. The meeting provided an excellent forum for the discussion of method development, validation and application and the meeting and we hope that these Proceedings will help the receptor modelers present their capabilities to the rest of the air pollution control community. In an effort to provide the best possible reports, the papers in this volume have been subjected to peer review. This marks the first time that an APCA specialty meeting proceedings has been so reviewed and has resulted in a slight delay in publication. However, we feel that the extra effort made by the reviewers and the authors results in a volume that will be of greater value because of the improvement it makes in the reports. We would urge future APCA meetings to follow this example.

1. M.S. Miller, S.K. Friedlander, and G.M. Hidy, J. Colloid Interface Sci. 39, 165 (1972).

2. T.J. Kneip and P.J. Lioy, eds, Aerosols: Anthropogenic and Natural, Sources and Transport, Annals of New York Academy of Sciences, volume 338, New York:1980.

3. E.S. Macias and P.K. Hopke, eds, Atmospheric Aerosols: Source/Air Quality Relationships, Symposium Series No. 167, American Chemical Society, Washington:1981.

4. J.G. Watson, Receptor Models Relating Ambient Suspended Particulate Matter to Sources, Report No. EPA-600/2-81-039, U.S. Environmental Protection Agency, March, 1981.

OVERVIEW OF RECEPTOR MODEL PRINCIPLES

John G. Watson

Air Resources Laboratory
Atmospheric Sciences Center
Desert Research Institute
Reno, Nevada

The objectives of this overview are: 1) to state the fundamental structure and assumptions of receptor models, 2) to identify the usefulness and limitations of these models in their present form and 3) to outline the potential future applications of receptor models to contemporary pollution problems. A division is made between conceptual models, the mathematical equations which relate variables to each other, and measurement models, which supply values to some of the conceptual model variables so that others may be calculated. Each model dictates, to a certain extent, the requirements of the other, and a logical evolution of receptor models to their present state-of-the-art is followed to illustrate this. In the process of this evolution, the limitations of receptor models become evident. The major limitation is that several basic assumptions are not generally applicable, and these can only be overcome by restricting the cases or by adding more complexity to the models. Future applications of receptor models are likely to be seen in the areas of standard attainment, visibility degradation, hazardous substances, and acid deposition.

Introduction

Several fine reviews of receptor models applied to suspended particulate matter have been previously published,[1,2,3,4,5,6] and this overview is not intended to repeat them. It is necessary, however, to introduce the topic of receptor models applied to contemporary air pollution problems to those who are unfamiliar with it and who rightfully expect the Proceedings of this specialty conference to tell them more about it. This overview, therefore, has three objectives:

- To state the fundamental structure and assumptions of receptor models.
- To identify the usefulness and limitations of these models in their present forms.
- To outline the potential future applications of receptor models to contemporary pollution problems.

The first objective will be met by stating what a model is and how it differs from the real world. A logical evolution of receptor models, covering advances which have occurred over the last ten years, will be presented, which will: 1) state the basic model assumptions, 2) present the mathematical relationships derived from those assumptions, 3) identify the measurement models required by those relationships and their basic assumptions, 4) recognize that certain of the assumptions of the conceptual and measurement models aren't met, and 5) modify the assumptions or the models to achieve greater consistency with reality. As assumptions are stated--then violated--then refined in this development, the powers and limitations of receptor models will become evident, thus meeting the second objective. A list of contemporary pollution situations, many of which will be dealt with by the presentations which follow, will be compiled to fulfill the third objective. The five step methodology can be applied to these situations in the years ahead. Throughout this discussion, reference will be made to other presentations in these Proceedings which are related to the topic at hand.

This brief report cannot hope to be the complete guide to receptor models and should not be viewed as such. It is meant only to show the process through which they have developed during the past in the hope that the same process will be of help to their future development.

The Conceptual Model and the Measurement Model

It must be recognized that a receptor model, like all models, is a representation of reality. We create models because we cannot effectively deal with all of the variables which affect the environment and because we want to see what the effects of changing some of those variables might be on others. Since some of the variables are always left out, our model is always an imperfect replica of the true situation. By improving our models, we strive to decrease the difference between our representations and the real world. In both source-oriented and receptor-oriented representations of reality, we really deal with two models: the conceptual model and the measurement model.

The conceptual model is based on our scientific understanding of the interaction of matter and energy. It consists of a set of mathematical relationships between variables. This conceptual model, no matter how complex, is still an inadequate representation of reality because we don't know what all of the important variables are or how they interact with each other. We also have to establish the values of some of the variables which we <u>have</u> identified as important in order to determine their effects on the other variables. As we shall soon see, the selection of which variables we assign values to and which ones we calculate is what differentiates a source from a receptor model; the conceptual model from which they are derived is the same.

Values are assigned to model variables through measurements, and since measurements are also imperfect representations of reality, any set of measurements constitutes a <u>measurement model</u>. The measurement model includes: 1) the observables measured, 2) the range of values into which the measurements are confined, 3) the frequency and duration of measurements, 4) the spatial density of measurements, and 5) the validity, precision and accuracy of the measurements.

Table I compares the number of significant physical variables and temporal and spatial scales for reality, measurement, source, and receptor models. These are only order-of-magnitude estimates, but reasonable changes could not reduce the vast difference between the number of variables for the real world and the number for any of the models.

Given our inability to even approach the conceptualization and measurement of so vast a number of variables, it might appear that there is no hope of ever coming to an understanding of what is happening in the atmosphere. However, if a full understanding cannot be reached, it can be approached by a careful interaction and refinement of conceptual and measurement models. The evolution of receptor models over the past ten years offers an excellent illustration of this process.

Receptor Model Evolution

The evolution of receptor models to their present state follows a logical, rather than a historical, route in this treatment. Much of the research encapsulized here did not occur in this order, though in hindsight it might have been carried out more efficiently if it had been. The group of receptor models considered here is that which is used to determine the direct contributions of various air pollution sources to ambient concentrations measured at a single receptor on an urban or smaller scale. Other receptor models have been created for the understanding of the inter-relationships of environmental variables and for larger scales, but they will not be treated here. The five steps outlined in the introduction will be followed in this manner:

1. Basic model assumptions are stated. They are labeled A1, A2, etc. Many minor assumptions must be considered as implicit in the basic assumptions and widely accepted physical laws are not stated explicitly.

2. The conceptual model which these assumptions imply is stated mathematically and/or verbally. These models are labeled CM1, CM2, etc.

3. The measurement models which are dictated by the conceptual model, or which have been developed because they are practical, are specified with their basic assumptions. These models are labeled MM1, MM2, etc.

4. Problems due to the violation of assumptions of the conceptual or measurement model are recognized. These are labeled P1, P2, etc.

5. The conceptual or measurement model is modified and given a new label.

Though the receptor model evolution described here has taken place within the context of suspended particulate matter studies on an urban scale, the conceptual model applies to all ambient pollution, both gaseous and particulate phases. The measurement models, however, have usually confined themselves to a single phase. The first basic assumptions are:

A1: Compositions of source emissions are constant.

A2: Components do not react with each other, i.e. they add linearly.

A3: p identified sources contribute to the receptor.

These assumptions lead to the fundamental conceptual model:

CM1: $C_{ik}=\sum_{j=1}^{p} a_{ij}D_{jk}E_{jk}$, The concentration of component i in the kth sample (C_{ik}) equals the sum of the products of the fractional amount of component i in the emissions from source j (a_{ij}), the atmospheric dispersion of emissions from source j for the kth sample (D_{jk}), and the total emission rate of all components from source j which corresponds to sample k (E_{jk}).

For a source model, the values of $a_{ij}E_{jk}$ and D_{jk} are obtained through measurements and the C_{ik} are calculated. For the receptor model, the values of the C_{ik} and a_{ij} are measured and the $D_{jk}E_{jk}$ products are calculated. The receptor approach results in:

CM2: $C_{ik}=\sum_{j=1}^{p} a_{ij}S_{jk}$ for i=1,n. The concentration of component i equals the sum of the products of the composition fractions (a_{ij}) and the total contribution of source j (S_{jk}) to the receptor. n is the total number of components measured.

The set of equations comprising CM2 is the chemical element balance[7] or the chemical mass balance[3] receptor model. It is probably more appropriate to call it a mass balance model since any property, e.g. crystalline structure optical appearance, isotope ratios, can be used as a component if it can be associated with a mass fractional composition of the source emissions. Drs. Houck, Liljestrand, and Davis demonstrate the universality of this mass balance model in these Proceedings. The mass balance model requires two additional assumptions:

A4: The number of sources, p, is less than or equal to the number of components, n.

A5: The compositions of all p sources (the set of a_{ij} for each S_j) are linearly independent of each other.

These assumptions provide a set of equations that can be solved by a least squares fit for the various source contributions (S_{jk}) if the source compositions (a_{ij}) and the receptor concentrations (C_{ik}) are known. The solution method is part of CM2. The measurement model required to provide these data is:

MM1: Measure p source compositions and m receptor concentrations for n components with routine measurement methods.

Now we find that several problems arise:

P1: CM1 needs exact values of a_{ij} and C_{ik}. Measurements are not exact they are really intervals.

By modifying CM2, this problem can be handled fairly easily:

CM3: Weight variables in the least squares fit so that less precise measurements have less influence than more precise measurements. Propagate errors through the least squares solution[6] with the following additional assumption.

A6: Measurement errors are random, uncorrelated and normally distributed.

Two more problems present themselves:

P2: There are more sources than components. p>n (A4 is untrue.)

P3: Several sources have the same compositions. (A5 is untrue.)

Two solutions exist. One is to modify the measurement model.

MM2: Measure more components and make sure they have different concentrations in different source emissions.

This model requires that new, non-routine measurements be developed and standardized. The presentations of Messrs. Johnson, Draftz and Jaklevic in these Proceedings explain several of the possibilities. In the absence of the capability to make these new measurements, CM3 must be further modified.

CM4: Group sources with similar compositions into source types. p then refers to the number of source types.

This gives up a lot, for now the model cannot distinguish between individual contributors. There are still other problems.

P4: Source compositions do not remain constant.

Friedlander[8] discusses the case of a linear reduction of a_{ij} with time and incorporates it into:

CM5: The source composition at the receptor equals $a_{ij}/(1+k_i\theta_i)$ where k_i is the decay rate and θ_i is the residence time of compoonent i in the atmosphere.

This model imposes two additional assumptions.

A7: Emissions reach a uniform concentration throughout the airshed immediately (i.e., a short time compared to the sample duration) upon introduction into the atmosphere.

A8: The decay rate is linear with time.

CM5 also modifies the measurement model.

MM3: Decay rates and residence times are measured in addition to the MM1 measurements.

Routine methods for MM3 do not yet exist. In fact, the inability to successfully carry out MM1 has resulted in several additional problems.

P5: It is not possible or too costly to measure the source compositions.

This has led to the development of models which make use of the variation of the ambient component concentrations (C_{ik}) over time (the number of consecutive samples, k=1,m). Mr. Hwang's presentation will expand on this. A linear regression model results from CM2 with A1, A2, A3, A6, and:

A8: Each source type j has one component t_j, or "tracer" in its composition such that $a_{t_j j}$ equals zero for all other sources.

The linear regression model is:

CM6: $S_{jk}=\sum_{j=1}^{p}(1/a_{t_j j})C_{t_j k}$. The total pollution level (S_k) equals the sum of the quotients of the ambient concentrations of tracer components ($C_{t_j k}$) and the fractional composition of the tracer for each source.

Dr. Daisey's presentation in these Proceedings illustrates this model. But other problems still exist.

P6: The number of source types, p, is unknown.

P7: Unique tracer components aren't available for all source types.

These are addressed by a factor analysis variational model.[10]

CM7: $C_{ik} = \sum_{j=1}^{p} \alpha_{ij} F_{jk}$, where the α_{ij} and F_{jk} are related to, but are not the same as the a_{ij} and S_{jk} of CM2.

This model is intended to determine p and the a_{ij} from the intercorrelations of concentrations. It implicitly includes the effects of measurement errors.[11] Dr. Hopke explains the concept and its applications in these Proceedings. The intimate relationship between the mass balance and variational models established here is important. While the variational models, under certain conditions, can specify the number of source types and their probable compositions, these results must be combined with the mass balance model to obtain source contributions.

Receptor Model Limitations

The interaction between assumptions, conceptual models, measurement models and challenges to assumptions may seem confusing, yet it is essential to the model development process. In fact, the previous presentation has been substantially simplified by the failure to state many less fundamental, but equally important, assumptions. For example, the variational models make certain assumptions about the statistical distributions from which measurements are drawn and minimum sample sizes which may or may not be true in typical applications. The problem is complicated, but it must be dealt with.

The limitations of a conceptual/measurement model combination depend upon which assumptions are true under a given situation. Exceptions can be found to every one: 1) source compositions are not constant, they vary with changes in process inputs, loads and cycles; 2) components _do_ react with each other and systems are _not_ linear; 3) one rarely knows exactly how many sources are contributing to a receptor; 4) there are many more sources than components which can be practically measured; 5) many sources have very similar compositions; 6) measurement errors are not necessarily random, uncorrelated, or normally distributed; 7) emissions definitely do _not_ instantaneously reach a uniform concentration throughout the atmosphere upon leaving a source; 8) nor do they exhibit decay constants which are linear with time; and 9) very few sources have their own unique tracer components.

It would seem that we have just refuted the very foundation upon which receptor models are based. Yet, for certain purposes they _do_ work. The fulfillment of an assumption is a question of degree rather than of yes or no. In many cases the deviations from basic assumptions can be tolerated by the goals of the application. It is important to quantify those deviations, however, and to consider them with the results of the models.

Receptor Model Advantages

The limitations of receptor models are to some extent offset by their advantages. They are relatively simple, mathematically, compared to source-oriented models of comparable accuracy and precision. Results of intercomparison studies and applications to simulated data sets, some of which are presented in these Proceedings by Messrs. Currie, DeCesar, Stevens, and Dzubay, show that the effects of reasonable deviations from basic assumptions can be tolerated. Dr. Henry develops in these Proceedings an analytical method of determining the effects of systematic, as opposed to random, measurement errors on the mass balance equations. He can now estimate the precisions required of measurement models to provide a specified precision for the output of the conceptual model. The models may seem to deviate from reality today, but they are much closer to it now than they were only a few years ago. There is every reason to believe that these improvements will continue.

New research into receptor models is already using CM1 to make use of more sophisticated measurement models. Drs. Yamartino and Cooper present hybrid approaches in these Proceedings, combining source and receptor oriented models by adding measurements of D_{jk} to ascertain E_{jk} in CM1. This composite approach shows great potential.

Receptor Model Applications to Contemporary Pollution Problems

To date, the areas and methods of application of receptor models have been as varied as the number of researchers using them. Messrs. Pace and Guertin discuss, in these Proceedings, receptor model applications for inhalable particulate matter and utility emissions. Four areas of application will present themselves during the 1980s: 1) criteria pollutant standard attainment, 2) western visibility degradation, 3) hazardous substances, and 4) acid deposition.

Standards Attainment

Receptor models have been used most widely in the past for determining the contributions of various sources to ambient concentrations which exceed National Ambient Air Quality Standards. In the United States, most air quality maintenance areas have created state implementation plans and are on their way to meeting standards. Messrs. Houck and Graf offer some examples in these Proceedings applied to lead and total suspended particulate matter. These standards may change in the future, as the suspended particulate matter standard is changing now,[12] and receptor model methods will be needed to devise new attainment strategies.

Countries other than the United States have other standards which must be met. Of particular concern is the air in rapidly industrializing countries of Latin America, Asia and Africa. Average TSP levels in factory towns in Taiwan, for instance, are over three times the U.S. standard.[13] The different sources and limited financial and technical resources of these countries will require creative applications of conceptual and measurement models.

Visibility Degradation

Visibility degradation is a big issue in the western United States where views extend for miles and small increases in pollution levels can reduce the visual range by tens of kilometers. Friedlander[8] describes a conceptual model and Heisler et al.[14] illustrate a measurement model appropriate for it. The National Parks Service, in cooperation with several western utilities, is presently designing a measurement model for a five year regional-scale receptor model application to characterize the frequency and magnitude of visibility degradation episodes in and around the Grand Canyon and to estimate the contributions of various sources to those episodes. These models involve an extra step, with its associated measurement and conceptual uncertainties, to determine the contributions of various pollutants to visibility degradation in addition to the contribution of various sources to the pollution levels.

Hazardous Substances

Communities which are in close proximity to major industries, hazardous waste incinerators, dumpsites, and chemical spills are expressing concerns about the potential contributions of these various sources to adverse health effects. They want to know the frequency and magnitude of their exposures and who is responsible for them. Because of the tremendous liabilities involved, the owners of these sources also want the answers. Because many of these substances are organic in nature, with a solid and vapor phase, and are highly reactive, both the conceptual and measurement models will be complex. As for the visibility problem, an extra step, the relationship between pollutant exposure and health effects, needs to be taken beyond the attribution of pollutant concentrations to sources. This will also add greater complexity and uncertainty to the receptor models.

Acid Deposition

Acid deposition, or "acid rain" in the popular parlance, is truly the "pollutant of the '80s," and the presentations by Messrs. Cooper, Thurston, Liljestrand, Stafford and Sampson in these Proceedings show that receptor models have a role to play in elucidating our understanding of acid deposition episodes and their sources.

This pollution problem presents the greatest challenge to the existing conceptual models in that the incorporation of industrial emissions (NO_x and SO_2 have been identified as the most important to date) into acid bearing vapors, particles and raindrops, and eventually into streams and onto plantlife, is an extremely complicated process. The conceptual models probably cannot accept the nonreactive assumption, A2, which means that CM1 probably does not apply; a more complex, nonlinear conceptual model is needed. Whereas the visibility degradation and hazardous substances models will require two sub-models, the acid deposition model will require multiple sub-models. For example, to determine the source contributions to the acidity of a lake will require the following relationships: 1) runoff from watershed to lake, 2) raindrops to watershed, 3) clouds to raindrops, 4) pollutants to clouds, 5) sources to pollutants. And these five steps do not include the potential of dry deposition! The systematic approach outlined in this overview needs to be applied very critically to these models to ensure their successful development.

Summary and Conclusions

Receptor models have followed a logical evolution which has derived conceptual models from basic assumptions, created measurement models intended to fulfill the requirements of the conceptual models, recognized that certain of the assumptions are not met, and modified the conceptual or measurement models to accommodate that recognition. Several assumptions of existing models are not met in real-world applications, and the evolution continues by determining the effects of deviations from them on model results and by modifying the assumptions and the models to better represent reality. This format of model evolution should continue as receptor models are applied to the contemporary pollution problems of the 1980s: standard attainment, visibility degradation, hazardous substances, and acid deposition.

References

1. G. E. Gordon, "Techniques for treating multielement particulate data to obtain information on sources: Overview," in Aerosols: Anthropogenic and Natural, Sources and Transport, edited by T. J. Kneip and P. J. Lioy, Annals of the New York Academy of Sciences, 338, p93 (1980).

2. G. E. Gordon, "Receptor models," Environmental Science and Technology, 14, p795 (1980).

3. J. A. Cooper, J. G. Watson, "Receptor-oriented methods of air particulate source apportionment," Journal of the Air Pollution Control Association, 30, p1116 (1980).

4. J. G. Watson, et al., "Receptor models relating ambient suspended particulate matter to sources," EPA-600/2-81-039, Research Triangle Park, N.C. (1981).

5. J. G. Watson, R. C. Henry, J. A. Cooper, E. S. Macias, "The state of the art of receptor models relating ambient suspended particulate matter to sources," Atmospheric Aerosol: Source/Air Quality Relationships, American Chemical Society Symposium Series #167, Washington, D.C., p89 (1981).

6. J. G. Watson, Chemical Element Balance Receptor Model Methodology for Assessing the Sources of Fine and Total Suspended Particulate Matter in Portland, Oregon, Ph.D. Dissertation, Oregon Graduate Center, Beaverton, Oregon (1979).

7. S. K. Friedlander, "Chemical element balances and identification of air pollution sources," Environmental Science and Technology, 7, p235 (1973).

8. S. K. Friedlander, "New developments in receptor modeling theory," Atmospheric Aerosol: Source/Air Quality Relationships, American Chemical Society Symposium Series #167, Washington, D.C., p1 (1981).

9. M. T. Kleinman, B. A. Pasternack, M. Eisenbud, T. J. Kneip, "Identifying and estimating the relative importance of sources of airborne particles," Environmental Science and Technology, 14, p62 (1980).

10. P. K. Hopke, "The application of factor analysis to urban aerosol source resolution," Atmospheric Aerosol: Source/Air Quality Relationships, American Chemical Society Symposium Series #167, Washington, D.C., p21 (1981).

11. B. A. Roscoe, P. K. Hopke, "Error estimates for factor loading and scores obtained with target transformation factor analysis," Analytical Acta, 132, p89 (1981).

12. J. G. Watson, J. C. Chow, J. J. Shah, "Analysis of inhalable particulate matter measurements," EPA-450/4-81-035, Research Triangle Park, N.C. (1981).

13. J. C. Chow, J. G. Watson, C. Y. Chaung, "Air pollution standards, monitoring, concentrations and control in the Republic of China (Taiwan)," presented at the 75th Annual Meeting of the Air Pollution Control Association, New Orleans, Louisiana (1982).

14. S. L. Heisler, R. C. Henry, J. G. Watson, "The sources of the Denver haze in November and December of 1978," presented at the 73rd Annual Meeting of the Air Pollution Control Association, Montreal, Quebec (1980).

TABLE I: Comparisons of measurement, source and receptor model components to reality on an urban scale.

	Order of Magnitude Estimates			
	Reality	Measurement	Source Model	Receptor Model
Air Volume (m^3)	10^{11} [a]	10^3 [b]	10^9 [c]	10^3
No. of variables	10^3 [d]	50	50	50
Range of pollutant concentrations (g)	10^{-15}--10^{-3}	10^{-9}--10^{-3}	10^{-6}--10^{-3}	10^{-9}--10^{-3}
No. of variable interactions	10^4 [d]	100 [e]	10	0
No. of pollutant sources	10^3	10	10^2	10
Range of pollutant emissions (g/sec)	10^{-2}--10^3	1--10^3	1--10^3	1--10^3
Emissions time scale resolution (sec)	10^2	10^7	10^7	10^7
Emissions spatial [f] scale resolution (m)	vert. 10 horz. 100	10 100	10 100	10 100
Dispersion time [g] scale resolution (sec)	10^2	10^2	10^3	10^4
Dispersion spatial scale resolution (m)	vert. 0--1000 horz. 100, micro 10^4, meso 10^6, macro	10, 100 1000 few few	10 1000 none none	10 1000 none none

[a] Volume of an urban area 10km x 10km with a 10^3m mixing height.

[b] Typical hi-vol sample volume.

[c] Gaussian plume models typically use wind data from 10m towers and calculate ground-level concentrations.

[d] Unknown: a rough estimate.

[e] This includes important chemical reactions.

[f] The precision with which emissions sources can be located.

[g] Time-scale of significant changes in wind speed and direction.

The Role of Receptor Methods for Revised Particulate Matter Standards

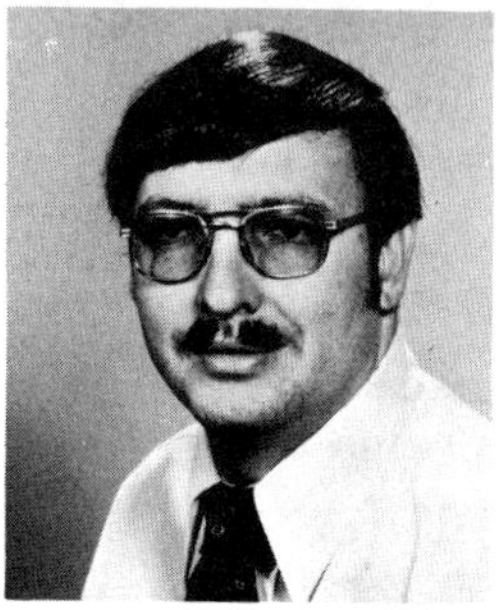

Thompson G. Pace, P.E.
U.S. Environmental Protection Agency
Office of Air Quality Planning and Standards
Research Triangle Park, North Carolina

This paper is directed to those persons responsible for developing control strategies for particulate matter under a revised PM standard. The selection of an appropriate source apportionment tool for development of control strategies for particulate matter depends on many factors, including the type of samplers and filter media available, study objectives, resources, and the availability of emissions and source characteristic data. The choices available include the use of receptor models alone, receptor and dispersion models together, and dispersion models alone. If receptor models are used, selection of the appropriate model depends upon filter type, particle size and chemical characteristics, prior knowledge of the sources, the objectives of the study, and resources available. All these are discussed and a rationale for selecting the appropriate models and/or combinations of models is suggested.

Mr. Pace is a senior environmental engineer serving in the Technolggy Development Section, Air Management Technology Branch, Office of Air Quality Planning and Standards, MD-14, U.S. EPA, Research Triangle Park, North Carolina

Introduction

The purposes of this discussion are to identify the receptor and dispersion models applicable to particulate matter analysis, discuss options for their use in control strategy development, selection of appropriate methods, and reference the technical documents describing the models and their uses. It is not intended as a reference on the technical aspects of the methods, as they have been described elsewhere in greater detail.[1]

Receptor methods, briefly stated, are a family of techniques which examine the ambient sample and the conditions of its collection to infer the types and relative mix of sources impacting on it during collection. Dispersion models, on the other hand, use information or source emission rates and meteorology to estimate the contribution of these sources to daily or annual average conditions. The terms "model" and "method" are used interchangeably, even though analysis methods such as scanning electron or optical microscopy are methods not models. The selection of appropriate receptor or dispersion models depends upon several considerations discussed below.

Types of Receptor Methods

There are several major categories of receptor methods which are potentially valuable for better understanding an area's ambient particulate problem and in formulating strategies for use in State Implementation Plans (SIP's). These include Mass Balance (MB), Factor Analysis (FA), Optical Microscopy (OM), and Automated Scanning Electron Microscopy (ASEM). While there are other techniques in addition to these, information and availability of the methods or analytical methods are limited. The major receptor methods are described briefly below and references are provided for more detailed information.[1,2,3,4,5,6]

Mass Balance

This method compares the chemical "fingerprints" or profiles of emissions from several source categories to the chemical composition of the sample. It is referred to in the literature by various names, including chemical mass balance, chemical element balance, species mass balance, and mass balance. Weighted least squares or some other statistical routine is used to find the relative proportions or mix of these sources which best "explains" or accounts for the composition of the sample. Species data (chemical compounds or elements) are required for both source and sample data.

Factor Analysis

Factor Analysis and other multivariate statistical methods, such as multiple linear regression, Target Transformation Factor Analysis (TTFA), and cluster analysis, are all variations of the least squares routine used in species mass balance. However, these methods (except TTFA) require no prior assumption of the impacting sources and can be useful where the types of sources are uncertain. Factor Analysis does, however, require at least 40 observations to give a reliable analysis.

Optical Microscopy

Particle identification by Optical Microscopy is one of the first and most widely used methods of source apportionment of coarse particles. The technique relies on identification of particles by their size, shape, color, surface properties, and birefringence (i.e., an optical property). It is generally used on particles larger than one micrometer. Several analysts have been asked to provide quality assurance of their analyses and the results have been mixed. Optical microscopy appears to be somewhat analyst dependent and semiquantitative. However, it is an excellent confirmatory or supplementary technique and is inherently "believable," since it requires no statistical, meteorological or mathematical manipulations or assumptions.

Automated Scanning Electron Microscopy

A computer-driven scanning electron microscope, equipped with x-ray fluorescence capability to identify the elemental composition, provides a particle-by-particle analysis of particle size, shape and elemental composition. This elemental composition and computer assisted sizing allows large numbers of particles to be analyzed and, thus, identified by source category in a relatively inexpensive manner.

Other Receptor Methods

X-ray Diffraction (XRD) - This method provides direct information on the crystalline nature of particles. This is particularly useful for distinguishing among minerals. It requires a heavy loading of coarse particulate matter on the filter and cannot be used on amorphous (i.e., non-crystalline) particles. The results of OM, ASEM and XRD may be used alone or in a mathematical procedure such as MB or FA.

Trajectory Analysis - This technique is used for tracing the history of an air mass to determine the possible area of origin of sources of ambient particulate. The simplest form is the pollution rose, where only average wind direction is considered. More sophisticated approaches account for the changing wind direction with time to trace air masses in time and space for as long as several days.

Microinventory - This technique is an orderly compilation of sources near a receptor. This is particularly useful as a finely gridded input to dispersion models, and in qualitatively estimating the potential sources of soil and other minerals to help interpret an observed mix of particulate matter on an ambient filter.

Source Apportionment Options for Control Strategy Development

There are three options for estimating source category contributions to observed levels of ambient particulate and applying this information in the development of control strategies. These are: (1) use of receptor models alone; (2) use of receptor and dispersion models in combination; and (3) use of dispersion models alone.

Use of Receptor Methods

A mass balance may be used for source apportionment studies. However, due to the uncertainty associated with any source apportionment approach, it is highly recommended that an additional analysis using at least one other receptor modeling approach be performed. This other approach could be selected from those discussed above, or from those discussed in Volume I of the Receptor Model Technical Series or the Digest of Ambient Particulate Analysis and Assessment Methods.[5] Such techniques as optical microscopy or ASEM may be used along with MB. They should be accompanied by an intensive chemical analysis of the samples. Other tools, such as x-ray diffraction, may also be used. The results of receptor method source apportionment may be used to estimate the degree of control required to demonstrate attainment at a given monitoring site.

Use of Receptor and Dispersion Models in Combination

Several demonstrations have been made of the use of receptor models to help validate the results of dispersion modeling.[7] This is especially useful when the emission inventory in a dispersion model is determined to be marginally adequate. The results of the receptor model can be used to carefully scrutinize the inventory assumed in the dispersion model to see whether emissions from certain source categories have been inadequately characterized. The use of a receptor model, such as MB in conjunction with dispersion modeling, is highly recommended in such situations.

Use of Dispersion Models

Dispersion models may be used in the development of control strategies for particulate matter. The guidelines and technical support for the use of dispersion models are well established and will not be discussed here.[8]

Selection of Appropriate Receptor Model(s)

Unfortunately, no one receptor model is best under all conditions. Several aspects of source characteristics and study objectives should be considered in selecting the most effective receptor methods for source identification. The filter availability and data requirements of the various models, the nature of the nonattainment situation and the nature of the suspected sources all should be considered in selecting source apportionment methods.

Filter Availability and Data Requirements

For some SIP's, time and resources will permit the design and execution of a nonattainment study, custom designed for the specific situation. For others, the available samples, and even the ambient and source species data, will be all that is available or feasible to obtain. The following is a discussion of the filter availability and data requirements for SIP development using receptor models.

Filter Availability. Several types of filters and filter materials are available and commonly used in air pollution sampling. Fiber filters include those made with glass or quartz fibers. Membrane filters include polycarbonate and teflon. Certain combinations such as teflon coated glass fibers are also available. Unfortunately, many SIP studies must be done using samples already collected and the filter material used may severely limit the types of receptor models which can be applied. For instance, glass fiber filters can contain a binder which interferes with carbon analyses. This is better accomplished on quartz fiber filters. OM and ASEM are best accomplished on a flat membrane surface. However, limited success has been reported using glass fiber filters. Table I summarizes the appropriate filter materials for use with the receptor models described above.

Ambient Chemical Data. The use of MB and factor analysis requires a data base of ambient chemical or species data. The particular elements or species required depend upon the suspected sources and their elemental composition. Generally, knowledge of the following elements is necessary to begin a MB or FA analysis: Aluminum, Potassium, Calcium, Iron, Nickel, Bromine, Copper, Lead, Silicon, and Sulfur. Other highly desirable species or elements, depending on nearby sources, include: Sodium, Magnesium, Chlorine, Vanadium, Zinc, Copper, Sulfate, Nitrate, Organic and Total Carbon, Arsenic, Tin, Chromium, Manganese, or Titanium. Additional species or elements may also be useful. For reasons discussed below, it is generally desirable that these data be available for both fine (< 2.5 μm) and coarse (2.5 to 15 μm) particle size samples individually.

Source Chemical Data. MB and Factor Analysis require the use of source "fingerprint" data. These data may be either the elemental or compound percent composition of the sources suspected of influencing ambient concentrations. Ideally, species data are collected at the specific sources in question, but in practice a set of literature derived fingerprints must frequently be used. Use of fingerprints reported in the literature is permissible only if good engineering judgment dictates that they are adequate for the specific application. A document is available which provides a compilation of much of the available literature on source profiles.[9] Considerable caution is needed in utilizing this information. As with the ambient data, size specific source data for both the fine and coarse fractions are highly desirable.

Nature of the Nonattainment Situation

The nonattainment situation can be governed by either the 24-hour National Ambient Air Quality Standard (NAAQS) for PM, the annual NAAQS or both. The governing situation, to some extent, dictates the most suitable type of nonattainment analysis.

24-hour Nonattainment. With an expected exceedance form of NAAQS, at least one observation above the NAAQS must occur if the area is nonattainment. The occasions on which these exceedances occurred should be reviewed to develop the control strategy. All samples above the PM standard should be analyzed. These samples must be determined to be representative of the types of days and, thus, the sources and situations contributing to the 24-hour nonattainment problem. It is likely to be quite difficult to devise an appropriate short term emission inventory needed for use in a dispersion model to characterize

24-hour episodes of high ambient particulates. Therefore, in 24-hour non-attainment situations, analysis of the observed monitoring data using a receptor model is likely to be particularly useful, either alone or as a check on dispersion model estimates. Of the models discussed, MB, OM, ASEM, XRD, and trajectory analysis are all appropriate for use with 24-hour observations.

Annual Nonattainment. For determining compliance with the annual arithmetic mean NAAQS, either size segregated particulate or TSP data should be available. Dispersion models, based on an annual emission inventory, and frequency distributions of observed meteorological conditions can be used. Also, the following receptor models and methods may be useful: MB, FA, and microinventory. OM, ASEM and XRD can be used on samples which have been carefully selected to represent the range of conditions comprising the annual average. Thus, the average source contributions would be the average from those samples. Once specific 24-hour or annual average source contribution factors are obtained, a proportional model may be used to estimate control requirements, if no dispersion model is appropriate for use in the analysis. In general, a dispersion model is more reliable for averaged concentrations than for 24-hour concentrations and is probably the method of choice in most annual nonattainment studies.

Nature of the Suspected Sources

Several considerations related to the nature of the suspected sources influence selection of the models selected for SIP analysis. The availability of particle size data and the size range of the predominant sources, prior knowledge of the sources, fingerprint similarity, and the need to identify specific sources vis-a-vis source categories are all important considerations in selecting a method. These are summarized in Table II.

Particle Size. Many researchers have discussed the bimodol distribution of particulate matter where source categories tend to emit either fine ($<$ 2-3 μm) or coarse ($>$ 2-3 μm) aerodynamic diameter particles. Sources of fine particles include combustion (motor vehicles, boilers, field burning, woodstoves, etc.) and industrial processes involving combustion, chemical reaction or condensation of vapors. Species mass balance, factor analysis, and dispersion models are well suited for either fine or coarse sources. Optical and scanning electron microscopy are suitable for coarse and fine particles down to 1 μm or less in size. Receptor models generally work best when the sample is segregated by size range (e.g., fine and coarse).

Prior Knowledge of Sources. In most instances, the sources suspected of contribution to PM concentrations are apparent but quantification is needed. Any of the methods discussed could be used if the sources are identifiable, provided the other requirements for their use are also met. However, two methods, the mass balance and the dispersion model, require knowledge of sources and their emission characteristics or emission rates.

Fingerprint Similarity. A "fingerprint" is the chemical or morphological pattern or characteristic of the source which is used to distinguish it from other sources. Some combustion source types are almost entirely carbon and there is very little to differentiate among them. Some help might be gained from optical properties. Also, carbon dating (C_{14}/C_{12} ratios) can be useful in distinguishing between modern or fossil carbon (e.g., woodsmoke versus fuel oil). Other common problems include distinguishing the sources of various soil components of a sample or between flyash and soil. Optical properties can be useful for some situations where fingerprints are similar and x-ray diffraction is very useful in separating various minerals by examining their crystalline structure. Other useful tools for identifying sources of combustion or soils are x-ray diffraction and the dispersion model, provided the emissions inventory can be estimated accurately.

Sources Versus Source Categories. Any of the receptor methods can be used to identify the impact of an isolated specific source unless its fingerprint is similar to that of background. Likewise, the techniques can be used to identify many of the source categories within an airshed consistent with the limitations identified in the preceding discussion. However, the impact of specific individual sources within an airshed may not be identifiable except by a dispersion model.

Establishing a Control Strategy Using Receptor Model Results

It is suggested that a proportional model similar to modified rollback be used to estimate the control requirements for SIP development. This model would first require that the Total Required Reduction in Ambient Concentration (TR) be calculated by equation (1):

(1) TR, $\mu g/m^3$ = Design value - NAAQS

where TR is the total required reduction in ambient concentrations

This TR may be accomplished by the reduction in the contribution of a single source but most likely it will be the sum of several individual sources or source category reductions (IR_i) as in equation (2).

(2) TR, $\mu g/m^3 = \Sigma$ IR $\mu g/m^3$

where IR is the reduction in ambient concentration desired from a source or source category (i)

These IR's are generally selected based on many considerations, including the technical feasibility of achieving a given emission reduction or additional reduction at that source. The percent reduction in emissions (%RE) for a source or source category is given in equation (3).

(3) $$\%RE_i = \frac{IR_i,\ \mu g/m^3}{AC_i,\ \mu g/m^3}$$

where AC_i is the ambient concentration due to the individual source (i)

For example, if the annual NAAQS for TSP were 75 μg/m^3 and the design value for the site were 100 μg/m^3, the total required reduction using equation (1) would be:

$$TR = 100 - 75 = 25\ \mu g/m^3$$

Assume the following ambient source contributions were estimated by the receptor model and a background analysis:

Source	Source Category Ambient Conc. (AC)
Steel mill roads	15 μg/m^3
Steel mill coke ovens	10 μg/m^3
Coal storage pile	5 μg/m^3
Urban paved roads	20 μg/m^3
Cement plant	10 μg/m^3
Background	30 μg/m^3
Unaccounted for	10 μg/m^3

Then, using equation (2), the TR would be the sums of the individual reductions in ambient concentration. Assume that the following reductions in source concentration are decided by the control agency:

Source	Individual Reduction (IR)
Steel mill roads	8 μg/m^3
Steel mill coke ovens	5 μg/m^3
Cement plant	8 μg/m^3
Coal storage pile	4 μg/m^3
TR =	25 μg/m^3

Since the TR sums to 25 μg/m^3, it appears that the individual reductions, if effected, would reduce the annual mean TSP to 75 μg/m^3. The required emission reductions (%RE) to accomplish this are calculated from equation (3) for each source category. For example:

$$\%RE_{cement} = \frac{IR_{cement}}{AC_{cement}} = \frac{8}{10} = 80\%$$

Therefore, the control requirements would be 80% for the cement plant. If this is not technologically feasible, the IR for cement would have to be decreased and the IR for another source increased to maintain a TR of 25 μg/m^3.

In using the proportional model, the background value should be estimated as that portion of PM which is transported into the area of jurisdiction from outside of the area. This estimate should be based on actual PM observations in nonurban areas near the boundary of the jurisdiction.

Resource Constraints and Study Adequacy

In a nonattainment situation, the resources must be allocated to perform an adequate source apportionment evaluation. Both the methods selected and the extensiveness of analysis will depend somewhat on the resources available.

For instance, if samples are available, but no current emission inventory is available, it may be less expensive overall to do a receptor study. However, if the skills are not available inhouse to do the filter and/or computer analysis, an emission inventory and dispersion model, using staff personnel, may be a more appropriate course of action. Likewise, the staff may be lacking to do the inventory but the funding is available for a receptor study. For whatever funding or staff resources are available, the analysis must be judged adequate to justify the changes in controls or emission limits which are prescribed.

Summary

Three options are available for source apportionment of particulate matter for use in control strategy development for PM. These are: receptor methods used in conjunction with a proportional model similar to modified linear rollback, receptor and dispersion models used together, and dispersion models used alone. (The EPA may choose to prioritize or otherwise constrain these three choices when guidance for PM SIP's is revised.) If receptor methods are selected, it is recommended that either mass balance or automated scanning electron microscopy be used as the primary method, and that one of several other receptor methods be used as supplementary approaches. The combined use of receptor and dispersion models offers an improved approach/methodology for use in developing control strategies to attain/maintain any PM standards that EPA might promulgate.

Acknowledgement

The Environmental Science Research Laboratory, including Dr. Thomas G. Dzubay, Dr. Charles W. Lewis and Robert K. Stevens have provided invaluable assistance in defining the technical limitations and utility of the various receptor methods. Dr. Edwin L. Meyer provided valuable advice and review of the manuscript. Mrs. Carole J. Mask provided typing and editorial assistance in a timely and highly efficient manner.

References

1. Receptor Model Technical Series, Volume I - Overview of Receptor Model Application to Particulate Source Apportionment. EPA-450/4-81-016a, U.S. Environmental Protection Agency, Research Triangle Park, North Carolina, July 1981.

2. Receptor Model Technical Series, Volume II - Chemical Mass Balance. EPA-450/81-016b, U.S. Environmental Protection Agency, Research Triangle Park, North Carolina, July 1981.

3. Receptor Model Technical Series, Volume III - User's Manual for Species Mass Balance Computer Program, Draft Report from Radian Corporation under Contract No. 68-02-3513 to U.S. Environmental Protection Agency, MD-14, Research Triangle Park, North Carolina, September 1982.

4. Receptor Model Technical Series, Volume IV - Technical Considerations in Source Apportionment by Particle Identification, Draft Final Report from Engineering Science to U.S. Environmental Protection Agency, Research Triangle Park, North Carolina, September 1982.

5. "Digest of Ambient Particulate Analysis and Assessment Methods," EPA-450/3-78-113, U.S. Environmental Protection Agency, Research Triangle Park, North Carolina, September 1978.

6. G. E. Gordon, "Receptor Models," Environmetal Science and Technology, 1980, 14, 792-800.

7. John E. Core and William M. Cox, "Particle Dispersion Model Evaluation: A New Approach Using Receptor Models," presented at the APCA 1982 Annual Meeting, New Orleans, Louisiana, June 1982.

8. Guidelines on Air Quality Models, EPA-450/2-78-027, U.S. Environmental Protection Agency, Research Triangle Park, North Carolina, April 1978.

9. Particulate Source Composition Library Development, Draft Report by PEDCo Environmental under Contract No. 68-02-3512, Work Assignment 20, to U.S. Environmental Protection Agency, MD-14, Research Triangle Park, North Carolina, October 1982.

TABLE I. APPROPRIATE FILTER MEDIA FOR USE WITH RECEPTOR MODEL STUDIES

	Glass Fiber	Quartz Fiber	Polycarbonate	Fluorocarbon
Elemental Analysis (Mass Balance) (Factor Analysis)	3	2	1	1
Optical Microscopy	2	2	2	2
Automated SEM	2	2	1	2
Carbon	2	1	3	3

1. Recommended
2. Useful under some circumstances or to some analysts
3. Not recommended

TABLE II. SELECTING SOURCE APPORTIONMENT METHODS TO FIT SOURCE TYPES AND STUDY OBJECTIVES

	Fine	Coarse	Sources Known	Sources Unknown	Fingerprints Similar	Fingerprints Dissimilar	Isolated Single Source	Airshed (Source Categories)	Airshed (Specific Sources Within Category)
MB	X	X	X		a,b	X	X	X	b
FA	X	X	X	X	b	X	X	X	
ASEM	c	X	X	X	b	X	x	x	b
OPTICAL	c	X	X	X	b	X	X	X	b
DISPERSION MODEL	X	X	X		X	X	X	X	X

a - Initial use of Factor Analysis may be helpful.

b - Useful if a source can be isolated from other similar sources by wind direction. Method usually cannot otherwise distinguish between sources in same category.

c - Particles larger than 0.5 to 1.0 μm.

X - Appropriate to use.

RECEPTOR MODELING CONCEPTS
APPLICABLE TO THE UTILITY INDUSTRY

Jacques P. Guertin
Peter K. Mueller
Electric Power Research Institute
Palo Alto, CA 94303

This paper outlines EPRI's approach to research on receptor modeling. The feasibility of applying receptor modeling to the problem of determining utility contributions to air quality, precipitation quality, and visibility impairment is being investigated. Intellectual and numerical analyses are to be conducted with the view of identifying usefulness and limitations. If these analyses of theoretical and observational aspects conclude that receptor modeling would be useful if some limitations are overcome, then a research plan would be developed and implemented.

Introduction

The purpose of this paper is to outline the approach of the Environmental Physics and Chemistry Program at EPRI to undertake a long term research project on receptor modeling with application to air quality, precipitation, and visibility.

Problem Statement

The understanding of utility contributions to airborne pollutants in the environment needs to be improved. Federal and state clean air regulations authorize the use of source-oriented plume models for deciding whether or not a source complies with air quality standards. A suitable regional scale model is being sought to establish source contributions to air quality and precipitation chemistry across the Canadian-United States Border.

The accuracy and reliability of models are important in pollution control decision which have significant economic consequences. The Sulfate Regional Experiment[1] has demonstrated the capability and limitation of source-oriented models for calculating the regional (100 to 1000 km) influences of SO_2 emissions. The Plume Model Validation Project[2] is designed to demonstrate capability and limitations for calculating near plume (0 to 50 km) air quality.

The results indicate that present source-oriented models are often not sufficiently precise or accurate for formulating control strategies reliably. The problems are with (a) uncertainties in mathematically simulating the dynamics of atmospheric mixing, transport, transformation and deposition, (b) uncertainty in meteorological and environmental data to represent starting conditions, (c) variations in the reliability of region wide emission inventories, and (d) fundamental inequalities between calculated and measured values. An alternative and complementary approach is a receptor model. In this approach, specific features of the atmospheric aerosols (particles and gases) are measured at the receptor and at the emission source. The contribution of a specific source is then calculated. This approach has been applied primarily to suspended particles. Owing to their complex characteristics, these particles carry much information. Attempts are being made to apply receptor modeling also to precipitation.

Receptor modeling is based on the premise that every process along the route from the source to the receptor can be represented statistically. Assuming conservation of mass, the total aerosol mass measured at a receptor is related to the linear sum of contributions from the individual sources. The mathematics are given in other papers of these Proceedings. The specificity of the source data can be increased for instance by classifying the measurements data at the receptor according to meteorological parameters, chemical classes, i.e., elements, and physical properties, i.e., particle size.

Thus, the receptor model relies on properties of material common to source and receptor but unique to specific source type. It requires no assumptions about dynamic atmospheric processes and emission rates. It does require statistical expressions accounting for all the processes operating along the path from source to receptor. A receptor-oriented model uses the ambient pollutant concentration measurements and the compositional characteristics of the emissions to identify and quantify the source contributions to those concentrations. Whereas a source-oriented model begins with estimates of emission rates and meteorological measurements to predict the contributions of a source at a receptor. Possibilities exist for combining receptor-oriented with source-oriented models. Such hybrids could predict future ambient pollutant concentrations with greater accuracy than is now attainable with source-oriented models alone.

The applications of receptor models have been heuristic rather than rigorously derived from basic principles. No application has been evaluated nor has a systematic protocol been established. Therefore, a study is needed to determine how practical receptor modeling could be with respect to tracing emissions from coal and oil fired power plants, and to establish whether or not research on receptor modeling could enhance the reliability of allocating and predicting the influence of utility emissions. There is a need to understand the usefulness and limitations of receptor modeling, in numerical terms, when applied to air quality, precipitation, dry deposition and visibility prior to establishing a research program on receptor modeling.

Study Design

The objectives of the initial EPRI research approach on receptor modeling are the following:

- to elucidate the fundamental basis for receptor models;
- to evaluate (preferably in quantitative terms) the extent to which receptor modeling could be a useful tool for ascribing materials in air and precipitation to coal and oil-fired power plants, and
- to establish which research topics in receptor modeling, if any, are likely to yield results useful for the utility industry.

The approach to this study is to describe and define all aspects of receptor models and identify needs for more knowledge on items such as:

- receptor modeling theory for applications to air quality, precipitation, dry deposition, and visibility.
- data treatment methods such as regression on mass and multivariate analyses.

- practical analytical methods for measuring the properties of emissions, air quality, precipitation, dry deposition, visibility, and perhaps meteorology. These methods could be limiting in the successful application and evaluation of receptor models.

A first step in this study is to generate a logic map or conceptual framework for examining the fundamental basis of receptor modeling and to compare the intrinsic aspects of receptor models with those of source-oriented models. It will also be necessary to establish approaches for relating the uncertainties in the input data to the uncertainties in the results from receptor modeling.

This conceptual framework will then lead to an intellectual and numerical analysis of current receptor modeling science, to identify its limitations and to determine if resolution of such limitations through research would make receptor modeling sufficiently reliable for application by the utility industry. From this study it should be possible to specify the activities and their costs for obtaining information that would be needed to operate receptor models with reliability.

It is anticipated that this analysis will include the following activities:

- delineating receptor modeling theory starting with the fundamentals and comparing the principles and assumptions with those of source-oriented models.

- moving model components and constructs from a heuristic basis.

- placing mathematical data handling methods (such as regression analysis on mass and statistical analysis on variability) in proper perspective and documenting their operational logic.

- establishing data requirements by processing simulated data sufficient to overdetermine a set of simultaneous equations. This should allow the reliability and resolution of output data to be related to those of input data.

- showing how atmospheric dynamics can be used to improve the specificity of the association between source and receptor models.

- identifying the methodologies necessary for obtaining relevant source and receptor characterization.

We believe there is sufficient expertise and knowledge available to undertake these activities through the involvement of a peer group of investigators. If the report of the detailed feasibility study justifies further needs for research for sponsorship by EPRI, then a plan for research will be formulated.

This plan would address the research necessary to evaluate the performance of receptor models on either, or both, local and regional scales by considering needed development of mathematical formulations, observational methods, and the establishment of variability in the characteristics of emissions from utilities. For application on local scales, EPRI would conduct studies which would establish a recommended receptor modeling methodology. For application on sub-regional or regional scales, EPRI would consider participating in studies that would be designed to determine utility contributions to air and precipitation quality at representative locations in the USA.

Summary and Conclusions

EPRI's approach to research on receptor modeling has been outlined. Based on an examination of the literature and consultation with leading receptor modeling practitioners, it is not clear that the application of receptor modeling would usefully quantify the contributions of coal and oil fired electricity generating stations to air quality, materials in precipitation and visibility. To assess the viability of receptor modeling in these regards, an intellectual and numerical analysis will be undertaken. It is anticipated that some aspects of receptor modeling will be useful. A plan for conducting research on these aspects is to be formulated. Follow-on studies may be undertaken with the goals of either or both establishing a recommended methodology for applying receptor modeling with regard to utility emissions on local scales and determining utility contributions at receptor locations on sub-regional or regional scales.

REFERENCES

1. P.K. Mueller and G.M. Hidy, "The Sulfate Regional Experiment. Report of Findings," Report EA-1901, Section 8, Electric Power Research Institute, 1983.

2. M.-K. Liu, G.E. Moore and H.-Y. Holman, "Survey of Plume Models for Atmospheric Applications," Report EA-2243, Electric Power Research Institute, 1982.

RECEPTOR MODEL SOURCE APPORTIONMENT OF LEAD
IN THREE AIRSHEDS WITH LEAD SMELTERS*

James E. Houck, John A. Cooper

Clifton A. Frazier, Richard T. DeCesar and James F. Mohan, all with NEA, Inc., 8310 S.W. Nimbus, Beaverton, Oregon 97005; David Maughan, with State of Montana, Helena, Montana and John Roberts, with Puget Sound Air Pollution Control Agency, Seattle, Washington.

Results of recent applications of chemical mass balance (CMB) and factor analysis (FA) to source apportionment of lead aerosol sources in Kellogg, Idaho, East Helena, Montana and Harbor Island, Washington are presented. Special emphasis was applied in these studies to separating the impact of fugitive dust sources such as windblown soils and road dust from smelter process emissions. A newly developed dilution size-segregating source sampler was used to characterize both fugitive and point sources in the East Helena study. The extensive source characterization of potential lead sources in East Helena provided sufficient source resolution to identify the contribution of specific smelter sources to ambient lead levels. CMB, FA and time series were used to identify major lead sources in Kellogg, Idaho. Blast furnace upset emissions were identified as the major source of lead in Kellogg. This was accomplished with a high level of confidence by using all three receptor model approaches despite the fact that direct source fingerprints were not obtained because of source sampling restrictions.

This paper presents the methodology used in these three studies, including sampling, analysis and interpretive approaches. The study results are discussed relative to control strategy development and their application to other airshed studies.

* This research was sponsored in part by funds provided by Regions X and V of the U.S. EPA and the States of Montana, Idaho and Washington.

INTRODUCTION

Atmospheric concentrations of suspended particulate material and lead have exceeded National Ambient Air Quality Standards in Kellogg, Idaho, East Helena, Montana and Harbor Island in Seattle, Washington where major lead smelters are located. Development of a cost-effective state implementation plan requires an accurate, quantitative understanding of the contributions that each source makes to the atmospheric lead concentrations. Attempts to apportion the contribution of potential lead sources with classical dispersion models have been generally unsuccessful. The objective of this paper is to summarize results of three receptor model studies (1-3) which have quantified the contributions of major lead sources in these airsheds.

BACKGROUND

Kellogg, Idaho is located in Silver Valley, an east-west trending valley which contains the south fork of the Coeur D' Alene River mining district. It is a narrow, deep valley in which flat regions of the valley floor rarely exceed one kilometer and mountain ridges rise on the average of 700 meters above the valley. The elevation of the valley floor rises from about 700 meters at it's west side to a little over 1,000 meters at the eastern end which promotes afternoon and evening drainage winds from the east. The nearly 100 years of mining and ore-processing in the Silver Valley have created large denuded areas from severe sulfur fume insults, and widespread contamination of soils due to erosion of solid wastes and deposition of airborne heavy metal contaminants. Previous studies of lead particulate sources in the Silver Valley suggested that these large area sources of lead contaminated fugitive dust may contribute substantially to the lead non-attainment problem.(4)

East Helena is located in the Helena Valley five kilometers east of Helena, the state capital. The Helena valley is bound on the north and east by the Big Belt Mountains and on the west and south by the main chain of the Continental Divide. The valley is approximately 40 kilometers in width from north to south and 56 kilometers east to west. The average height of the surrounding mountains above the valley floor (elevation, 1128 meters) is about 900 meters. The valley can best be described as broad and flat with limited topographic features.

Harbor Island is a man-made island located in Elliott Bay south of the main business district of Seattle. It is located in a heavily industrialized portion of the city near the north end of the Duwamish Valley. A secondary lead smelter is located on the island along with a large variety of other industry and nearby roads with high traffic densities.

Primary lead smelters are the only major industry in the Idaho and Montana valleys and traffic volumes are relatively low.

EXPERIMENTAL

Ambient Sampling

Three types of samplers were used to collect ambient samples in Silver Valley and East Helena. A standard high-volume sampler with glass fiber was used as a standard reference while low-volume TSP and dichotomous samplers were used with teflon filters for chemical analysis. Samples were collected at six locations in the Silver Valley on an every third or sixth day basis and every day at Silver King School over about a 9 month period except for a 6 day intensive sampling period in which samples were collected every four hours at Silver King School and the Doctors Clinic.

East Helena ambient samples were collected from January, 1981 through April, 1982 with the most extensive sampling program being conducted during the fourth quarter of 1981. Nearly 700 valid dichotomous and low-volume teflon filters were collected and analyzed for mass and elemental concentrations during this period.

Samples were collected at one location on Harbor Island on an every sixth day basis during the first half of 1982. High-volume samplers with teflon filters and University of Washington designed fine particle samplers were used.(5)

Source Sampling

The only extensive source sampling was conducted during the East Helena study after completion of the Kellogg study in Silver Valley, Idaho. Although sampling of the smelter sources in Kellogg was originally planned, legal restrictions prevented on-site source sampling. An extensive fugitive road and soil dust sampling program was implemented and factor and time variate analysis along with photographic techniques were used to deduce smelter source characteristics.

A size segregating dilution sampler developed by NEA, Inc. was used to sample major point sources in East Helena. Process fugitive dust samples were collected with a variety of special samplers designed to yield uniformly deposited fine and coarse samples. Passive fugitive dust sources were sampled by collecting grab samples, sampling the resuspended dust samples with dichotomous and low-volume TSP samplers to simulate the size fraction collected with ambient samplers.

Stack samples were collected at the Harbor Island secondary lead smelter using a Radar high-volume source sampler while process fugitive and passive fugitive samples were collected by methods similar to those used in East Helena.

Analysis

The State of Idaho measured the concentration of Pb and Cd deposits on high-volume filters by atomic absorption spectrophotometry (AA) while the States of Montana and Washington measured only lead by AA. Fifty elements were measured on low-volume TSP and dichotomous filters by NEA using x-ray fluorescence (XRF) and neutron activation analysis (NAA). The use of three independent analytical methods and four independent laboratories provided excellent quality assurance.

Over 950 filters were analyzed by XRF and 380 by NAA during the Kellogg study, while over 700 filters were analyzed in the East Helena study. Only 50 filters were analyzed in the Harbor Island study.

Interpretive Procedures

An effective variance chemical mass balance (CMB) method (6) was used in all three studies. This CMB procedure provides the highest level of quantification and confidence when source chemical fingerprints are available. Application of this method to the Kellogg airshed, however, was restricted half way through the program and alternative approaches such as time variate and target transformation factor analysis (7) were used.

RESULTS AND DISCUSSION

Kellogg, Idaho

The Kellogg, Idaho aerosol lead study represented the first application of a receptor model to source apportionment of lead sources in an airshed with a lead smelter. This particular study, however, was conducted during a period in which the smelter company, Bunker Hill, was involved in intense legal activity and they refused to allow source sampling on their property. The effect of this restriction was minimized with a mid-program modification which shifted emphasis to alternate methods of obtaining chemical fingerprints for the industrial sources, and focused more sharply on the contributions of fugitive area sources such as road and windblown dust and automotive exhaust. Since there are only three major lead source categories in the Silver Valley, precise characterization of two categories (automotive exhaust and area dust) defines the contribution of the third (industrial) category by difference. Factor analysis, intensive four-hour sampling, combined with video taping, as well as literature values were used to develop chemical fingerprints indirectly to subdivide the industrial contribution.

A crude size distribution of the Pb aerosol was constructed by comparing the results from the three samplers used in this study. Although the lead was enriched in the fine fraction, it contributed less than 23% of the lead collected with a high-volume sampler at the Silver King site while the fraction greater than about 15 μm accounted for more than 50% of the lead. It was clear from the size distribution that coarse particles are a major source of lead and their control would be essential in any SIP.

The analytical results for some of the major elemental species of interest are listed in Table 1. The crustal components Al, Si and Ti were highly correlated in the coarse fraction even after minimizing the influence of meteorology by normalizing the results to total mass. Although the coarse fraction Cd and Pb concentrations were highly correlated (normalized correlation coefficient of 0.947), Si and lead exhibited a strong negative correlation.

The average low-volume Pb concentration over the entire study period exceeded 9 $\mu g/m^3$ at the Doctors Clinic site but approached a maximum of 60 $\mu g/m^3$ during 4 hour averaging periods in September as shown in the time series plot in Figure 2. These periods of peak Pb concentrations at the Doctors Clinic were characterized by low wind speeds (generally less than 5 miles per hour) persistently out of the west which established the Doctors Clinic clearly in a position downwind from the ore-processing operations. The one deviation from this pattern during the six-day intensive study was September 28 when the wind speed exceeded 15 miles per hour and the daily maximum lead concentration was the lowest of the period. Sulfur and zinc also exhibited a similar concentration variability pattern and were correlated with lead. The strongest correlation during this period was between Al and Si (0.966) with the next strongest between fine and coarse Cd and Pb (0.947 and 0.906). A clear and strong negative correlation was exhibited between Pb and Si in the coarse fraction. This latter observation strongly suggests that road and windblown dust are not the dominant source of coarse particulate lead at the Doctors Clinic.

Source contributions to hi-vol levels were based on CMB analysis results for lo-vol TSP and dichotomous samples by assuming that the relative source contributions to the "giant" hi-vol TSP lead particles were the same as the coarse dichotomous particles. The source contributions to hi-vol TSP lead levels at the Silver King School and the Doctors Clinic sites calculated on this assumption are summarized below:

- Silver King School; Winter-Spring and Summer
    - Direct Industrial — 5.6 ± 0.5 μg/m^3 (74%)
    - Road and Windblown Dust — 1.8 ± 0.1 μg/m^3 (24%)
    - Transportation — 0.17 ± 0.06 μg/m^3 (2%)
- Silver King School; Intensive Sampling Period
    - Direct Industrial — 8.6 ± 0.9 μg/m^3 (83%)
    - Road and Windblown Dust — 1.6 ± 0.1 μg/m^3 (16%)
    - Transportation — 0.14 ± 0.05 μg/m^3 (1%)
- Doctors Clinic; Winter-Spring and Summer
    - Direct Industrial — 3.1 ± 0.3 μg/m^3 (79%)
    - Road and Windblown Dust — 0.66 ± 0.03 μg/m^3 (17%)
    - Transportation — 0.16 ± 0.05 μg/m^3 (4%)
- Doctors Clinic; Intensive Sampling Period
    - Direct Industrial — 9.3 ± 0.9 μg/m^3 (90%)
    - Road and Windblown Dust — 0.84 ± 0.04 μg/m^3 (8%)
    - Transportation — 0.22 ± 0.07 μg/m^3 (2%)

These source contributions represent the average of 24 hour-samples collected during the winter-spring and summar sampling periods and the four hour samples collected during the intensive sampling period.

The road and windblown dust contribution includes both current and historical indirect industrial contributions from dust fall, spillage, trackout, etc. followed by resuspension. Although these specific sub-categories were not quantified, current industrial activities are thought to contribute a significant portion of the lead associated with road and windblown dust. Road and windblown dust contributions at the Doctors Clinic site were strongly supported by the close agreement between the elemental fingerprint used for this source and the ambient fingerprint measured during a known high road and windblown dust impact. The lead content of the Silver King road dust (4%), however, was not in good agreement with the lead in the ambient aerosol during the period of high road and windblown dust impact. It was over tenfold greater than the ambient lead during this period and almost threefold greater than the surrounding soil. Thus, the lead attributed to this source at the school is likely biased substantially to the high side or contains a large fraction of current indirect smelter contribution by way of paved road dust.

Up to five subgroups were resolved from the industrial category with factor analysis. These subgroups included a

- High sulfate source,
- Lead plant,
- Zinc plant,
- Arsenic/antimony source, and
- Possibly a weak phosphate source.

Ambient lead levels were, however, strongly dominated by a single lead plant emission which has the following characteristics:

- Lead accounts for more than 28% of the mass.
- Contains both fine and coarse lead particles.
- The lead to cadmium ratio in both the fine and coarse particles is about 10.
- The fine and coarse particle emissions are correlated.
- The ambient impact from this source's coarse particles is inversely correlated with wind speed and ambient silicon levels.

The source which best fit this description was the blast furnace and its emissions during upset conditions. Control of the blast furnace would have been essential in any SIP. The Bunker Hill lead smelter closed its Kellogg operations late in 1981 and a dramatic reduction in Pb levels has been observed as noted in Table 2.

East Helena, Montana

The East Helena lead study was the first one in which extensive source sampling was conducted within a lead smelter using size-segregating dilution sampling for receptor modeling. In this case, 79 source fingerprints for each size category were determined for point, process fugitive and passive fugitive sources. Three sources which had easily traceable fingerprints and were identified in ambient aerosols in the near-plant East Helena ambient samples are: (1) blast furnace upsets, (2) dross building fugitive emissions and (3) a copper kiln stack. The blast furnace upset had a high cadmium and lead content, 10.2% and 42.7%, respectively, which is consistent with the characteristics implied for this source category from the Kellogg study. The dross building had a high arsenic content and a modestly high lead value, 24.0% and 13.4%, respectively. The copper kiln stack produces particles made up of almost entirely a mixture of elemental Cu, CuO and Cu_2O as determined by x-ray diffraction measurements. X-ray fluorescence analysis revealed that the total copper content was 54% by weight. There is also about 0.2% of Ni by weight in the particles. While this is not a high concentration in an absolute sense, it is one of the highest nickel contents seen among the sources measured (except for speiss) and it was a useful feature when the copper kiln was fit by the CMB model. The only source with a fingerprint similar to the blast furnace upset with its high cadmium content was the baghouse stack. The high (and pure) copper content of copper kiln stack emissions and the high arsenic content produced by the dross building made them quite distinct, although the dross building emissions had some copper (2.9%). The other sources had less distinct characteristics and are described in detail in the final report.(2)

The East Helena source matrix represents perhaps the most extensive and well defined compilation of source data used in any CMB study to date. Both the number of sources from which samples have been collected and analyzed, and the number of elements which can be quantified due to the metallurgical nature of most of the sources combine to provide more than adequate information for CMB source apportionment. Source apportionment for both mass and lead concentrations have been conducted for each of the key mean quarterly data sets. The source contributions to ambient mass (TSP) and lead concentrations vary with site and quarter, however, it is clear which sources are the overall major contributors. The overall major sources of TSP in the East Helena airshed are: (1) road and soil dust, (2) fugitive ore concentrate and residue emissions, (3) carbonaceous sources (residential wood combustion, coal dust, and unexplained H, O, N and C probably from miscellaneous combustion sources, agricultural materials and pollen), (4) a zinc oxide rich material and (5) a copper kiln. The overall major sources of lead are: (1) road and soil dust, (2) fugitive ore concentrate and residue emissions, (3) blast furnace upsets and (4) a zinc oxide rich material. The blast furnace upset is a major source of lead even though it is a minor source of TSP due to its high lead content (43%). The copper kiln, on the other hand, is often a major (or at least intermediate) contributor to TSP but its contribution to ambient lead concentrations is very small due to its low lead content (0.2%).

Minor TSP sources are: (1) vehicular exhaust, (2) dross building fugitive emissions, (3) blast furnace upsets, (4) slag pouring, (5) secondary sulfate, (6) fugitive slag dust and (7) limestone. Minor but measurable lead sources are (1) vehicular exhaust, (2) dross building fugitive emissions, (3) slag pouring, (4) fugitive slag dust and (5) the copper kiln (very small).

An example of the results for one of the eleven sites is illustrated in Figure 2. Although the largest single source of the 79 $\mu g/m^3$ of total suspended mass was road and soil dust at 36.6 $\mu g/m^3$, eight other sources were clearly identified as substantial contributors. The road and soil dust category, however, contributed less than 20% of the lead (1.04 $\mu g/m^3$) while the blast furnace contributed 26%, and fugitive ore and residue handling contributed 35%. Fugitive emissions from the dross building contributed less than 5% of the lead.

For the road and soil dust source category, it is worth noting that road dust sources provided the best CMB fits for the key mean quarterly data sets. Soil was left in the title, "Road and Soil Dust", because small contributions of soil dust would not be separatable from the larger road dust contribution due to their similar chemical fingerprints. When CMB source apportionment was conducted on individual 24 hour samples during the quarters from which the quarterly means were determined, it was found that in a significant minority of the days, soil dust provided a better fit than did road dust. It is also interesting to note that the road dust source fingerprint determined from city streets provided the best mean quarterly fit at each of the key sites during the second and third quarters, whereas the road dust source fingerprints for Highway 12 provided the best mean quarterly fit during the first and fourth quarters. Again it must be emphasized the CMB source apportionment on quarterly mean values will assign the contribution to the dominate source among sources with very similar chemical compositions and some fine detail is lost, e.g., there was some city street dust impact on the ambient samplers during the first and second quarters and there was some highway dust impact during the second and third quarter.

Harbor Island

The total particulate mass at the Harbor Island sampling site which was less than 300 feet from the secondary lead smelter was apportioned between 12 major source categories while the fine mass was apportioned into 13 major categories. The largest contributor to the TSP was soil and/or road dust at 31.4%. Secondary sulfate was the largest source category in the fine particulate fraction at 30.2%.

Most of the lead was in the coarse particle fraction with only about 20% occurring in the less than 2.5μm size range. Transportation sources contributed less than 1% of both the fine and coarse particulate lead. Although soil and road dust contributed less than 10% of the lead in both size fractions, this source was directly related to the smelter and reduction of the smelter emissions would indirectly reduce the soil and road dust contributions. Yard dust was the largest single source (47%) of total particulate lead. The lead pot baghouse was the second largest contributor to the TSP lead and the largest contributor to the fine particle lead.

REFERENCES

1. Cooper, J.A., R.T. DeCesar, C.A. Frazier, J.E. Houck and J.F. Mohan, "Determination of Source Contributions to Air Particulate Lead and Cadmium Levels in Kellogg, Idaho Using the Receptor Model," final report to U.S. EPA No. 40523, NEA, Inc., Beaverton, Oregon.

2. Houck, J.E., J.A. Cooper, C.A. Frazier and R.T. DeCesar, "East Helena Source Apportionment Study", final report to the Montana State Dept. of Health and Environmental Sciences, September, 1982, NEA, Inc., Beaverton, Oregon.

3. Cooper, J.A., C.A. Frazier, J.F. Mohan and J.E. Core, "Harbor Island Lead Aerosol Source Apportionment Study," final report to Puget Sound Air Pollution Control Agency, 12/20/82 , NEA, Inc., Beaverton, Oregon.

4. von Lindern, I.H., "An Empirical Air Quality Analysis Utilizing Computer-Assisted Cartographic Modeling Techniques," Ph.D. Dissertation, Yale University, New Haven, Connecticut, 1980.

5. Waggonner, A., University of Washington, Seattle, Washington.

6. Watson, J.G., Jr., "Chemical Element Balance Receptor Model Methodology for Assessing the Sources of Fine and Total Suspended Particulate Matter in Portland, Oregon," Ph.D. Dissertation, Oregon Graduate Center, Beaverton, Oregon.

7. Malinowski, E.R., D.G. Howery, P.H. Weiner, J.M. Soroka, P.T. Funke, R.B. Selzer and A.R. Levinstone, "FACTANL - Target Transformation Factor Analysis," Quantum Chemistry Program Exchange, Indiana University, Bloomington, Indiana, 1976, Program 320.

Table 1

AVERAGE AMBIENT MASS, Si, Pb, Cd, Zn, S and As CONCENTRATIONS AND INDUSTRIAL Pb, Cd, Zn and S CONTRIBUTIONS* (FEBRUARY TO SEPTEMBER, 1980)

($\mu g/m^3$)

Site	Data** Set Type	N	Sample	Mass	Si	Pb (total)	Pb (Indus.)	Cd (total)	Cd (Indus.)	Zn (total)	Zn (Indus.)	S (total)	S (Indus.)	As
Silver King	1	27	fine	14.8	.200	1.73	1.61	.154	.153	1.12	1.10	1.34	1.29	.062
		27	coarse	23.7	3.08	1.49	.951	.094	.084	1.35	1.16	.608	.461	.021
		27	lo-vol	80.2	5.37	6.40	4.85	.422	.396	4.40	3.88	2.50	2.08	.078
		27	hi-vol	92.9	-	7.51	-	.466	-	-	-	-	-	-
Doctor's Clinic	1	28	lo-vol	74.5	6.11	3.42	2.68	.254	.242	2.47	2.30	1.88	1.49	0.71
		28	hi-vol	82.4	-	3.89	-	.274	-	-	-	-	-	-
Silver King	2	84	fine	16.4	.208	1.92	1.77	.179	.178	1.22	1.20	1.56	1.50	.081
		84	coarse	24.6	2.79	1.79	1.20	.111	.100	1.42	1.21	.779	.616	.030
Silver King	3	30	fine	32.0	.259	4.26	4.08	.589	.588	2.85	2.83	3.84	3.76	.150
		30	coarse	40.0	4.03	2.03	1.42	.180	.168	3.43	3.20	1.74	1.56	.031
		30	lo-vol	94.5	6.68	8.81	7.46	.894	.872	8.53	8.09	6.54	6.12	.172
Doctor's Clinic	3	29	fine	39.0	.241	5.40	5.18	.754	.753	3.40	3.39	5.04	4.96	.244
		29	coarse	43.1	4.86	2.38	1.98	.189	.181	3.29	3.18	1.70	1.46	.031
		29	lo-vol	98.6	7.23	9.10	8.25	.945	.931	7.91	7.73	6.78	6.34	.237
Smelter-ville	1	30	fine	17.0	.224	1.65	1.56	.149	.149	.97	.97	1.34	1.3	.063
		30	coarse	41.8	6.74	1.13	.91	.067	.064	1.00	.93	.44	.29	.017
		30	hi-vol	135	-	5.48	-	.338	-	-	-	-	-	-
Kingston	1	33	lo-vol	41.1	3.71	.59	.50	.044	.045	.38	.345	.76	.64	.016
		33	hi-vol	51.3	-	.60	-	.049	-	-	-	-	-	-

* All data in $\mu g/m^3$, data from samples collected from May 18 through June 30 voided and not included in means due to Mt. St. Helens eruption.

** Type 1 - hi-vol samples collected at site simultaneously with dichotomous and/or lo-vol samples.
Type 2 - all valid dichotomous pairs.
Type 3 - intensive sampling period.

Table 2

Comparison of Quarterly Lead Levels Before And After Closure of the Bunker Hill Lead Smelter ($\mu g/m^3$)

Quarter	Silver King		Doctors Clinic	
	1981	1982	1981	1982
1	11.8	(1.8)	6.7	(0.4)
2	3.7	(1.0)	2.4	(0.4)
3	6.5	-	2.2	-
4	7.9	-	5.0	-

Numbers in parenthesis are preliminary.

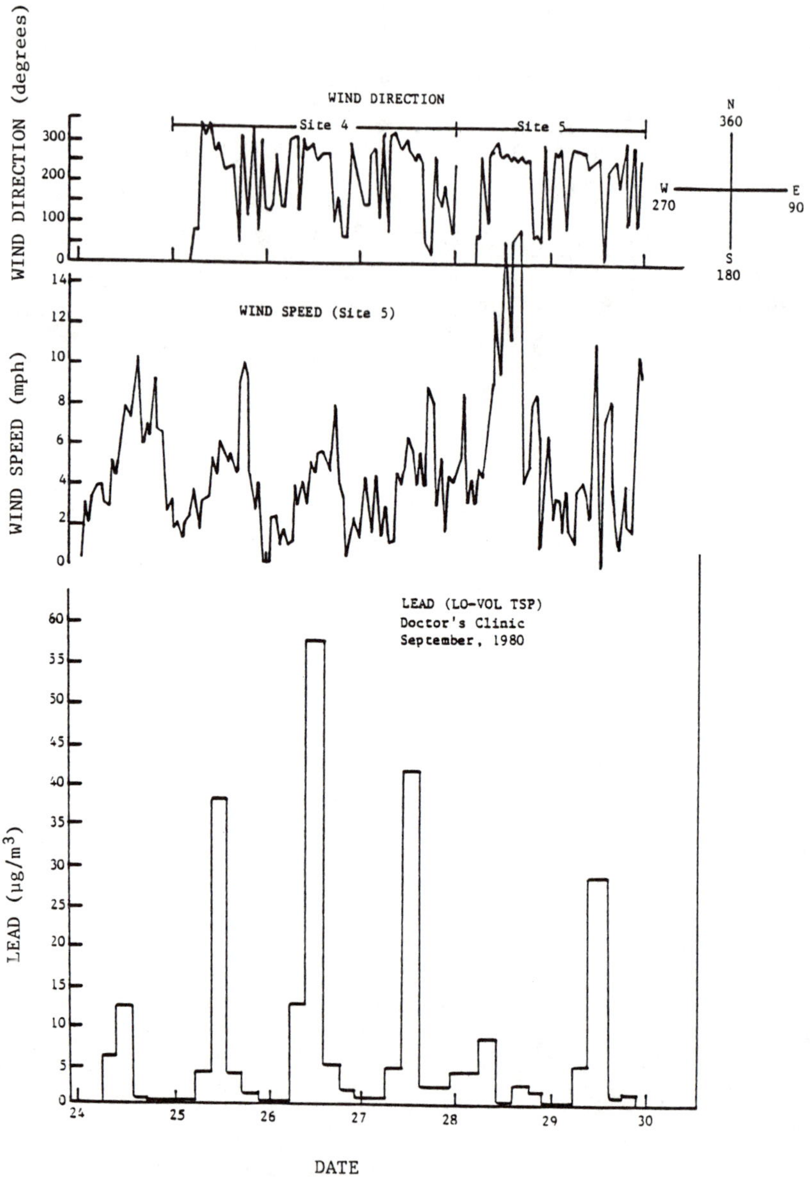

Figure 1. Comparison of the Peak Lead Concentrations With Wind Speed and Wind Direction During September, 1980 Intensive Sampling Study.

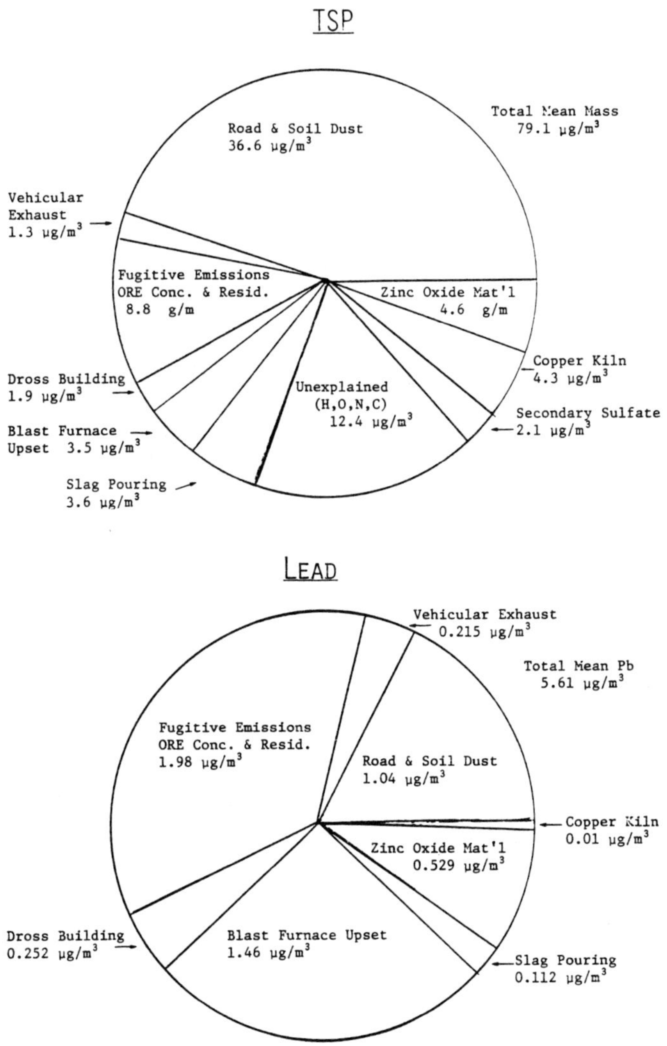

Figure 2. CMB Source Apportionment - Firehall Site, First Quarter

STATUS OF SOURCE APPORTIONMENT METHODS:
QUAIL ROOST II

Robert K. Stevens,→

Thompson Pace,
U. S. Environmental Protection Agency
Research Triangle Park, North Carolina

A source apportionment workshop sponsored by the United States Environmental Protection Agency was held at Quail Roost Conference Center in Rougemont, N.C. Thirty-five scientists from academic, governmental, and private institutions participated. The objective was to compare and document results of source apportionment analyses of simulated and real aerosol data sets. The simulated set was developed by scientists from the National Bureau of Standards and consisted of mass and elemental concentrations generated using a dispersion model that simulated transport of aerosols from a variety of sources to a receptor site. The real set contained the results of x-ray fluorescence, combustion, ion chromatography, x-ray diffraction, and scanning electron and optical microscopy analyses of aerosol samples collected in 18 12-hr sampling periods in Houston, TX. Ten groups analyzed these data sets using a variety of receptor models to apportion the simulated and real ambient measurements to potential contributing sources. The following recommendations and conclusions were derived from these workshop exercises: (1) Chemical mass balance and factor analysis methods could not easily resolve the data into more than eight categories; (2) chemical mass balance and ridge regression receptor models were able to resolve sources with similar elemental signatures; however, the effects of similarities in source signatures were manifest in the large uncertainties calculated for those source contributions; (3) source apportionment methods may need to be modified to incorporate chemical transformation of SO_2 and NO to sulfates and nitrates because current models tend to resolve particles into chemical species, rather than deducing their sources; and (4) scanning electron microscopy, optical microscopy, and X-ray diffraction methods are often needed to complement elemental-based receptor models in order to resolve sources having similar signatures.

STATUS OF SOURCE APPORTIONMENT METHODS: QUAIL ROOST II

Introduction

On March 14-17, 1982, the "Mathematical and Empirical Receptor Models: Quail Roost II" Workshop was held in Rougemont, NC. Thirty-five scientists participated. The results of receptor modeling applied to real and simulated data sets were discussed, evaluated, and structured into components that would permit intercomparisons of the models. The participants also were asked to prepare a U.S. Environmental Protection Agency (EPA) report on the state of the art of receptor modeling.

The predecessor of this Quail Roost II Workshop was a March 1980 workshop also held at the Quail Roost Conference Center. At that 2½-day meeting (the "Quail Roost I Receptor Modeling Workshop"), scientists met to discuss the status of mathematical and analytical procedures to determine the natural and man-made sources that contribute to atmospheric particles. The objective was to obtain data and information for an EPA report that would describe the state of the art of receptor modeling (sometimes referred to as "source apportionment modeling"). This report was to contain recommendations for research projects that would advance knowledge in this relatively new area of air pollution science.

The EPA report that resulted from Quail Roost I[1] presented three major recommendations:

1. Develop a simulated air quality and source contribution data set perturbed by random error in order to test predictions of several different types of receptor models against known source contributions.

2. Obtain a real air quality data set to compare the consistency of receptor models and source apportionment procedures currently used by various groups in the United States.

3. Standardize validated models so that they can be readily implemented by Federal, state, and local agencies.

Following the Quail Roost I Workshop, several EPA and National Bureau of Standards (NBS) scientists worked to generate simulated and real air quality data sets consistent with Recommendations 1 and 2. The simulated set consisted of lists of mass and elemental concentrations for 40 sequential simulated fine particle aerosol observations at a single receptor site, and elemental composition profiles for 13 source types from which the observations were generated.

The real data set was derived from an EPA visibility study conducted in Houston, TX, during summer 1980. The data set was based on 18 sequential fine and coarse particle observations at a single field site, and included mass, elemental, and ionic species concentrations from beta gauge, x-ray fluorescence, neutron activation, and wet chemical analyses of the collection filters. In addition, filter samples from the Houston field study were subjected to x-ray diffraction (XRD), light microscopy (LM), and combined automated scanning electron microscopy (SEM)/x-ray fluorescence (XRF) analysis by selected participants.

Although the Houston aerosol was collected on three different filter media in order to permit a wide range of chemical analyses, the sampling protocol was not specifically designed for receptor modeling. For example, the filter mass loadings were a factor of ~2 less than adequate for XRD analysis. Davis of the South Dakota School of Mines and Technology (SDSM) (this volume) addresses this problem. Also, no special effort was made to collect and preserve the ambient nitrates. Ambient nitrates can be reliably measured with a denuder difference sampling system.[2] However, if particle sampling is conducted without these precautions, nitrate concentrations can be underestimated by a factor of 10. In spite of these sampling limitations, the Houston data set was considered to be an adequate and cost-effective basis for the Quail Roost II model intercomparison.

To complement the Houston ambient data, participants were provided with emissions inventories for Houston.[3] Unfortunately, emissions inventories are generally not adequate for receptor modeling as they seldom contain information on the elemental composition of the primary particle emissions or of the soil in the airshed under study. Source profiles specific to Houston, TX were not available. To redress this deficiency, source signatures from other cities[4,5] were provided. The participants were also permitted to use any other available source signatures, although this was not encouraged.

The availability of these data sets thus provided a basis for a second Quail Roost receptor model workshop. The principal objective of the Quail Roost II Workshop was to document the current status of source apportionment methodology by determining, for the simulated data set, how well the different receptor models were able to reproduce the true source contributions, and by comparing, for the real (Houston) data set, the consistency of results from different source apportionment methods when the true source contributions were unknown. A second objective was to determine the number of sources that could be reliably determined using only elemental compositional data and whether sources of similar emission characteristics could be resolved. A third objective was to delineate the weaknesses of receptor models and make recommendations to improve their performance.

The simulated and real data sets and associated source profiles were distributed in advance of the Quail Roost II Workshop on nine-track magnetic tape and in hard copy form. Between the time of receipt and the beginning of the Workshop (5 wk for the real data set, but less than 2 wk for the simulated set), the participating laboratory was expected to apply its particular source apportionment method to either or both data sets and prepare a report that would contain the following information:

1. Method(s) of source apportionment, including any special means for making judgments.

2. Estimates of the numbers and identities of source types.

3. For each source thought to be present, an estimate of its contribution ($\mu g/m^3$) to the individual periods.

4. Uncertainty estimates for each component, together with limits for components believed to be present but not detectable.

Discussion of Results

The key Quail Roost II participants who submitted results for the simulated and real data sets are listed in Table I.

For both simulated and real data sets, the University of Illinois and Harvard University submitted factor analysis results, the Texas Air Control Board submitted multiple linear regression results, and Environmental Research and Technology, Inc., Radian Corporation, and Nuclear/Environmental/Analysis, Inc. submitted results based on chemical mass balance models. EPA submitted factor analysis results for one of the simulated data sets. Table I lists these and other contributions to the analysis.

Simulated Data Set

NBS sent to the participants three simulated data sets. However, only sets I and II were analyzed by the participants, since they had only 2 wk prior to the Workshop to prepare their reports.

Data set I was made up of 40 observations along with elemental source signatures for 13 potential contributing sources. (Only eight of these were in fact active.) Unknown to the participants, an additional unspecified source also contributed to the set I observations. Data set II also consisted of 40 observations. (All 13 known sources contributed.) Each data set contained at least four source groups within which signatures were similar. Sets I and II were designed to determine whether any of the receptor models could resolve sources emitting particles of similar elemental composition, and whether any model could detect and calculate the mass contributed to the receptor site by an unknown source.

None of the conventional chemical mass balance (CMB) methods detected the unknown source, but two of the factor analysis methods and a multilinear regression model were able to deduce the mass contribution of the unknown source. All of the receptor models had difficulty resolving sources with similar signatures. These difficulties created substantial uncertainties in the apportionment of those sources. In a companion paper (this volume), Currie of NBS presents additional details on the development of the simulated data and the participants' results for those data.

Real Data Set

Figure 1 shows that the conventional receptor or factor analysis models could only apportion the fine and coarse into three chemical categories (carbon, nitrate, and sulfate) and four general sources (vehicle exhaust, steel, marine, and crustal). On the other hand, LM and automatic SEM/XRF analysis were able to resolve 18 sources. Even though the sample mass loadings were inadequate for XRD, participants from Lawrence Berkeley Laboratory and SDSM were able to determine that most of the calcium in the coarse particles was of local soil origin based on their identification of a form of $CaCO_3$ indigenous to Houston.

Dzubay et al. (this volume) detail the results of the receptor model intercomparison based on the Houston data.

Considerations in Design of Source Apportionment Studies

A group of participants headed by Gordon of the University of Maryland examined and evaluated the preliminary results presented at the Workshop and prepared reports that contained (1) an assessment of the value of the real and simulated sets, and (2) a series of recommendations that serve as a basis for design of source apportionment studies by private or governmental institutions. Their findings and recommendations, prepared principally by Gordon with input from other Quail Roost II participants, appear in forthcoming report on the Workshop.[7] One of the key issues addressed by this Workshop subgroup was the development of protocols for data collection and analysis appropriate to different levels of source apportionment. The various levels of effort listed below assumes the source apportionment study is being implemented to address the issue of developing air pollution control strategies to reduce particulate mass loadings in an airshed to be in compliance with EPA's air quality standard for particulate matter. Therefore, the site where samples are to be collected and analyzed for receptor modeling are located where particulate mass has been observed to be in violation of the standard. The elements (and/or chemical species) that would be measured as input to the receptor model would vary from airshed to airshed but some species that are typically determined are listed in Table III. The Quail Roost II participants generally recommended that source apportionment studies be divided into three levels of intensity based principally on resource availability:

Level I: Minimal Resources. This effort uses existing source signature libraries, existing ambient air quality data (mass and elemental composition of fine and coarse particles), local emissions inventories, and literature values for soil composition. These data suffice for an initial apportionment by CMB and for factor analysis to determine if there are contributing sources that are not included in the emissions inventory. The cost for this level of effort is between $10-25,000, depending on whether or not additional sample analysis is required. Most of the effort will be spent gathering emission data and inputting data to the receptor model.

Level II: Moderate Resources. This level of effort includes all of the Level I work plus source elemental profiles from grab samples of fugitive sources, soil, and major primary particle emitters in the airshed. These grab samples are resuspended and recollected in the same manner and on the same filter material used to collect particles at the receptor site(s). Most of the expenditures for level II recptor modeling study is in the collection, redeposition, and analysis of grab samples. Level II costs could range from $25-100,000, depending on the size of the airshed under study.

Level III: Maximum Resources. This level of effort calls for (1) collection of aerosols at receptor sites where maximum plume fumigation may occur; (2) collection of particle emissions from the major emission sources with a dilution sampler that is coupled with a sampler similar to those used at the receptor sites; (3) analysis of filters from the sources and receptor sites by XRF and instrumental neutron activation analysis; (4) use of XRD, SEM, and LM to resolve sources with similar emission profiles. This level of effort could cost between $100-500,000, again depending on the size of the airshed under investigation. Most of the costs are in the collection and analyses of source and ambient samples.

The subgroup compiled a list (Table II) of aerosol sampling and gas measurement methods that should be part of any station that collects ambient data for receptor modeling purposes. Figure 2 depicts the aerosol sampling and analysis protocol that would be followed in a complete aerosol apportionment study. Note that at least four different sampling procedures may be necessary to perform reliable source apportionments. Gordon's subgroup further recommended that multivariate and wind trajectory receptor models be used whenever there is a need to obtain source signatures not readily available by conventional source sampling procedures. These models can provide source signatures by their unique treatment of air quality data. However, multivariate statistical models require large data sets (40 or more observations), and wind trajectory methods require short sampling periods (6 hr or less) and appropriate meteorological measurements.

Recommendations for Future Research

Table III lists the marker elements used by the Quail Roost II participants in their various receptor models to report the emission sources contributing to ambient particles impacting our Houston receptor site. The sulfate and nitrate listed in Table III do not, in the absence of appropriate source signatures, have marker elements that reliably indicate their origins. Even analysis of the Houston filters by SEM and LM failed to deduce the origins of the sulfate and nitrate. In most airsheds in the United States, sulfate and nitrate and their associated cations represent a substantial portion of the particle mass below 10 μm. Procedures must be developed to determine the origins of sulfate and nitrate that impact a given receptor site if source apportionment methodology is to adequately address urban and rural particulate problems.

At Quail Roost I and II, no participant presented methods to incorporate within the receptor modeling framework procedures to deal with nonconserved species, particularly sulfates and nitrates. In view of the dominance of these species in urban and rural aerosols, this issue requires resolution. Since Quail Roost II Lewis et al. (1983) have proposed a receptor model theory of secondary species which has the general form:

$$M_j = S_j \cdot R_j \cdot T \qquad (1)$$

where: M_j = mass concentration of secondary particles (e.g., $SO_4^=$) from source j impacting receptor site.

S_j = mass concentration of primary fine particles from source j impacting receptor site. This term is derived from CMB calculations.

R_j = ratio of SO_2 to fine particle emissions measured at source j.

T = transformation/deposition function [i.e., loss of SO_2 enroute from source j to receptor site, due to SO_2 dry deposition and $SO_4^=$ conversion]. T requires only average rate values for these processes.

A key assumption underlying equation (1) is that SO_2, primary fine particles, and secondary sulfate have identical behavior with regard to dispersion, but that SO_2 may experience additional dry deposition loss beyond that for the other two aerosol species. The application of equation (1) requires that back trajectory meteorological data are available.

Conclusions

The Quail Roost II Workshop exercises led to a number of conclusions that enhance our understanding of source apportionment methodologies. Some of the most important are as follows:

1. Multivariate procedures isolated at most seven of the nine sources active in simulated data set I. For simulated data set II, target transformation factor analysis isolated only four of the 13 sources.

2. The simulated data set exercises indicate that neither a multivariate nor CMB approach to source apportionment can be considered the method of choice for all circumstances. While a CMB approach is mathematically easier to implement and does not require a large number of samples, it lacks credibility in circumstances where the sources and their compositions are not well known. On the other hand, while multivariate analysis does not require comprehensive information on source signatures in advance, it does require a very large number of samples, and it appears that only four to six classes of sources can be resolved in a multisource airshed.

3. The unavailability of source profiles from the major emission sources in Houston and the absence of adequate soil samples limited the resolving power of the source apportionment models as applied to real data set.

4. LM and SEM successfully resolved the general categories of the Houston samples into several specific source related categories. The integration of these methods into a mathematical receptor model has not been demonstrated.

5. XRD appears to be a viable alternative to other analytical techniques for identifying certain crustal and combustion sources. However, this alternative was not adequately demonstrated with the Houston data set because special sampling techniques tailored to XRD were not used in collecting those samples.

6. Current models, which tend to resolve particles into chemical categories and not to deduce their sources, need to be modified to incorporate chemical transformation of SO_2 and NO into secondary products.

Recommendations

The major recommendations listed below are based on the Quail Roost II Workshop participants' evaluation of the receptor model intercomparisons and of their past experience with air quality source apportionment studies conducted for state and local agencies.

1. EPA should develop a well documented source apportionment package to assist Federal, state, and local agencies involved in air quality management. This package should include multiple mathematical options for performing source apportionments. The selection of a given option would depend on the resources and ambient and source data that are available. One option, for example, would be to perform factor analysis on receptor samples, followed by source sampling (as indicated by the factor analysis), followed in turn by CMB analysis. Selected ambient samples might then be analyzed by LM as a quality assurance check on the apportionment of the coarse particles.

2. The categories for reporting results from receptor modeling need to be standardized. The source profiles should be given (or referenced to the literature) along with the source intensities derived from them. One of the difficulties in intercomparing the Quail Roost II results was the ambiguity of some of the source categories used by different participants in their analyses of the Houston data set.

3. A source signature library needs to be established for receptor modeling studies. Each signature must be obtained in a receptor-compatible manner. That is, the source sample must be collected in such a way that the properties of the sample at the source and at the receptor are as similar as possible, and the analytical techniques used on the source and receptor samples must be essentially identical. These criteria may be met by dilution sampling of source emissions in which the aerosol is collected on filter media and with collection devices equivalent to those used at the receptor site.

4. Conduct a Quail Roost III Workshop in which the various source apportionment methods are applied to air quality and source signature data obtained by EPA in 1982 for the Philadelphia airshed. The Philadelphia source signature and air quality measurements were obtained by sampling and analysis procedures recommended in this report. Participants of Quail Roost III should be provided with the analytical data at least 9 mo prior to the workshop to allow sufficient time for analysis and report preparation.

5. The results of receptor and dispersion modeling should be assessed in concert to refine and improve the accuracy of both approaches.

Disclaimer

Although the research described in this article has been funded in part by the United States Environmental Protection Agency, it has not been subjected to Agency review and therefore does not necessarily reflect the views of the Agency and no official endorsement should be inferred.

Acknowledgments

The authors wish to thank Northrop Services, Incorporated, for their help, specifically Leslie Silkworth, for coordinating the Quail Roost II Workshop, and Slate Raymond, for the editing and typing of reports prepared by the Workshop attendees. Also thanks to the participants themselves who contributed so freely of their time to the preparation of the various receptor modeling reports cited in our manuscript.

References

1. J. G. Watson et al. "Receptor Models Relating Ambient Suspended Particulate Matter to Sources" EPA-600/2-81-039, U. S. Environmental Protection Agency, Research Triangle Park, NC (1981).

2. R. W. Shaw, R. K. Stevens, J. Bowermaster, J. W. Tesch, E. Tew. "Measurements of atmospheric nitrate and nitric acid: the denuder difference experiment," Atmos. Environ., 16:845 (1982).

3. S. L. Heisler, J. G. Watson, J. J. Shah, J. C. Chow, J. C. Collins, J. Whitney. "Recommendations for the Design of Aerosol Characterization Studies in Houston and El Paso, Texas," P-A716-110, Environmental Research and Technology, Inc., Concord, MA, 1981.

4. J. G. Watson, "Chemical element balance receptor model methodology for assessing the sources of fine and total suspended particulate matter in Portland, Oregon," Ph.D. Dissertation, Oregon Graduate Center, Beaverton, OR (1979).

5. G. R. Cass, G. J. McRae, "Source-Receptor reconciliation of South Coast Air Basin Particulate Air Quality Data," California Institute of Technology, Pasadena, CA, 1977.

6. W. E. Full, R. Ehrlich, J. E. Klovan "Extended Q-model -- objective definition of external end members in the analysis of mixture," Math Geol., 13:331 (1981).

7. R. Stevens, T. Dzubay, C. Lewis, T. Pace, L. Currie, D. Johnson, R. Henry, G. Gordon, B. Davis, et al., "Mathematical and empirical receptor models: Quail Roost II." Atmos. Environ., (In Review).

8. C. W. Lewis (1983) private communication.

TABLE I. PARTICIPANTS WHO SUBMITTED RESULTS FOR THE SIMULATED AND REAL AIR QUALITY DATA SETS

Participant	Method	DATA SET	
		Simulated	Real
Environmental Research and Technology, Inc. (ERT)	Effective variance CMB	X	X
Nuclear/Environmental/Analysis, Inc. (NEA)	Effective variance CMB	X	X
U.S. Environmental Protection Agency (EPA) (Environmental Sciences Research Laboratory (ESRL)	Effective variance CMB/factor analysis	X	
Radian Corporation (RC)	Weighted least-squares CMB (real set) amd weighted ridge regression CMB (both sets)	X	X
Texas Air Control Board (TACB)	Multiple linear regression by tracers/factor analysis	X	X
South Dakota School of Mines and Technology (SDSM)	X-ray diffraction-effective variance CMB		X
State University of New York at Syracuse (SUNY)	Q-mode factor analysis[6] of SEM/XRF data		X
Boeing Aerospace Company, Seattle, WA (BC)	Light microscopy of coarse fraction		X
Lawrence Berkeley Laboratory (LBL)	X-ray diffraction		X
Harvard University (HU)	Weighted CMB/factor analysis	X	
University of Illinois (UI)	Target transformation factor analysis	X	X

TABLE II. SUGGESTED INSTRUMENTATION TO COLLECT AMBIENT DATA FOR RECEPTOR MODELING

Particulate Sampler	Flow Rate (liters/min)	Filter Medium	Species Measured	Method
Dichotomous	17	Teflon	Mass	β gauge
			Elements	XRF, INAA
		Quartz	Carbon	Combustion
			Ionic species (SO_4^{2-}, NO_3^-, etc.)	Ion chromatography/chemistry
		Nuclepore	Individual particles	SEM/XRF
Dichotomous	50	Teflon	Crystals, minerals	XRD, LM
Hi-Vol (with cyclone) 2.5 μm cutpoint	500-1000	Quartz	Organic compounds	gas chromatography/ mass spectrometry, Liquid Liquid Chromatography, etc.

Gas Sampler	Gases Measured	Output
Total sulfur	SO_2 and other sulfur-containing gases	Real time, need data-logging system
NO_x	NO, NO_2	Real time, need data-logging system
O_3	O_3	Real time, need data-logging system
Polyurethane	Organic vapors	Later extraction and analysis in laboratory

Other Instrument	Parameter Measured	Output
Meteorological tower	Wind speed, wind directions, temperature, dew point	Real-time, need data-logging system

TABLE III. SOURCE CATEGORIES AND MARKER ELEMENTS USED IN SOURCE APPORTIONMENT OF HOUSTON AEROSOL

Marker Elements	General Source Category	Specific Source Categories
Na, Cl	Marine	Marine
Al, Si, K, Ca, Mn, Fe	Crustal	Wind erosion of soil Paved and unpaved roads Construction Limestone crushing and handling Cement Lime kiln Fly ash Slag
C, Br, Pb	Transportation	Noncatalyst vehicles Catalyst cars Diesels
Mn, Fe	Steel	Iron and steel production Steel finishing and handling
Cu, Zn, As, Sm, Sb	Nonferrous metals	
NH_4^+, SO_4^{2-}	Sulfate	Primary emissions in Houston airshed Conversion of SO_2 in Houston airshed Regional background
C	Carbonaceous	Refineries Botanical Vegetative burning Tire wear
NO_3^-	Nitrate	Photochemical

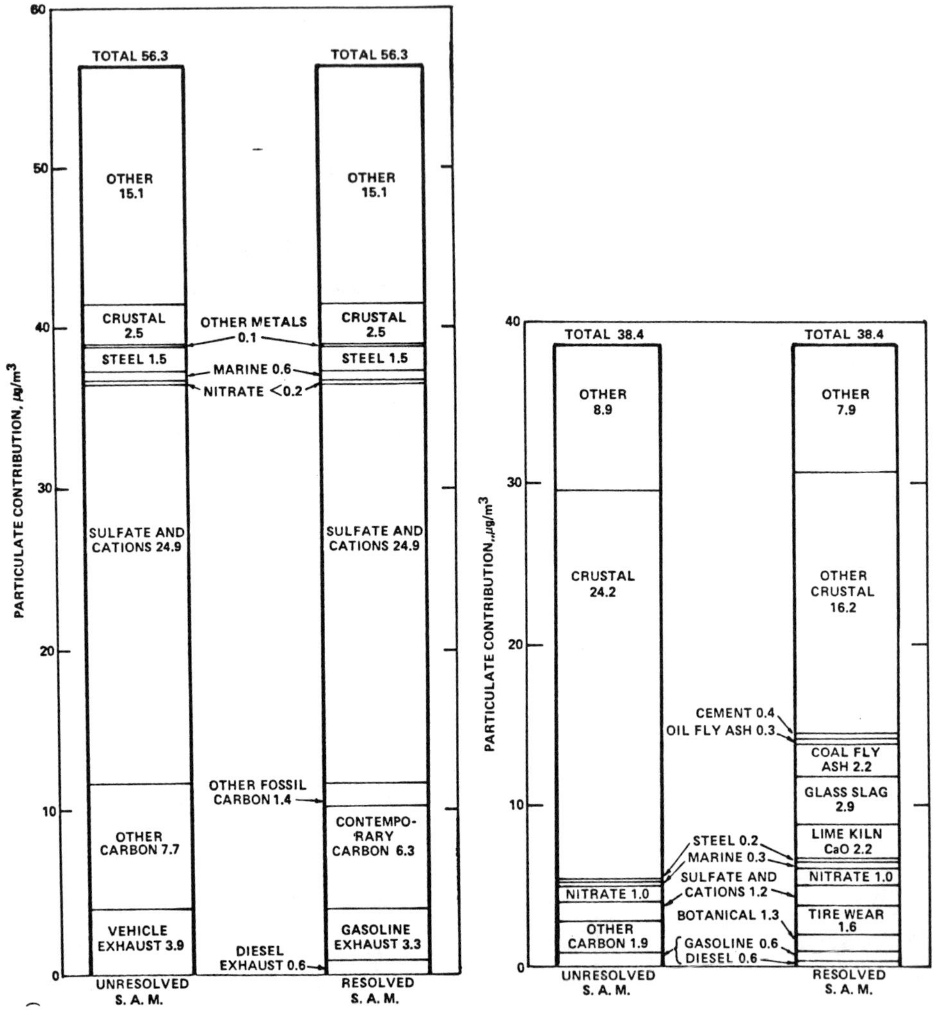

Figure 1. Results of source apportionment of real (Houston) data set. Left graph: fine fraction. Right graph: coarse fraction. S.A.M. = source apportionment method

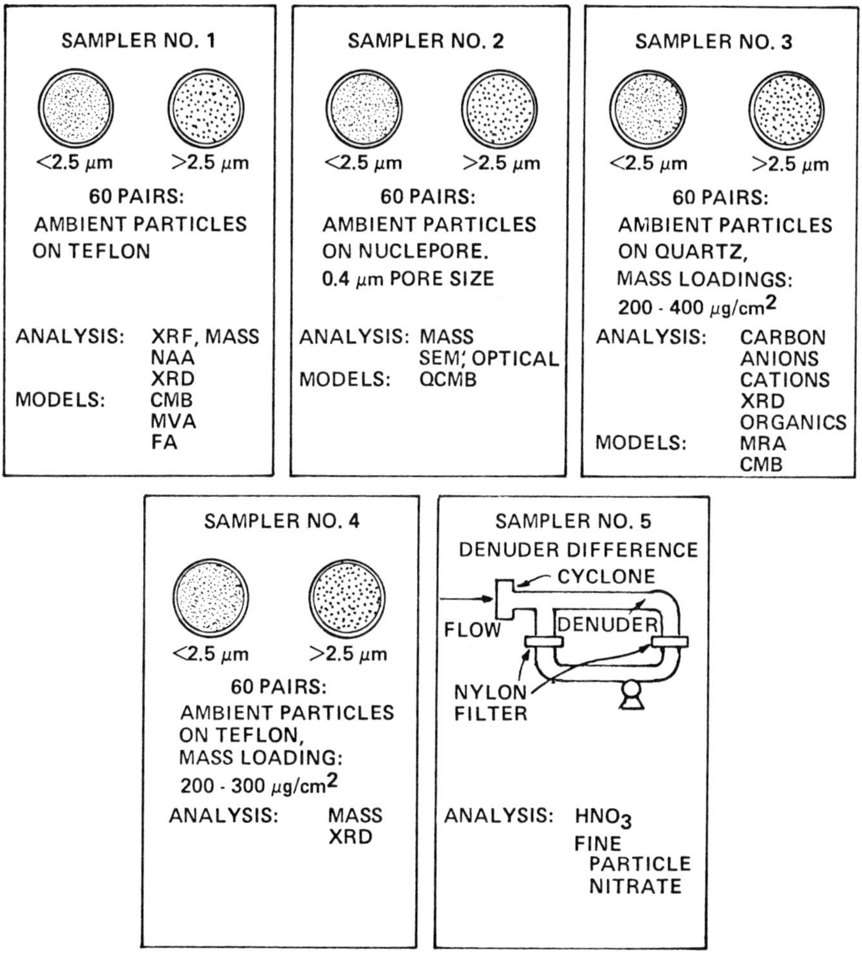

Figure 2. Ambient sampling and analysis protocol for a complete aerosol apportionment study.

INTERCOMPARISON OF RESULTS OF SEVERAL RECEPTOR MODELS FOR APPORTIONING HOUSTON AEROSOL

Thomas G. Dzubay,→

Robert K. Stevens,
Environmental Sciences Research Laboratory,
U. S. Environmental Protection Agency,
Research Triangle Park, North Carolina,

and

J.A. Cooper, R.T. De Cesar, and J.E. Core, NEA, Inc., Beaverton, OR; E.R. Crutcher, Northwest Microlab, Redmond, WA; S. Dattner, Texas Air Control Board, Austin, TX; B.L. Davis, South Dakota School of Mines, Rapid City, SD; P.K. Hopke, University of Illinois, Urbana, IL; D.L. Johnson, SUNY, Syracuse, NY; H.J. Williamson and W.D. Balfour, Radian Corp., Austin, TX; S.L. Heisler and J.J. Shah, Environmental Research and Technology, Concord, MA.

Using a set of data on aerosol composition in Houston, TX, nine participants of the Quail Roost II receptor modeling workshop estimated impacts of major classes of emission sources. The data set included mass, elemental composition, and ionic composition of fine- and coarse- fraction aerosol collected at the University of Houston during 18 12-hr periods from Sept 10 to 19, 1980. Methods used to analyze this data set included chemical mass balance, ridge regression, multiple linear regression, and target transformation factor analysis. Some participants performed additional analyses using x-ray diffraction, scanning electron microscopy, and light microscopy. The number of classes detected ranged from 4 (by multiple linear regression) to 23 (by light microscopy). The results were condensed into eight general categories, intercompared, and found to agree favorably. Differences in results were associated with the use of different source profiles.

INTERCOMPARISON OF RESULTS OF SEVERAL RECEPTOR MODELS FOR APPORTIONING HOUSTON AEROSOL

Introduction

At the recent Quail Roost II Workshop, participants used synthetic and real data sets to evaluate the comparability of several types of receptor models.[1] Here we report the intercomparison based upon the real data set, which consisted of analytical results for ambient aerosol samples collected at the University of Houston in Houston, TX.[2] The sampling site was near grassy areas and campus buildings to the west and south, a large paved parking lot 60 m to the north, and a major freeway 0.9 km to the northeast. Refineries were located to the east 5 km and beyond. Industrial districts were located 5 km to the northeast, east, south, and southwest.

Four dichotomous samplers were operated simultaneously from Sept 10 to 19, 1980, to collect ambient aerosol in the fine (0 to 2.5 μm) and coarse (2.5 to 15 μm aerodynamic equivalent diameter) size ranges. The filters were changed twice daily (at 0600 and 1800 CST). Table I summarizes the analyses of each type of filter. Five weeks before the workshop, participants were given data on mass, carbon, and ionic and elemental composition for 18 consecutive sampling periods. Table II shows the average concentration of each species. For particles smaller than 4 μm, the ratio $^{14}C/^{12}C$ was measured at the National Bureau of Standards and used to deduce the fraction of carbon from fossil fuel combustion.[3]

Filter samples were sent to some participants who performed further analyses. Crystalline species were determined by x-ray diffraction (XRD) analysis at Lawrence Berkeley Laboratory (LBL),[4] and the resulting XRD spectra were also analyzed by Davis at the South Dakota School of Mines (SDSM).[5] Individual particles from each coarse-fraction sample were analyzed by scanning electron microscopy with automated image analysis and x-ray spectrometry (SAX) at the State University of New York (SUNY) at Syracuse.[6] Individual particles from one coarse-fraction sample were analyzed by light microscopy (LM) by Crutcher of Northwest Microlab.[7] The compositions of particles from emission sources in Houston were not available. Thus, it was necessary to use source signatures measured elsewhere; participants were provided with libraries compiled by Watson[8] and Cass and McRae.[9] Also provided were the compositions of three Houston soil and road dust samples, and two Houston emission inventories.

Receptor Model Methods and Results

Except as noted below, participants analyzed the coarse and fine fractions separately. Six participants based their receptor model calculations on elemental and ionic compositional data; one participant used carbon and XRD data; one used SAX data; and one used LM data. No specifications were given for the categories to be used in reporting results. Consequently, the various participants used about 40 different ones. To facilitate comparisons, the participants' results were condensed into eight general categories: marine, crustal, vehicle exhaust, steel, nonferrous metals, sulfate, carbon, and nitrate. The latter three were defined to consist of sulfate, carbon, and nitrate in excess of their amounts in the other five categories. The sulfate category included the related cation, ammonium. Table III presents the participants' results in the eight categories for aerosol collected during daytime, Sept 10.

Source Apportionment Using Ionic and Elemental Composition

Effective Variance Chemical Mass Balance [CMB(EV)]. Shah of Environmental Research and Technology (ERT-A) used the CMB(EV) method developed by Watson.[1,8] A library of about 25 source profiles from Watson[8] and Cass and McRae[9] was used. Included were a profile for noncatalyst car emissions consisting of 21% lead and another for a weighted mixture of car and truck emissions containing 13% lead. For some samples, the weighted mixture was selected to represent the vehicle exhaust category; for others, the noncatalyst car profile was used. Solutions were obtained using six to nine profiles in each of the fine- and coarse-fraction samples. For each sample, fine-fraction sulfate was designated to be either ammonium sulfate or ammonium bisulfate (but not both).

To test for effects of subjectivity in the choices of marker elements and source profiles, Heisler of Environmental Research and Technology (ERT-B), independently followed a procedure similar to that of Shah. For any particular sample, ERT-A and ERT-B did not always use the same set of marker elements and source profiles. For example, ERT-B always used both ammonium sulfate and ammonium bisulfate as profiles for sulfate in the fine fraction, whereas ERT-A always used only one of them. The consequence of this difference was very small. However, the meaning of the vehicle exhaust category is ambiguous because a source profile describing a single vehicle type (noncatalyst cars) was used interchangeably with a profile describing a mixed population of cars and trucks.

De Cesar, Cooper, and Core of NEA, Inc. used a CMB(EV) method similar to that of ERT-A and ERT-B. A library of 33 source profiles, including the three Houston road dust and soil profiles, was used. Most of the other profiles were taken from the Portland Aerosol Characterization Study. Sulfate and ammonium were fit with single-species profiles. The profile selected to represent the vehicle exhaust category in the fine fraction pertained to a weighted mixture of cars and trucks; its lead content was 14%. The number of profiles used in the least-squares fits ranged from six to eight for the fine fraction and from six to nine for the coarse fraction.

Weighted Ridge Regression Chemical Mass Balance [CMB(RR)]. Williamson, Balfour, and Schmidt of Radian Corp. applied both weighted least-squares and weighted ridge regression methods to the Houston data set in a manner described in References 1 and 10. Computations were made using the average daytime and nighttime concentrations measured between 0700 Sept 11 and 0700 Sept 19. The source profiles used in the analysis were taken from Watson[8] and Cass and McRae.[9] The weighting coefficients consisted of uncertainties in the ambient concentrations but not in the source profiles. For both the least-squares and CMB(RR) methods, 18 profiles were used.

Radian's CMB(RR) results for 18 components were less affected by multicollinearity than their least-squares results. The concentrations predicted by CMB(RR) were positive numbers for almost every source, whereas about 35% of the concentrations predicted by the least-squares method were negative. Moreover, the predicted standard errors σ were smaller with CMB(RR) by factors of typically 2 to 5 or more. Nonetheless, the case of 18 profiles was such a severe test of CMB(RR) in terms of multicollinearity that the reported concentrations exceeded 2σ for only two sources in each size fraction. This suggests that optimal performance would have been obtained if fewer than 10 profiles had been used.

Multiple Linear Regression (MLR). Dattner of the Texas Air Control Board (TACB) analyzed Houston aerosol using a backward stepwise unweighted multiple linear regression (MLR) method.[1] Factor analysis indicated four significant multielement components in each size fraction. In each size fraction, two components were determined to be significant by MLR. In the fine fraction, the tracer species for those components were sulfur and bromine; in the coarse fraction, they were silicon and calcium. It was assumed that sulfur and bromine were representative of the sulfate and vehicle exhaust categories, respectively, and that silicon and calcium together represented the general crustal category. The small number of samples in the data set prevented the detection of additional components that significantly affected the variance in mass. At least 40 samples are needed for MLR analysis, but the Houston data set contained only 18.

Target Transformation Factor Analysis (TTFA). Hopke, Severin, and Chang of the University of Illinois (UI) applied TTFA to Houston aerosol.[1,11] Statistical tests suggested the presence of four detectable components in the fine fraction. Their mean concentrations during daytime on Sept 10 and the names given them are: 17.6 $\mu g/m^3$ vehicle exhaust, 35.2 $\mu g/m^3$ sulfate, 0.04 $\mu g/m^3$ smog, and 1.0 $\mu g/m^3$ refinery. Because the sulfate, smog, and refinery components consisted of mixtures of sulfate, nitrate, and carbon, in Table III we present these components as a sum of all three mixtures. In the coarse fraction, five components were detected: their mean concentrations on Sept 10 are shown in Table III with the soil and limestone components combined in the general crustal category. These results do not reveal the full capabilities of TTFA. As in the case of MLR, at least 40 samples are needed, but only 18 samples were provided to workshop participants.

Source Apportionment Using XRD, SAX, and LM Data

Effective Variance Chemical Mass Balance Using XRD Data [CMB(XRD)]. Davis of SDSM used a CMB(EV) method similar to that used by investigators from ERT and NEA.[1,8] However, the only elemental data used were the carbon concentrations. The other data used were XRD results for the following crystalline species: ammonium sulfate, lead ammonium sulfate, quartz, calcite, dolomite, and halite. These XRD results consisted of Davis' analysis of XRD spectra provided by LBL. Davis determined the relative abundances of the crystalline species by applying a self-normalization procedure[5] that requires that the results accounted for the total measured mass minus carbon. Profiles consisting of the abundances of seven species were constructed for the following five sources: vehicle exhaust, combustion, ocean aerosol, soil, and road dust. The combustion profile contained the equivalent of 12.3% lead (expressed as 12.7% lead oxide). The soil and road dust profiles were obtained by XRD analyses at SDSM of the road and soil samples from Houston. The fine and coarse ambient concentrations were combined into a single fraction for which the CMB computations were performed. Results for the combustion category are listed in Table III as the sum of the sulfate and carbon categories. Subsequent to the CMB calculations, the contributions of transportation and combustion to the fine and coarse fractions were made on the basis of their major components (e.g., sulfate in combustion) present in both size fractions.[12] The results reported here do not reveal the full capabilities of CMB(XRD) because the samples were too lightly loaded for complete XRD analysis to be performed.

Q-Mode Analysis of SAX Data. All coarse-fraction ambient samples, as well as source materials used in the apportionment, were characterized by SAX under the same operating conditions to produce "signatures." These signatures consisted of the weight percent contribution by each of 25 particle types. A fixed classification scheme was used for all particles regardless of particle size. The source signatures used in the apportionment covered a variety of area and point source types. Most were obtained in and around Syracuse, NY, but the three Houston soil and road dust samples were included. Using the supplied Houston emissions inventories, 30 source signatures were chosen to represent the Houston area.

All 18 ambient and all 30 source sample signatures were combined in a matrix and subjected to a Q-mode factor analysis[13] with a VARIMAX rotation. The extremes in variable space (25 particle classes) were identified. No Houston ambient samples represented extreme points in variable space after the VARIMAX rotation. This indicated that all important source types were in the data set. "Extreme" samples were those closest to the rotated factor axes and at the outer boundary of the hyperellipsoid. The source apportionment consisted of a vector solution for ambient samples contained within the boundaries of the hyperellipsoid defined by the 16 most extreme signatures. The composition score on the 16 axes was reported as the fractional contribution of each source type to each ambient sample. Products of the fractional contributions and the masses measured by β-gauge were then used to represent the results in units of $\mu g/m^3$.

Light Microscopy (LM). One coarse-fraction ambient sample on a Nuclepore filter was analyzed using LM.[1,7] Particles were first identified in 23 different classes. Then, on the basis of densities assigned to the classes and measured particle dimensions, the sample's composition was constructed. The mass concentration of each particle type was determined by multiplying the sample's total mass concentration by the particle type's fractional mass abundance. Table IV shows the results for ten specific crustal categories; in Table III, these specific categories are combined into the general crustal category. The remaining 13 detected categories were condensed into general categories in the following way: the steel category was used for industrial hematite, rust, iron scale, and magnetite spheres; the sulfate and carbon categories were used for soot, tire wear, charred wood, plant parts, pollen, and secondary industrial particles.

Discussion

Total Mass

When the Houston data set was first obtained, the measured mass was apportioned to the measured species plus the oxygen and hydrogen assumed to relate to them.[2] The amounts that could be accounted for in the fine and coarse fraction were 78 ± 10% and 98 ± 20%, respectively. Because measured species plus assumed oxygen and hydrogen do not account for all fine-fraction mass, the way that mass is used in the calculations will affect the prediction of mass. Three of the investigators (ERT-A, ERT-B, and NEA) did not use mass as an explicit parameter in their CMB(EV) computations. Their results for the fine fraction agree well with each other but account for only 70 to 80% of the measured fine-fraction mass, which is consistent with the fraction accounted for by the measured species.

Both MLR and TTFA used mass explicitly as a parameter fit by least-squares procedures. Consequently, the TTFA results for average mass are in almost exact agreement with the average measured values for both the fine and coarse fractions, and the TTFA results agree fairly closely with the measured value for the particular sampling period shown in Table III. MLR permitted an intercept; its values were -0.1 and -2.5 $\mu g/m^3$ for the fine and coarse fractions, respectively. The intercepts plus predicted component values agree closely with the measured values. CMB(XRD) used mass implicitly in normalizing the abundances of seven species so that their sums in the fine and coarse fractions equaled the total mass. Thus, the CMB(XRD) results agree fairly closely with the total mass.

Vehicle Exhaust Category

Because lead is the most important marker for the vehicle exhaust category, the concentration predicted for this category will be related closely to the abundance of lead in the source profile. The abundances of lead in source profiles used by different participants varied by a factor of 4, causing estimates for the vehicle exhaust category (Table III) to vary by a similar factor. Such variations partly reflect the use of different definitions of the category. For example, for the results shown in Table III ERT-B used a profile for noncatalyst cars and ERT-A used a profile for a mixed population of vehicles, which caused their results to differ by 44%.

In the CMB(EV) analysis performed by NEA and ERT-A, the abundance of lead in the vehicle exhaust profile was 13-14%; this profile was assumed to represent a mixed population of vehicles. Such values differ greatly from the 4 to 5% deduced by MLR and TTFA. The former is based on source measurements made before 1979 and may overestimate the lead content in 1980. Also, the former includes primary emissions only. Because the MLR and TTFA profiles are based on ambient measurements, they may include the actual lead content plus secondary aerosol associated with vehicles. If the vehicle exhaust profile deduced by TTFA could be proven to be correct, then substantial increases in the CMB(EV) estimates for the contributions of vehicles to ambient air would be necessary.

According to Table III, the CMB(XRD) result for the vehicle exhaust category is much lower than the results by other methods. However, such a large difference pertains only to a single 12-hour sampling period on September 10, 1980 and may be anomalous. When all sampling periods between September 10 and 18 were considered, the average result for vehicle exhaust obtained by CMB(XRD) agreed well with the results by other methods.[1]

Crustal Component

Table IV shows how the crustal component, which constitutes at least 70% of the coarse-fraction mass concentration, was resolved into specific components by XRD, SAX, and LM. For quartz and calcite, the XRD data by LBL are in fairly good agreement with the LM results. Because the sample loadings were light, XRD could not detect the clay component reported by LM. The numbers of crustal components determined by the various methods are: ten by LM, eight by SAX, five by XRD, and two by CMB(EV), MLR, TTFA, and CMB(XRD). It is encouraging to note that SAX and LM together resolved about 12 $\mu g/m^3$ or 40% of the crustal component into the following specific source classes: cement, slag, fly ash, industrial calcium oxide and carbonate, lime kiln, and glass. Good agreement existed between LM and SAX for the first three of these specific classes but not for the last three. Only SAX could resolve fly ash into components from coal and oil.

Conclusions

For the eight general categories, the source apportionment results obtained by the different methods show fairly good agreement. Although apportioning aerosols into eight categories is of great value, obtaining categories that are more specific to emission source types would be desirable. The sulfate, nitrate, carbon, and crustal categories were defined by their chemical nature rather than by emission source type. Thus, it is encouraging to note that LM and SAX did successfully resolve the general categories into several specific source related categories. By combining the CMB results with the LM and SAX results, it is possible to resolve the coarse fraction mass into 15 specific categories.

Although the present study provides valuable insight into the relative performances of several receptor models, certain limitations in the study prevented a demonstration of the models' full capabilities. One limitation was the lack of source profiles specific to Houston. Most were obtained in other cities, making their applicability to Houston only an assumption. Three soil and road dust profiles for Houston were provided, but it is not known if these were representative of the mineral dust suspended in Houston's atmosphere. A second limitation was the lack of a protocol for reporting the vehicle exhaust category. This led some participants to include only primary emissions from noncatalyst cars and others to include primary emissions from a weighted mixture of vehicles. A third limitation was that the 18 sampling periods were too few for MLR and TTFA, which require at least 40 samples. A fourth limitation was that the sample loadings were too low for the full capability of XRD to be demonstrated.

Overall, this intercomparison was valuable because it was the first of its kind and provided impetus to better define the data requirements of receptor models. Recently, a program of ambient and source sampling designed to overcome the limitations of the Houston data set was completed in Philadelphia. Ambient samples were collected during 60 12-hr sampling periods between Jul 15 and Aug 13, 1982. Also, several local source signatures and numerous soil samples were obtained. The data will be available by Sept 1983 and may be used in a comprehensive test of receptor models.

Acknowledgment and Disclaimer

We thank W. J. Courtney, J. W. Tesch, M. A. Mason, C. Pressley, C. Owen, and D. Branton of Northrop Services Inc. for providing the mass and compositional data, and O. Radcliff of Texas Air Control Board for obtaining the soil samples. Although the research described in this article has been funded in part by the United States Environmental Protection Agency, it has not been subjected to Agency review and therefore does not necessarily reflect the views of the Agency and no official endorsement should be inferred.

References

1. R. Stevens, T. Dzubay, C. Lewis, L. Currie, D. Johnson, R. Henry, G. Gordon, B. Davis, et al., "Mathematical and Empirical Receptor Models: Quail Roost II," (to be submitted to Atmospheric Environment).

2. T. G. Dzubay, R. K. Stevens, C. W. Lewis, D. Hern, W. J. Courtney, J. W. Tesch, and M. A. Mason, "Visibility and aerosol composition in Houston, Texas," Environ. Sci. Technol., 16: 514 (1982).

3. J. A. Cooper, L. A. Currie, G. A. Klouda, "Assessment of contemporary carbon combustion source contributions to urban air particulate levels using carbon-14 measurements," Environ. Sci. Technol., 15: 1045 (1981).

4. A. C. Thompson, J. M. Jaklevic, B. H. O'Conner, C. M. Morris, "X-ray powder diffraction system for chemical speciation of particulate aerosol samples," Nuclear Instrum. Methods 198:539 (1982).

5. B. L. Davis, "Standardless X-ray diffraction quantitative analysis," Atmos. Environ., 14: 217 (1980).

6. D. L. Johnson, B. McIntyre, L. Fortmann, R. K. Stevens, R. B. Hanna, "A chemical comparison of individual particle analysis and bulk chemical analysis," Scanning Electron Microsc., I: 469 (1981).

7. E. R. Crutcher, "Quail Roost" II: Quantitative Optical Microscopic Analysis," Northwest Microlab Unpublished Report, Redmond Washington, March, 1982.

8. J. G. Watson, "Chemical element balance receptor model methodology for assessing the sources of fine and total suspended particulate matter in Portland, Oregon," Ph.D. Thesis, Oregon Graduate Center, Beaverton, OR (1979).

9. G. R. Cass G. J. McRae, "Source-Receptor Reconciliation of South Coast Air Basin Particulate Air Quality Data," report for California Air Resources Board, 1981.

10. H. J. Williamson, W. D. Balfour, C. E. Schmidt, "Weighted ridge regression a statistical technique for increased source resolution in chemical element balance analysis," Atmos. Environ. (submitted)

11. P. K. Hopke, "The application of factor analysis to urban aerosol source resolution," Atmospheric Aerosol: Source/Air Quality Relationships, ACS Symposium No. 167, Am. Chem. Soc., Washington, DC, 1981, pp 21-49.

12. B. L. Davis, "Source Characterization and Apportionment of the Houston Aerosol by Means of X-Ray Diffraction Techniques," in Proceedings of APCA Specialty Conference on Receptor Models Applied to Contemporary Pollution Problems, Danvers, Mass., October 17-20, 1982.

13. W. E. Full, R. Ehrlich, J. E. Klovan, "Extended Q-model -- objective definition of external end members in the analysis of mixtures," Int. J. Math. Geol., 13: 331 (1981).

TABLE I. SUMMARY OF SAMPLING DEVICES AND ANALYTICAL PROCEDURES USED TO DETERMINE AEROSOL COMPOSITION IN HOUSTON FROM SEPTEMBER 10 TO 19, 1980

Sampler Type	Manufacturer (Model)	Height of Inlet, m	Filter Medium	Species Measured	Analytical Procedures
Dichotomous	Sierra (244)	2	Teflon[a]	NO_3^-, SO_4^{2-} H+ NH_4^+	Ion chromatography Gran titration Colorimetry
Dichotomous	Beckman	4	Teflon[a]	Mass Elemental composition Crystalline species	β-gauge X-ray fluorescence X-ray diffraction
Dichotomous	Beckman	4	Nuclepore[b]	Elemental composition Individual particles	X-ray fluorescence Light microscopy (LM) SAX[d]
Dichotomous	Beckman	4	Quartz[c]	C	Combustion
Hi vol (<4 μm)	Prototype	2	Quartz[c]	C ^{14}C	Combustion β-counter

[a] 1 μm pore size (Membrana Corp.).

[b] 0.3 μm pore size (Nuclepore Corp.).

[c] QAST 2500 (Pallflex Corp.).

[d] Scanning electron microscopy with automated x-ray analysis.

TABLE II. AVERAGE COMPOSITION (ng/m^3) OF HOUSTON AEROSOL IN FINE AND COARSE FRACTIONS FOR 0700 SEPT 10 TO 0700 SEPT 19, 1980[a]

Species	Fine	Coarse
Mass	38600 ± 300	24800 ± 200
Al	123 ± 60	1093 ± 366
Si	210 ± 32	2990 ± 965
S	4834 ± 279	397 ± 430
Cl	32 ± 6	366 ± 111
K	119 ± 6	170 ± 20
Ca	155 ± 8	2775 ± 144
Ti	-8 ± 9	36 ± 12
V	-2 ± 5	-5 ± 6
Cr	4 ± 3	7 ± 3
Mn	14 ± 2	21 ± 3
Fe	162 ± 9	604 ± 34
Co	1 ± 2	0 ± 2
Ni	4 ± 1	4 ± 1
Cu	28 ± 3	18 ± 3
Zn	84 ± 5	58 ± 4
As	2 ± 3	0 ± 2
Se	1 ± 1	0 ± 0
Br	55 ± 3	36 ± 3
Rb	0 ± 1	1 ± 1
Sr	1 ± 1	8 ± 1
Cd	0 ± 2	0 ± 2
Sn	2 ± 3	-1 ± 3
Sb	6 ± 3	-2 ± 3
Ba	48 ± 13	91 ± 16
Pb	465 ± 23	124 ± 10
SO_4^{2-}	14610 ± 1207	907 ± 165
NO_3^-	593 ± 617	1629 ± 235
NH_4^+	4180 ± 379	<190
Na	317 ± 254	278 ± 238
C	7100 ± 467	3100 ± 36

[a] The tabulated results are based upon final β-gauge analysis in Sept 1980, ionic analyses in Sept and Oct 1980, carbon analyses in Oct and Nov 1980, and XRF analyses in Oct 1981.

TABLE III. INTERCOMPARISON OF SOURCE APPORTIONMENT RESULTS ($\mu g/m^3$) FOR DAYTIME, SEPT 10, 1980[a]

Size Range	Category	ERT-A CMB(EV)	ERT-B CMB(EV)	NEA CMB(EV)	TACB MLR	UI TTFA	SDSM CMB(XRD)	SUNY SAX	Microlab LM
Fine	Marine	0.58	0.53						
Fine	Crustal	3.9	1.9	1.7					
Fine	Vehicle Exhaust	3.8	2.3	3.7	11.5	17.6	0.46		
Fine	Steel	1.9	1.8	0.74					
Fine	Metals[b]			0.11					
Fine	Sulfate	23.4	25.6	25.8	41.1		50.8		
Fine	Carbon	6.8	8.8	7.7		36.3			
Fine	Nitrate								
Fine	Total	40.5	40.9	39.8	52.6	53.9	51.3		
Coarse	Marine	0.26	0.23	0.27		0.33	<0.01	0.0	
Coarse	Crustal	17.0	24.3	31.4	39.9	40.3	42.1	28.9	30.6
Coarse	Vehicle Exhaust	0.69	0.42	0.95			0.05	0.3	
Coarse	Steel		0.12	0.33				3.1	3.0
Coarse	Metals[b]			0.04				0.9	
Coarse	Sulfate	1.4	1.0	1.1		2.5		0.6	4.7
Coarse	Carbon	1.1	1.6	3.1			1.1	0.3	
Coarse	Nitrate	1.3	0.57	1.0		0.57			
Coarse	Total	21.6	28.2	38.3	39.9	43.7	43.3	34.1	38.4

[a] The mass concentrations measured in the fine and coarse fractions were 56.3 ± 2.8 and 38.4 ± 2.8 $\mu g/m^3$, respectively.

[b] Nonferrous metals.

TABLE IV. SOURCE APPORTIONMENT OF CRUSTAL MATTER ($\mu g/m^3$) IN THE COARSE FRACTION FOR DAYTIME, SEPT. 10, 1980

Crustal Matter	LBL XRD	SUNY SAX	Microlab LM
Road dust	a	14.5	a
Construction	a	3.9	a
Clay	a	a	10.6
Quartz	4.9	a	4.4
Ankerite	0.75	a	a
Gypsum	0.91	a	0.19
Calcite	4.7	2.0	3.5
Industrial calcium oxide and carbonate	a	a	2.1
Cement	a	0.3	0.46
Lime kiln	a	2.4	a
Slag	a	3.3	2.5
Coal fly ash	a	2.2	a
Oil fly ash	a	0.3	a
Fused fly ash	a	a	2.1
Glass	a	a	1.4
Minerals, misc.	0.19	a	0.38
Total crustal	11.5	28.9	27.6

[a] Not reported.

QUANTITATIVE X-RAY POWDER DIFFRACTION ANALYSIS OF AIR PARTICULATE SAMPLES

Albert C. Thompson,→

L. R. Johnson* and J. M. Jaklevic
Lawrence Berkeley Laboratory
University of California
Berkeley, California 94720 U.S.A.

The design, operation and calibration of an x-ray powder diffraction system for the analysis of air particulate samples are described. The instrument analyzes a sample non-destructively for its major crystalline components. The system has been optimized for the non-destructive analysis of major crystalline species deposited on samples collected with dichotomous air samplers. It includes a position-sensitive detector for increased sensitivity and a small computer for rapid on-line data analysis. A calibration procedure has been developed to give direct quantitative measurements of the primary chemical compounds on the sample.

This system was used to analyze 18 pairs of samples of the Houston aerosol. The results of this analysis when combined with the measurement of the elemental concentrations by XRF analysis give additional discrimination in source receptor models.

*On leave from the Institute of Atmospheric Sciences, South Dakota School of Mines and Technology.

QUANTITATIVE X-RAY POWDER DIFFRACTION ANALYSIS OF AIR PARTICULATE SAMPLES

Introduction

X-ray powder diffraction analysis (XRD) has long been recognized as a powerful method for the chemical characterization of crystalline compounds[1]. This technique is a promising method for the characterization of atmospheric aerosol particles and there are several reports of such applications[2-5]. However the method has not been widely applied principally because the procedures involved in conventional XRD are poorly adapted to the analysis of aerosol deposits on filter substrates. Furthermore, because of the minute quantities of materials available for analysis, measurement times were excessively long and the results were more qualititative than quantitative.

We have recently developed an automated XRD instrument with which it is possible to non-destructively analyze atmospheric aerosol samples for the concentrations of the major crystalline compounds. The system is unique in its use of a position-sensitive wire chamber counter and an on-line data processing system for rapid compound identification and quantification[6].

The instrument is optimized for the analysis of particulate aerosol samples collected with dichotomous air samplers. These air samplers separate the particles into two distinct particle size ranges, 2-15 μm diameter (coarse) and 0-2 μm (fine) particles, before deposition onto two filter substrates. Total mass loadings of up to 300 $\mu g/cm^2$ on the fine fraction filter are possible with a cellulose ester filter substrate before it becomes clogged. The coarse fraction can have somewhat more mass deposited on it. Teflon* substrate material can have mass loadings of up to 200 $\mu g/cm^2$.

The Houston aerosol was monitored at several locations during the second week of September, 1981 and the samples were analyzed by a variety of techniques. As part of the analysis of this aerosol, we have measured the crystalline material on 18 pairs of samples taken during this time on teflon substrates with a dichotomous sampler. Although these samples were acquired primarily for elemental analysis and as a consequence were too lightly loaded ($\simeq$30 $\mu g/cm^2$ total mass loading) for optimum XRD analysis, the results do illustrate the usefulness of this approach in source receptor modeling of the urban aerosol.

Description of Instrument

Figure 1 shows a schematic of the XRD instrument. It consists of a conventional 800 watt copper anode x-ray diffraction tube mounted on a θ-2θ goniometer in a Bragg-Bretano diffractometer geometry. The samples are processed as they are received from the dichotomous sampler with no remounting. Computer controlled slits are used to collimate the x-ray beam onto the sample and to limit the field of view of the x-ray detector to that part of the sample which is irradiated by the x-ray beam.

The usual single slit detector is replaced with a position-sensitive proportional wire chamber which enables the parallel acquisition of data over an extended angular range[7]. The sealed xenon filled detector is sensitive over an angular range of up to 7° in 2θ with a position resolution of 0.24°(FWHM), comparable to that achieved in normal slit diffractometers.

The obvious advantage gained from such a detector is the increased data acquisition rate due to the equivalent multiple detecting element. We observe a 50-fold increase in counting rates compared to a single slit system. A secondary advantage is that the use of the extended detector causes data from the same diffraction angles to be acquired over a range of sample orientations. This results in an averaging process in which many microcrystals are oriented in the proper direction for diffractions resulting in increased sensitivity and some averaging over possible preferred orientations. This can be especially significant for the case of lightly loaded coarse particle filters where the total number of particles contributing to the diffraction signal might be statistically small.

The positions of the detected x-rays are converted to digital form using a time-to-amplitude convertor, transformed into equivalent 2θ values and then stored in the computer memory of a Texas Instruments 9900* microprocessor. This microprocessor also controls the goniometer and slits and inserts samples from a sample tray. The entire system is under the control of a Digital Equipment Corporation LSI II* computer which is also used to perform data analysis on previously acquired spectra while the microprocessor accumulates data from the current sample. This feature increases analytical output since a typical data acquisition interval for an aerosol sample is approximately two to twelve hours.

Data Analysis

The use of an on-line computer greatly facilitates the data analysis by providing rapid and convenient interpretation of the diffraction spectrum which is then followed by immediate conversion into compound concentrations in the aerosol. Since most filter substrates employed in typical air sampling programs contain both interfering lines and continuous scattered intensities[8,9], we subtract from the unknown spectrum a normalized background profile obtained from a blank filter. This both improves the signal to noise and reduces the curvature of the background. If the filter substrate is teflon, the sample spectrum is first shifted (usually less than 0.1 channel) to make the peak centroids of the strong teflon peaks the same as the centroids in the background spectrum. The remaining spectrum is then analyzed for diffraction peaks using a peak fitting subroutine. The computer software has been designed to utilize a color graphics terminal to overlay the peak analysis with the raw spectrum to verify the peak analysis. The resulting list of peak positions and intensities is used as the input data set to a chemical compound search-match routine.

The search-match procedure used in the present system incorporates certain assumptions regarding aerosol composition which greatly simplifies the algorithm. In particular, the library of possible compounds is currently limited to approximately 77 of those most likely encountered in typical aerosols. Many of these compounds are minerals which are commonly measured in aerosol samples. The search-match procedure proceeds by comparing the most prominent lines in the unknown spectrum to the most prominent lines in the stored spectra associated with the standard compounds. A figure of

*Reference to a company or product name does not imply approval or recommendation of the product by the University of California, the U.S. Environmental Protection Agency or the U.S. Department of Energy to the exclusion of others that may be suitable.

merit which incorporates both the quality of the match-up in angle and in the intensities of the respective lines is calculated for each identified compound. The lines corresponding to the compound with the highest figure of merit are subtracted proportionately from the data set and the process is then iterated until no more candidate compounds are matched. Residual diffraction peaks not attributed to the library of compounds are listed in order that additional search-match procedures can be performed if necessary.

Because of the large number of lines associated with a given compound, the relatively poor peak-to-background ratio exhibited by the lightly loaded filters and the possibility of peak overlap from the various compounds on the filter, some ambiguity in compound identification can occur for the less abundant compounds. For this reason, the list of possible compounds must be examined carefully and some judgment must be exercised by the analyst in assessing the likelihood of certain compound identifications. These judgments are based on the number of prominent lines which are identified out of the total possible and the compatibility of the assignment with measured elemental abundances as measured by x-ray fluorescence. The graphics terminal and a digital plotter are used to facilitate this evaluation. The final analysis gives an estimate of the concentrations of compounds found and their associated error. An estimate of the elemental concentrations is also produced for convenient comparison with XRF analysis. A list of the instrumental detectability for compounds which were not observed can also be produced.

Calibration of Instrument

The absolute concentrations of the compounds extracted from the peak list are assigned on the basis of experimentally measured calibration coefficients which relate the counting rate in selected peaks in a spectrum of a prepared standard to the mass concentration of that compound. These standards were prepared from either laboratory reagents or mineral samples. The particles were ground to a fine powder with a ball mill and then suspended in an air flow and collected on a filter substrate using a conventional dichotomous sampler equipped with an inlet which limited the particle size to less than 20 μm. The masses of the prepared standards were determined either by beta gauge or by x-ray fluorescence analysis assuming a stoichiometric ratio to some easily measured element. The fine fraction sample was then analyzed in the same way as the aerosol samples. Asssignment of concentrations to compounds in the unknown spectrum then assumes that the same linear relationship exists between the total diffracted intensity associated with one or more prominent lines in the diffraction and the amount of material present.

Sources of Error

A quantitative analysis of the most probable errors associated with XRD analysis of aerosol samples is a complicated problem. The subtraction of substrate background and the extraction of peak locations and intensities are relatively straightforward procedures and their precision can be estimated assuming Poisson counting statistics and applying well-developed mathematical models. In calculating both the intensity error and the detectability for individual compounds, the background in the original spectrum is used. Similarly, the accuracy of prepared standards and the related calibration factor can be estimated from fundamental principles.

However, significant systematic errors in the analysis can result from the variability of the diffraction patterns obtained from different samples of the same chemical compound or mineral. Although the locations of diffraction peaks are fixed, the relative intensities of peaks within a diffractogram for a particular compound can vary in response to several unpredictable factors which can be associated with differences in source form or sampling methods. The most serious problems occur when the distribution of particle orientations is not completely random, due either to preferential orientation of particles with respect to some axis on the sample, or when the number of particles present in the lightly loaded sample does not represent a statistically random distribution. The potential lack of consistency in relative peak intensities can be compensated by basing the diffraction intensity for a given compound as a sum over several diffraction lines. However, in cases where the quantitative determination of minor constituents depends critically on the subtraction of more intense and possibly interfering patterns, the analytical error associated with this measurement must reflect this uncertainty.

The preferential orientation problem is illustrated in Figure 2. Figure 2a shows the spectrum obtained for a standard sample of mascagnite $(NH_4)_2SO_4$. Figure 2b shows the spectrum obtained from sample number 224302F in the Houston data set. It exhibits a spectrum dominated by components similar to 2a. Figure 2c shows the spectrum for sample 224307F which demonstrates a predominance of lines overlapping the lines of 2a but with totally different relative intensities. The simplest explanation for this effect in terms of XRD invokes a mechanism whereby the mascagnite crystals are in some manner aligned preferentially in a given direction on the filter. This could be explained by transformations which occur after the particles have been collected. One observes that the cases of preferential alignment did tend to predominate in the overnight samples where relative humidity effects could be significant. The search-match algorithum uses two calibration coefficients for regular and aligned components of mascagnite and adds them in the final analysis. Although it is still possible to obtain quantitative estimates for the amount of mascagnite even in the case of preferential alignment, the error is correspondingly increased due to this alignment artifact.

An additional problem can occur for systems in which the individual crystalline domains are too small a size to generate a significant coherent x-ray reflection. In these cases the diffraction peaks can be broadened even to the point where a given compound is not observed.

An additional variance of about 30% arises for some minerals which have different calibration coefficients for samples from different widely separated sites. Some reduction in the error associated with this variance can be achieved for the case of aerosol analysis by obtaining standards representing local sources of the material of interest. This will be especially useful in source receptor modeling where samples can often be collected of sources which will be important in receptor models. However, since this is not always possible, this variance represents a limitation of this technique.

The results of XRD analysis of the 18 coarse fraction Houston samples are presented in Table I. These samples were collected sequentially with a sampling time of 12 hours. The average calcite concentration in these samples is 7.3 $\mu g/m^3$ with very little diurnal variation while the total mass loading varies considerably. The average quartz concentration, however, is about 15% of the total mass loading and varies with it. Although large concentrations of mascagnite were measured on the samples, when the concentrations were corrected for the 10% mixture of fine particles on the sample, the average coarse concentration of mascagnite is only 0.3 $\mu g/m^3$. Figure 3a shows a plot of the concentration of silicon calculated from the concentration of quartz from XRD as a function of the elemental concentration of silicon determined by XRF. Figure 3b is a similar plot for the calcium concentration from XRD calcite as a function of elemental calcium from XRF. This figure shows that in the Houston coarse aerosol, both calcium and silicon are highly correlated with specific mineral species.

The results of XRD analysis of the 18 fine fraction Houston samples are presented in Table II. These samples contained predominantly mascagnite (49%) which unfortunately has a large number of lines which makes identification of other components difficult. The samples were also lightly loaded ($\simeq$ 100 $\mu g/cm^2$ total mass) which also makes further compound identification difficult. There is a large diurnal variation in the mascagnite concentration as well as a large variation in the percentage of mascagnite that exhibits preferred orientation. The reasons for this large diurnal variation are still being studied. The double salt $PbSO_4$-NH_4SO_4 was the other compound found on many samples. This salt accounts for about 70% of the elemental lead concentration. Whether this salt is produced in the aerosol or is the result of conversion on the filter has not yet been determined.

These samples provide interesting information on the Houston aerosol which will need further analysis to determine their significance. The combination of this data with results from other analytical methods performed during the same sampling period of the Houston study should improve the source receptor modeling of this aerosol.

Summary

Although the samples used in the present study had mass loadings less than the 200 $\mu g/cm^2$ required for optimal XRD analysis, we were none the less able to observe measureable concentrations of several compounds of interest in source receptor modeling. Detectabilities of better than 5 $\mu g/cm^2$ can be achieved for many environmentally interesting compounds using this instrument. We have demonstrated that it is possible to achieve convenient, quantitative and non-destructive analysis of particular aerosol samples which require detailed analysis for evaluation of specific pollution sources and their health effects. Although x-ray powder diffraction is not as precise as x-ray fluorescence and is more difficult to perform, the information it provides on the chemical composition of aerosol samples may lead to more accurate apportionment of particulate material in aerosol samples to specific sources and lead to improved pollution control strategies.

Acknowledgments

W. Searles and R. Gatti assisted in the development and calibration of this instrument.

Although the research described in this article has been funded wholly or in part by the U.S. Environmental Protection Agency through an Interagency Agreement No. IAG-AO-89-F-0-013-0 with the U.S. Department of Energy under Contract No. DE-AC03-76SF00098, it has not been subjected to Agency review and therefore does not necessarily reflect the views of the Agency and no official endorsement should be inferred.

References

1. H. P. Klug and L. E. Alexander, X-Ray Diffraction Procedures, John Wiley and Sons, New York. 1974.

2. P. D. E. Biggins and R. M. Harrison, "The Identification of specific chemical compounds in size-fractionated atmospheric particulates collected at roadside sites", Atmos Env. 13: 1213-1216 (1979).

3. B. H. O'Connor and J. M Jaklevic, "Characterization of ambient aerosol particulate samples from the St. Louis area by x-ray powder diffractometry", Atmos. Env. 15: 1681-1690 (1981).

4. B. L. Davis, "Use of x-ray diffraction quantitative analysis in air quality source studies", in: Electron Microscopy and X-Ray Applications to Environmental and Occupational Health Analysis, D. A. Russell, ed., Ann Arbor Science. 1981, Vol. 2, Ch. 7.

5. B. L. Davis, L. R. Johnson, D. T. Griffen, W. R. Phillips, R. K. Stevens and D. Maughan, "Quantitative analysis of Mt. St. Helens ash by x-ray diffraction and x-ray fluorescence spectrometry", J. of App. Meteorology 20: No. 8, 922-933 (1981).

6. A. C. Thompson, J. M. Jaklevic, B. H. O'Connor and C. M Morris, "X-Ray powder diffraction system for chemical speciation of particulate aerosol samples", Nuclear Instruments and Methods 198: 539-546 (1982).

7. H. E. Gobel, "A new method for fast XRPD using a position-sensitive detector", Adv. X-Ray Anal. 22: 255 (1979).

8. B. H. O'Connor and J. M. Jaklevic, "X-Ray diffractometry of airborne particulates deposited on membrane filters", X-Ray Spectrometry 9: 60-65 (1980).

9. B. L. Davis and L. R. Johnson, "On the use of various filter substrates for quantitative particulate analysis by x-ray diffraction", Atmos. Envir. 16: No. 2, 273-282 (1982).

TABLE I XRD ANALYSES FOR COARSE FRACTION ($\mu g/m^3$) HOUSTON SAMPLES

Date/Time	Calcite	Quartz	Mascagnite Meas[a]	Mascagnite Net[b]	Dolomite	Other[d]	Total[c] Mass
9/10/0700	5.7	6.3	2.2	0.0	2.9	--	38.4
9/10/1900	4.8	4.5	2.6	0.0	--	0.3 Ps	22.6
9/11/0700	4.9	4.7	3.6	0.2	--	--	23.2
9/11/1900	3.1	2.2	3.0	1.4	--	--	13.7
9/12/0700	4.0	3.4	4.3	0.0	--	--	17.6
9/12/1900	4.7	3.0	2.4	1.0	1.7	--	18.2
9/13/0700	9.0	5.0	3.6	0.0	1.3	0.6 Mg,0.5 Ru	34.3
9/13/1900	5.8	1.8	1.7	1.1	--	--	9.6
9/14/0700	4.9	2.0	2.1	0.5	--	0.3 Mg	21.2
9/14/1900	9.2	1.6	0.2	0.0	0.5	0.3 He,0.9 Ha	14.5
9/15/0700	9.4	5.4	1.2	0.0	--	0.5 Ps	29.6
9/15/1900	10.1	5.8	1.6	0.7	--	--	29.8
9/16/0700	9.4	4.6	1.6	0.0	--	--	34.8
9/16/1900	6.6	1.2	0.2	0.0	2.2	0.5 Ha	11.9
9/17/0700	11.4	4.2	0.8	0.2	1.0	1.0 Th,0.5 Ps	30.3
9/17/1900	12.9	4.3	--	--	--	1.5 Ha	24.7
9/18/0700	11.5	4.5	1.3	1.3	--	1.2 Ha	26.8
9/18/1900	13.5	4.5	0.3	0.3	--	0.6 Ha	45.3
AVERAGE	7.3	3.8	1.9	0.3			24.8

[a]Measured concentration on filter.

[b]Net concentration after subtraction of 10 fine fraction on filter.

[c]Not measured by XRD; provided by EPA.

[d]Ha = Halite; He = Hematite; Mg = Magnetite; Ps = Ammonium Lead Sulfate
Ru = Rutile; Th = Thenardite.

TABLE II - XRD ANALYSES FOR FINE ($\mu g/m^3$) FRACTION HOUSTON SAMPLES

Date/Time	Mascagnite	% Mascagnite in Preferred Orientation	Lead Ammonium Sulfate	Total[a] Mass
9/10/0700	23.0 ± 7.0	0	1.1 ± .4	56.3
9/10/1900	12.0 ± 4.0	8	1.6 ± .5	60.8
9/11/0700	34.0 ± 10.0	3	--	55.3
9/11/1900	16.0 ± 5.0	54	--	52.9
9/12/0700	46.0 ± 16.0	9	--	46.3
9/12/1900	14.0 ± 5.0	34	0.8 ± .3	45.1
9/13/0700	39.0 ± 11.0	2	0.9 ± .3	73.4
9/13/1900	6.0 ± 2.0	54	0.8 ± .3	28.3
9/14/0700	16.0 ± 5.0	6	1.0 ± .3	47.5
9/14/1900	3.0 ± 1.0	18	--	17.4
9/15/0700	12.0 ± 4.0	0	0.5 ± .2	28.9
9/15/1900	9.0 ± 3.0	24	1.3 ± .4	53.5
9/16/0700	16.0 ± 5.0	3	1.1 ± .4	47.5
9/16/1900	2.0 ± 0.7	40	0.4 ± .2	10.8
9/17/0700	6.0 ± 2.0	9	0.8 ± .3	27.1
9/17/1900	1.5 ± 0.5	29	1.3 ± .4	18.0
9/18/0700	0.5 ± 0.3	100	0.5 ± .2	13.8
9/18/1900	0.3 ± 0.2	100	--	12.6

[a]Not measured by XRD; provided by EPA.

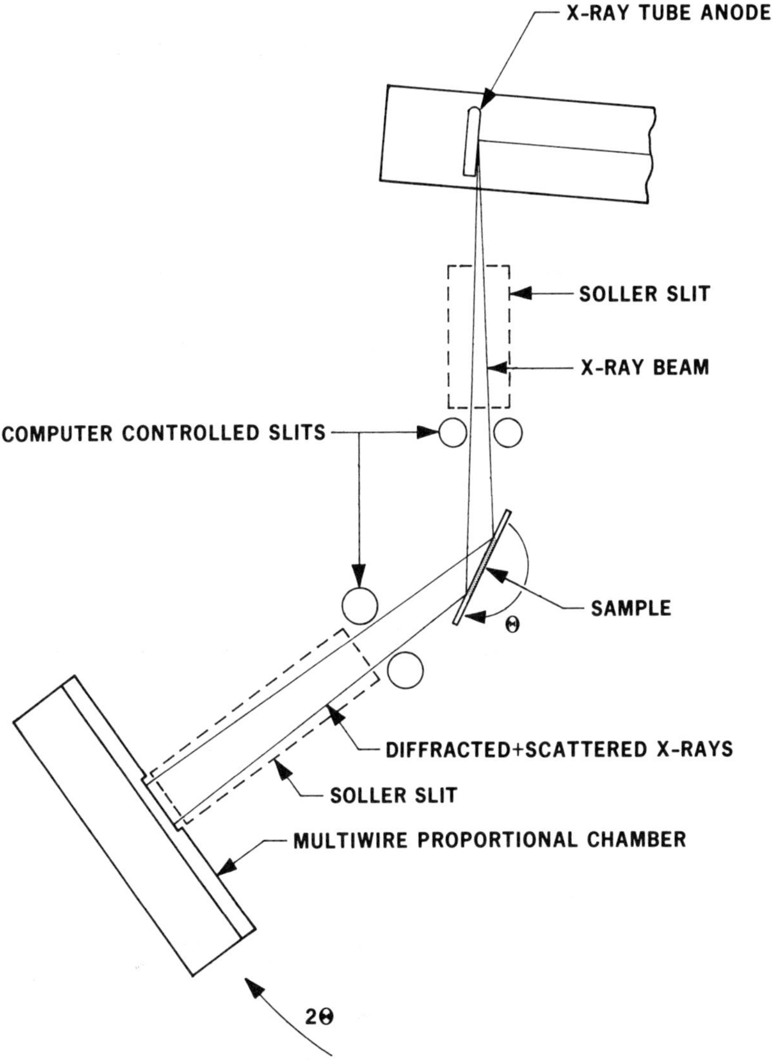

XBL 8110-7197

Fig. 1. General layout of powder diffraction system.

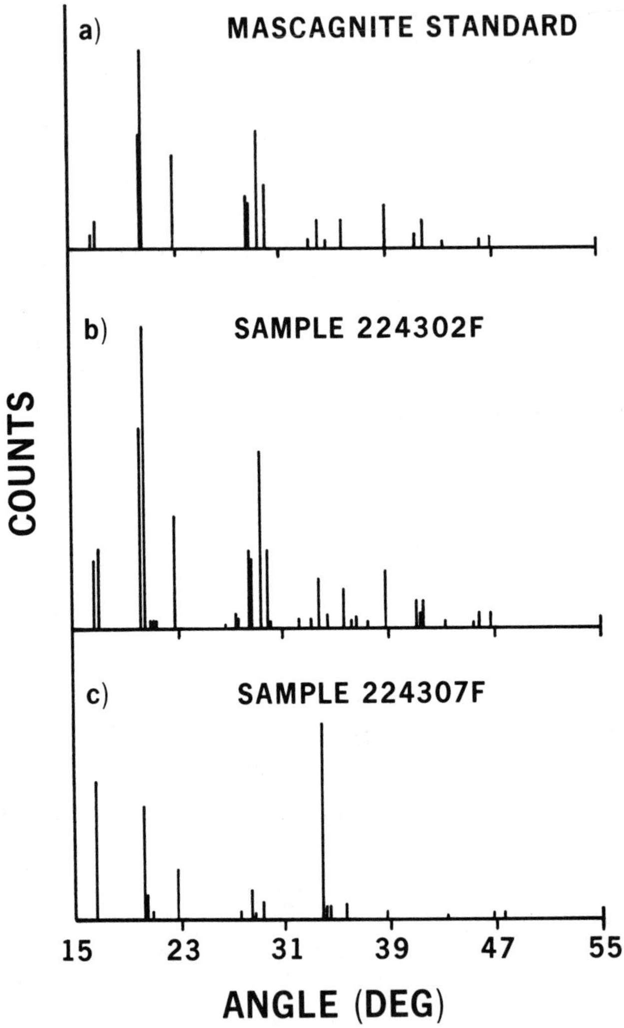

XBL 831-7967

Fig. 2. Analyzed peaks from XRD analysis for a) Mascagnite standard, b) Sample 224302 fine fraction from Houston aerosol, and c) Sample 224307 fine fraction from Houston aerosol.

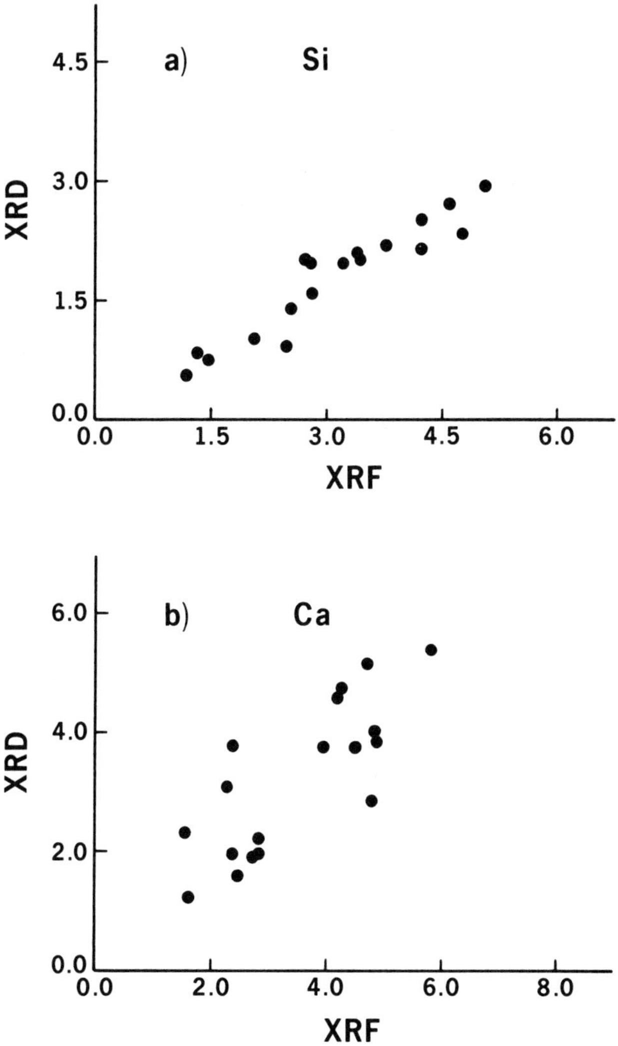

XBL 831-7968

Fig. 3. Concentration in μg/m^3 of coarse fraction elements determined by XRD as a function of concentration as measured by XRF. Fig. 3a is of XRD silicon from quartz and Fig. 3b is of XRD calcium from calcite and dolomite.

SOURCE CHARACTERIZATION AND APPORTIONMENT OF THE HOUSTON AEROSOL BY MEANS OF X-RAY DIFFRACTION TECHNIQUES

Briant L. Davis
Senior Scientist, Cloud Physics Laboratory
Institute of Atmospheric Sciences
South Dakota School of Mines and Technology
Rapid City, South Dakota

Eighteen pairs of dichotomous sampler filters (teflon) used in collecting ambient aerosol at Houston, Texas, from 10-18 September 1980, have been studied by x-ray diffractometry. Calcium carbonate from shell fragments constitute the dominating single component of the crustal source material. Lesser amounts of clays, quartz, feldspars, and organic matter are also present. For the ambient filter analyses the coarse fraction was dominated by calcite and quartz, whereas the fine fraction was composed almost exclusively of ammonium sulfate and lead ammonium sulfate. Sea salt aerosol was observed at times in the coarse fraction. Carbon was observed in both coarse and fine fractions.

A source apportionment of the aerosol was completed by least squares minimization. Source composition data were compiled from emission inventories as well as from the analyzed crustal material. Considering average data for the 9-day sampling period, the apportionment revealed that approximately 1/3 of the total aerosol at the receptor was of crustal origin (soil and street dust), fuel oil combustion contributed about 1/2 of the mass (as fine particles) but only 2.0% as coarse particles, transportation contributed about 6% as fine particles and 1% as coarse particles, and ocean aerosol (coarse NaCl) contributed just over 1% to the total aerosol.

Although none of the filters analyzed contained aerosol loads sufficiently adequate for optimum XRD analysis, these results compared favorably in many aspects with results obtained by scanning electron microscopy, additional x-ray diffraction work at the Lawrence Berkeley Laboratory, and elemental analyses completed by EPA and other groups.

SOURCE CHARACTERIZATION AND APPORTIONMENT OF THE HOUSTON AEROSOL BY MEANS OF X-RAY DIFFRACTION TECHNIQUES

1. Introduction

In this paper we report results which are part of a multi-laboratory cooperative effort to develop source apportionment techniques based on several approaches, including elemental analysis, x-ray diffraction analysis (XRD), scanning electron microscopy, and analysis of simulated data. This program is directed by the U.S. Environmental Protection Agency. The analyses were completed on a set of 18 dichotomous filters taken during the period 10-18 September 1980 on the campus of the University of Houston, Houston, Texas. Although the field protocol was not developed specifically with x-ray diffraction included as an analytical tool, recent improvements in XRD quantitative analysis procedures prompted the U.S. Environmental Protection Agency to include this laboratory (IAS) and the Lawrence Berkeley Laboratories (LBL), as participants in the study. For accurate analysis by x-ray diffraction, approximately 200 $\mu g\ cm^{-2}$ aerosol load should be present on the filter. None of the filters satisfied this condition, although sufficient load was present to justify proceeding with the study so that the capabilities of the technique and future needs could be assessed.

2. Experimental Techniques

Actual x-ray diffraction analysis was completed at the Lawrence Berkeley Laboratories using a position-sensitive x-ray detector with computerized instrument control and data acquisition (Thompson et al.[1]). Analyses of the data were completed at the IAS laboratories (as well as at LBL -- to be reported elsewhere) using the data compilation contained by the LBL instrument. Analyses of the soil materials were made using the IAS x-ray diffractometer, which employs copper radiation, and theta compensator, but which requires manual intensity measurements. Component intensities and appropriate physical data were combined in the reference intensity analysis procedure which has been described elsewhere (Davis[2] and Davis and Johnson[3]). Source apportionment calculations included least squares minimization, according to procedures developed by Watson[4] and applied recently by many other investigators (see the review by Gordon[5]).

The analytical and computational aspects of the source apportionment study has resulted in the accumulation of a large volume of tabular data which cannot be presented here, but which may be reviewed in more detail in the report by Davis[6]. In this paper we present only a portion of the total data set with brief comments on the procedure and results.

3. Results

a. Analysis of ambient filters

Although procedures have now been developed to combine x-ray transmission results with x-ray diffraction to obtain a complete amorphous and crystalline component analysis, x-ray transmission data were not available for this set of samples. Therefore, only direct analysis of the crystalline components could be made. Table I and Table II present the results of the quantitative multi-component analysis of coarse and fine fractions, respectively. The tables also include the analyses for total carbon made by classical combustion procedures at the EPA laboratories. The crystalline components (excluding carbon) sum to 100%. For comparison with other analytical techniques and for source apportionment purposes we have assumed that the carbon material represents the only noncrystalline substance in the sample. This is a reasonable assumption for the Houston study area, although small amounts of fly ash of a siliceous glassy nature, or slag glass, could be present. (These materials have been observed by R. Crutcher for the 9/10/80-0700 samples, using optical microscopy.)

Observation of the magnitude of values in the tables demonstrate that calcite and quartz constitute the major "crustal" components in the coarse fraction, although some ammonium sulfate is also present. It should be noted here that in these analyses a correction for contamination of fine materials in the coarse fraction has been made. Approximately 10% of the fine aerosol may be present in the coarse fraction, which accounts for at least part of the observed "coarse" ammonium sulfate and carbon. These corrections have been made in the data used for the source apportionment calculations described below.

Table II presents the components observed by XRD in the fine fraction of the 18 filter set. Only two components were observed, ammonium sulfate $(NH_4)_2SO_4$ and ammonium lead sulfate $(NH_4)_2 \cdot Pb(SO_4)_2$. One cannot be certain that all of the observed sulfate compounds represent original atmospheric reaction; some portion may represent artifact reaction on the filter. The carbon values in Table II represent measurements by EPA, and may be combined with the components observed by XRD into a total component analysis by simple renormalization.

The EPA steering group for the Houston study has selected two cases, Sep. 10, 0700 and Sep. 13, 0700, along with the average data set for making comparisons between analytical techniques. Table III presents selected elemental ratios for these cases based on analyses completed by x-ray diffraction at the IAS and LBL laboratories, by scanning electron microscopy combined with x-ray energy spectroscopy (SAX) at the State University of New York at Syracuse (SUNY-CESF), and by x-ray fluorescence at EPA. Table IV presents these ratios for the 18 filter set average. Rather large differences, as well as close comparisons, occur in the two individual case days, whereas more general agreement in magnitude occurs for the average set. Most obvious is the low aluminum reduced from the XRD compounds, apparently the result of insensitivity of XRD to the low concentrations of poorly diffracting clay and mica minerals. The low iron content obtained by XRD is also apparent. Fly ash and minor iron oxides are probably the sources in this case. The XRD analysis completed at the LBL was completed by a calibration technique and therefore the differences observed with IAS are not merely the differences in pattern interpretation or intensity measurement but are differences in calibration factors, and reference intensity constants. A comparative study dealing with these difficulties is presently being conducted by the two laboratories.

b. Soil and street analyses

Table V presents the analyses of the three soil and street dust samples collected in the vicinity of the receptor site at the University of Houston. The major component observed here, which is not observed in the ambient filters, is the clay montmorillonite. This is not surprising when considering the structural characteristics of this material, since it is well known that individual montmorillonite crystallites tend to swell and agglomerate in the presence of moisture and resist saltation into the atmosphere. The high calcite component in the Shell road and Wheeler road analyses account for the major amounts of this component observed in the ambient analyses. Also included in Table V are element concentrations in weight percent obtained by stoichiometric reduction of the known compounds in the multi-component analysis.

4. Source Apportionment Computations

On-site sampling of the large number of point sources in the vicinity of the receptor was not available for the apportionment analysis. Because of the inability of the x-ray diffraction analyses to quantitatively evaluate minor components at such light filter loadings, only a few major source categories were included in the apportionment study. These are transportation (F_1), fuel oil combustion (F_2), soil (F_3), streets (F_4), and ocean aerosol (F_5). Table VI presents the source matrix used in the apportionment calculations. These source compositions were obtained from various published inventory source data including a NEDS[7] compilation for Harris County, Texas. The composition for the ocean aerosol was taken from Deer et al.[8], and Holmes[9], and represents the nominal composition for sea water. The following factors and assumptions are noted in the development of the source matrix.

1. Twenty percent of the gaseous hydrocarbon emissions were assumed to be converted to carbon bearing particulate matter;

2. The weight ratio of lead to carbonaceous matter (particulates) of 0.166 was taken as an average of one value calculated directly from carbon and lead emission data and one value quoted in the literature (Dzubay et al.[10]); and

3. Emission factors of 0.66 gram lead and 13.91 gram carbon per liter of gasoline consumed are obtained from the lead and carbon emission data.

The lead sulphates were recalculated to lead oxide for both the ambient and source composition, although in actuality the lead may occur as PbBrCl, or $PbBr_2$, etc. in the atmosphere. The nature of the lead species is irrelevant in the source apportionment calculations providing one is consistent in both source and filter matrices.

The error quantities used in the least source apportionment were obtained from the variance error calculations (Davis[11]) for both the ambient and soil/street dust analyses, and were estimated for the source inventory data by the several methods of computation or of the differing values described in the literature.

Table VII presents the contribution factors (in $\mu g\ m^{-3}$) determined for the two case days and for the 36-filter weighted mean. All data represent the fourth iteration in the effective variance least squares procedure. Table VIII presents the observed component compositions (as used in the source apportionment matrix) and those values calculated from the contribution factors obtained at the fourth iteration.

Contributions from soils (F_3) and streets (F_4) were computed separately, as implied in Table VI; similarities in source compositions resulted in negative values for F_3 in over half of the 18 cases run. This colinearity problem was solved by adding the F_3 and F_4 factors into a single "crustal" source, the resultant of which was a positive number in all cases.

The most dominant feature of these results is the crustal contribution to the coarse fraction and the significant fuel oil combustion contribution to the fine fraction. The analyses for the ambient filters for fine and coarse fractions were combined when completing the source apportionment calculations and then reapportioned into coarse and fine components on the basis of the sulfate for combustion and carbon for transportation. This was necessitated by the fact that only two components were present in the fine fraction of the filter analyses which would not readily lend itself to source apportionment computations. From the average data we observe that approximately 38% of the aerosol observed at the receptor is of crustal origin, whereas 55%, the dominant species overall, derives from combustion sources. A small amount of sea salt is present in a number of samples collected at the end of the sampling period. If one examines the trend of the composition factors from beginning to end of the Houston sampling period, one notices two major segments of aerosol transport which are related to air mass and wind conditions. This is displayed in Figure 1 where wind velocity and resultant 6 hour average wind direction parameter are plotted against sampling time. Also superimposed on these data are the 12-hour values of the ratio F_4/F_2, or the ratio of street contributions to combustion contributions. Also plotted is the extinction coefficient measured during the study as reported by Dzubay _et al._[12]. The hourly data set was provided by Dr. C. W. Lewis, EPA/RTP. The change in air mass correlates with beginning of an upward trend in F_4/F_2 as shown by a change in the wind speed maximum during the afternoons, and shift to a more southeasterly wind direction. The increase in wind velocity signals the dilution of the combustion aerosol trapped over the city and creates a corresponding traffic-induced increase in fugitive dust sources at the sampling site.

5. Conclusions

X-ray diffraction analysis has been applied to the Houston aerosol for compound characterization and source apportionment based on broad source categories. Results show that the major contribution to the aerosol in the coarse fraction is of crustal origin, and the major contribution in the fine fraction from fuel oil combustion. Even though the ambient filters were not loaded with sufficient aerosol mass to achieve optimum XRD sensitivity these results are in reasonable agreement with those of companion studies completed by other techniques. A good correlation of source contribution trends with observed meteorological conditions may be seen over the 9-day sampling interval at the University of Houston sampling site.

6. Acknowledgments

X-ray diffraction analyses for the 18 filter set was completed at the Lawrence Berkeley Laboratories by Dr. J. Jaklevic and Mr. L. R. Johnson. Dr. Stuart Dattner of Texas Air Control Board provided soil and street dust samples. Helpful suggestions from R. K. Stevens, T. Dzubay, and C. W. Lewis are acknowledged. This work was completed under Cooperative Agreement CR806769020 of the U.S. Environmental Protection Agency.

REFERENCES

1. A. C. Thompson, J. M. Jaklevic, B. H. O'Connor, and C. M. Morris, "X-ray powder diffraction system for chemical speciation of particulate aerosol samples," Lawrence Berkeley Laboratory, Preprint 12827, 22 pp. (1981).

2. B. L. Davis, "'Standardless' x-ray diffraction quantitative analysis," Atmos. Environ., 14: 217-220 (1980).

3. B. L. Davis, L. R. Johnson, "Sample preparation and methodology for x-ray quantitative analysis of thin aerosol layers deposited on glass fiber and membrane filters," Advances in X-ray Analysis, 25: 295-300 (1982).

4. J. G. Watson, "Chemical element balance receptor model methodology for assessing the source of fine and total suspended particulate matter in Portland, Oregon," Ph.D. dissertation, Oregon Graduate Center, Beaverton, Oregon. (1979).

5. G. E. Gordon, "Receptor models," Environ. Sci. and Tech., 14: 792-800 (1980).

6. B. L. Davis, "A study of the errors in x-ray quantitative analysis procedures for aerosols collected on filter media. Atmos. Environ., 15: 291-296 (1981).

7. National Emission Data System, "U.S. Environmental Protection Agency, report for Harris County, Texas, 21 July 1981 (1981).

8. W. A. Deer, R. A. Howie, J. Zussman, Rock Forming Minerals Volume 5 (Non-silicates). John Wiley and Sons, New York. 371 pp. (1963).

9. A. Holmes, Principles of Physical Geology. The Ronald Press Company, New York. 1288 pp. (1965).

10. T. G. Dzubay, R. K. Stevens, W. J. Courtney, E. A. Drane, "Chemical element balance analysis of Denver aerosol," In Electron Microscopy and X-ray Applications to Environmental and Occupational Health Analyses, Volume 2. P. A. Russell, ed. Ann Arbor Science Press, Ann Arbor, pp. 23-42. (1981).

11. B. L. Davis, "A study of the errors in x-ray quantitative analysis procedures for aerosols collected on filter media," Atmos. Environ., 15: 291-296 (1981).

12. T. G. Dzubay, R. K. Stevens, C. W. Lewis, B. Hern, W. J. Courtney, J. W. Tesch, M. A. Mason, "Visibility and aerosol composition in Houston, Texas," unpublished manuscript (1982).

TABLE I

Quantitative Component Analyses-Coarse Fraction*

Weight Percent (Variance Error)

	SEP 10		SEP 11		SEP 12		SEP 13	
	0700	1900	0700	1900	0700	1900	0700	1900
Mascagnite $(NH_4)_2SO_4$	8.5 (2.2)	22.6 (4.9)	24.8 (5.2)	41.7 (6.8)	34.0 (6.3)	42.4 (6.8)	23.2 (5.0)	24.6 (5.2)
Quartz SiO_2	43.2 (3.3)	38.1 (3.7)	35.1 (3.7)	27.2 (3.9)	32.0 (4.0)	21.2 (3.2)	38.0 (3.8)	20.8 (2.5)
Calcite $CaCO_3$	48.3 (3.1)	28.1 (2.5)	21.1 (2.1)	25.5 (3.3)	34.0 (3.7)	26.6 (3.4)	38.8 (3.4)	54.6 (4.2)
Dolomite $(Ca,Mg)CO_3$	--	11.3 (1.6)	19.0 (2.6)	5.6 (1.8)	--	9.8 (1.7)	--	--
Halite $NaC\ell$	--	--	--	--	--	--	--	--
Carbon (TOTAL)	10.2 (1.4)	15.5 (2.4)	12.5 (2.2)	16.8 (3.8)	14.2 (3.0)	17.0 (2.9)	10.2 (1.5)	20.8 (5.4)
$\check{M}_B$ (μgm cm^{-2})	80	52.9	52.9	34.9	40.9	42.0	76.9	22.9
$\check{M}_V^B$ (μg m^{-3})	38.4	22.6	23.2	13.7	17.6	18.2	34.3	9.6

Weight Percent (Variance Error)

	SEP 14		SEP 15		SEP 16		SEP 17		SEP 18	
	0700	1900	0700	1900	0700	1900	0700	1900	0700	1900
Mascagnite $(NH_4)_2SO_4$=	12.8 (3.1)	19.6 (4.4)	--	10.3 (2.6)	6.8 (1.8)	--	--	--	--	--
Quartz SiO_2	27.9 (2.8)	26.3 (3.0)	42.1 (3.2)	36.9 (3.1)	34.9 (3.0)	8.6 (1.0)	35.0 (3.0)	19.2 (2.0)	23.1 (2.3)	22.4 (2.3)
Calcite $CaCO_3$	59.3 (3.2)	17.8 (1.7)	57.9 (3.2)	48.4 (3.0)	58.3 (3.0)	48.9 (2.4)	65.0 (3.0)	53.2 (2.3)	46.8 (2.4)	60.5 (2.6)
Dolomite $(Ca,Mg)CO_3$	--	36.3 (3.9)	--	4.4 (0.6)	--	18.4 (2.2)	--	9.4 (1.2)	6.9 (0.9)	17.2 (2.1)
Halite $NaC\ell$	--	--	--	--	--	24.0 (2.1)	--	18.2 (1.7)	23.2 (2.0)	--
Carbon (TOTAL)	11.9 (2.5)	16.6 (3.6)	10.8 (1.8)	12.8 (1.7)	10.3 (1.5)	20.2 (4.4)	9.2 (1.7)	14.2 (2.1)	13.4 (1.9)	9.7 (1.2)
$\check{M}_B$ (μgm cm^{-2})	48.0	30.0	60.0	64.9	72.9	24.0	60.9	48.9	52.0	86.0
$\check{M}_V^B$ (μg m^{-3})	21.2	14.5	29.6	29.8	34.8	11.9	30.3	24.7	26.8	45.3

*Analysis of all components (except carbon) completed by x-ray diffraction (RIM). Sum of all components except carbon equals 100%. Carbon analyses provided by EPA (performed by classical combustion procedures) are given as percent of total mass.

TABLE II

Quantitative Component Analysis - Fine Fraction*

		Weight Percent (Variance Error)					
		$(NH_4)_2SO_4$		$(NH_4)_2Pb(SO_4)_2$		Carbon	$\breve{M}_B$ ($\mu g\ cm^{-2}$)
SEP 10	0700	100		--		18.3	92.0
	1900	85.5	(13.2)	14.5	(13.2)	18.1	100.9
SEP 11	0700	89.3	(7.7)	10.7	(7.7)	15.6	92.0
	1900	99.5	(0.4)	0.5	(0.4)	11.5	88.0
SEP 12	0700	80.9	(13.4)	19.1	(13.4)	11.2	76.9
	1900	100		--		14.2	74.9
SEP 13	0700	83.5	(10.3)	16.5	(10.3)	14.9	122.0
	1900	78.8	(14.9)	21.2	(14.9)	19.1	46.9
SEP 14	0700	98.0	(1.5)	2.0	(1.5)	16.0	78.9
	1900	100		--		27.6	28.9
SEP 15	0700	100		--		27.0	48.0
	1900	99.3	(0.5)	0.7	(0.5)	20.9	89.0
SEP 16	0700	94.8	(3.8)	5.2	(3.8)	20.0	78.9
	1900	94.8	(11.2)	5.2	(11.2)	35.8	18.0
SEP 17	0700	90.4	(6.9)	9.6	(6.9)	25.5	44.9
	1900	100		--		24.4	30.0
SEP 18	0700	100		--		30.4	22.9
	1900	100		--		23.0	20.9

*Analysis of all components (except carbon) completed by x-ray diffraction (RIM). Sum of all components except carbon equals 100%. Carbon analyses provided by EPA (performed by classical combustion procedures) given as percent of total mass.

TABLE III

Selected Elemental Ratios Calculated for Two Case Days of the Houston Study, A, 9/10-0700; B, 9/13-0700, Coarse Fraction

METHOD	$\frac{Ca}{Si}$		$\frac{Al}{Si}$		$\frac{Fe}{Si}$		$\frac{Mg}{Si}$		$\frac{S}{Si}$	
	A	B	A	B	A	B	A	B	A	B
SAX	1.06	2.02	0.13	0.25	0.06	0.36	0.03	0.01	0.10	0.15
XRD (IAS)	0.92	0.84	0	0	0.02	0.02	0	0.003	0.03	0.18
XRD (LBL)	0.93	1.59	0	0	0.10	0.23	0.02	0	0.07	0
XRF	0.78	0.84	0.35	0.37	0.27	0.24	--	--	0.09	0.14

TABLE IV

Selected Elemental Ratios, averaged for the Houston 18-filter set, coarse fraction

METHOD	$\frac{Ca}{Si}$	$\frac{Al}{Si}$	$\frac{Fe}{Si}$	$\frac{Mg}{Si}$	$\frac{S}{Si}$
SAX	1.77	0.19	0.30	0.02	0.10
XRD (IAS)	1.52	0.04	0.04	0.08	0.13
XRD (LBL)	1.94	0.02	0.20	0.03	0.11
XRF	0.93	0.37	0.20	--	0.13

TABLE V

Receptor Site Soil and Street Dust Analyses
(Compounds and Reduction to Elements)

Component	Weight Percent		
	Shell Rd. (UHS-1)	Library (UHS-2)	Wheeler Rd. (UHS-3)
Kaolinite	0	0.8	2.9
Montmorillonite	13.6	34.4	19.0
Quartz	33.8	41.5	37.7
Microcline	0.9	2.3	3.0
Andesine	0	1.7	1.2
Calcite	50.5	18.8	33.9
Dolomite	0	0.7	2.4
Organic Matter	1.2	0	0
ELEMENT:			
Si	20.0	30.6	24.9
Al	1.7	4.7	3.2
Na	0.3	0.8	0.5
K	0.1	0.6	0.6
Ca	19.5	7.7	13.7
Fe	0.7	1.0	0.8
Ni	0.5	0.2	0.4
Mg	0.3	0.7	0.6
Mn	0.4	0.2	0.3
C(CO_3)	6.1	2.3	4.4
C(Free)	1.2	0	0
H_2O	0.8	2.1	1.5

TABLE VI

Source Compositions

Component	F_1 Transportation	F_2 Combustion	F_3 Soil	F_4 Streets	F_5 Ocean Aerosol
	Weight Percent (Variance Error)				
Quartz SiO_2	0	0	68.6 (4.7)	45.5 (4.7)	0
Calcite $CaCO_3$	0	0	30.4 (2.5)	52.6 (4.7)	0.7 (1.1)
Dolomite $(Ca,Mg)CO_3$	0	0	1.0 (0.1)	1.4 (0.2)	0
Halite NaCl	0	0	0	0	90.4 (9.0)
PbO	12.7 (4.1)	0	0	0	0
$SO_4^{=}$	6.4 (6.2)	72.3 (8.7)	0	0	8.9 (8.9)
Carbon	80.9 (2.2)	27.8 (8.7)	0	0.5 (5.6)	0

TABLE VII

Contribution Factors for September 10 (0700), September 13 (0700), and for the 9-day (36-filter) Sampling Period

Date/ Time	Units	F_1 Transportation Coarse	F_1 Transportation Fine	F_2 Combustion Coarse	F_2 Combustion Fine	F_3+F_4 Crustal Coarse	F_5 Ocean Coarse	Total
	%	0.05	0.5	1.2	53.7	44.5	0	100.0
9/10/80	μg cm^{-2}	0.07	1.0	1.6	71.4	59.1	0	133.2
0700	μg m^{-3}	0.05	0.5	1.1	50.8	42.1	0	94.6
	%	1.1	7.2	3.5	57.4	31.6	<0.01	100.8
9/13/80	μg cm^{-2}	1.1	7.2	3.5	57.3	31.5	<0.01	100.6
0700	μg m^{-3}	1.2	7.7	3.8	61.8	34.1	<0.01	108.6
Mean	%	0.8	5.6	2.0	55.4	38.3	1.3	103.4
36	μg cm^{-2}	1.0	6.5	2.4	64.8	44.7	1.5	120.9
Filters	μg m^{-3}	0.5	3.6	1.3	35.2	24.3	0.8	65.7

TABLE VIII

Observed and Calculated Component Concentrations for Case-day Filters and 36-filter Mean

$\mu g\ m^{-3}$

Component	9/10/80-0700		9/13/80-0700		Mean 36-filters	
	Obs.	Calc.	Obs.	Calc.	Obs.	Calc.
Quartz	19.6	19.6	16.6	16.6	8.5	9.3
Calcite	21.9	21.8	17.0	16.9	13.4	14.4
Dolomite	0	0.6	0	0.5	2.4	0.4
Halite	0	0	0	0.004	1.3	0.7
PbO	0	0.06	5.6	1.1	1.1	0.5
$SO_4^=$	37.4	37.6	52.1	48.0	24.7	26.7
Carbon	15.8	15.1	16.5	25.6	12.0	13.2
Totals	94.7	94.8	107.8	108.7	63.4	65.2

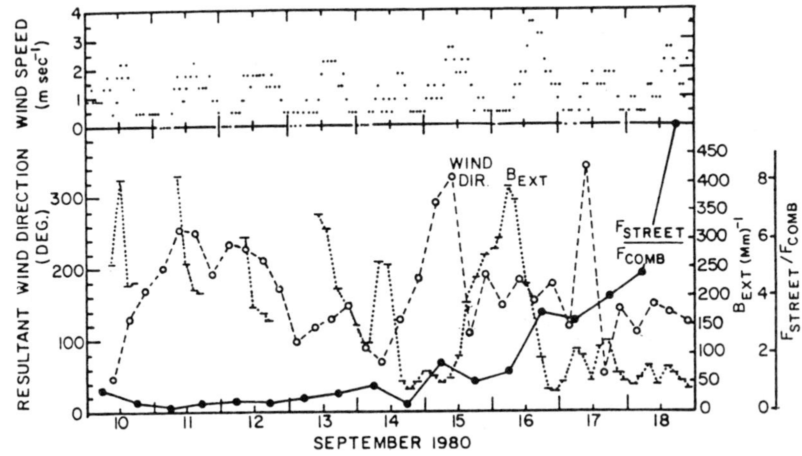

Figure 1. Variation of wind speed, resultant wind direction, extinction coefficient, and source contribution ratio F_{STREET}/F_{COMB} through the sampling period Sep 10-18, 1980. Site is University of Houston campus. B_{EXT} values earlier than 0900, Sep 14, contain ±15% normalization uncertainty.

REVIEW OF THE QUAIL ROOST II RECEPTOR MODEL SIMULATION EXERCISE

R. W. Gerlach and L. A. Currie
National Bureau of Standards, Washington, D.C. 20234

C. W. Lewis
Environmental Sciences Research Lab, U.S. Environmental Protection Agency, Research Triangle Park, N.C. 27711

As a principal component of the Quail Roost II Receptor Model Workshop, 3 sets of ambient aerosol compositional data were generated by computer simulation and distributed to several participants as a preliminary exercise in receptor model intercomparison and validation. Objectives of the exercise were to judge: (a) the resolving power and accuracy of alternative source apportionment methods, (b) the meaningfulness of predicted vs. actual uncertainties, and (c) the adequacy of the preliminary data set for defining the performance of candidate methods. Generation of the data was accomplished with the RAM dispersion model using real meteorological data, reasonable chemical profiles for up to 13 source types, and random profile and measurement errors for 19 elements and one isotope ratio ($^{14}C/^{12}C$). Geographic source placement and emissions rates were adjusted to yield an interesting level of complexity at the receptor site. The exercise proved useful in indicating strengths and weaknesses of both the simulation process and the various deconvolution methods. Factor Analysis and Chemical Mass Balance approaches proved complementary, and indicated for these data sets that 6 to 9 sources could be resolved with results within about a factor of two of the true values. Limitations in the analysis of the data related primarily to resolving power and the statement of uncertainty. One important outcome to the exercise was a set of 10 recommendations for future simulations.

REVIEW OF THE QUAIL ROOST II RECEPTOR MODEL SIMULATION EXERCISE

INTRODUCTION

Within the context of Source Apportionment Analysis a simulated data set is a collection of mathematically generated numbers which are intended to represent a set of ambient air quality measurements, typically aerosol elemental concentrations. In contrast to a "real" data set resulting from actual field measurements, however, in a simulated data set the impacts of the contributing sources are known exactly from a knowledge of the manner in which the simulated data set was generated. It is the existence of this absolute standard of truth against which the predictions of various receptor models may be compared that makes the use of simulated data sets such a potentially powerful (though little used) method to evaluate the performance of alternative mathematical source apportionment techniques. In the following text, a summary is given of the generation of the simulated data sets and the results of the intercomparison. A more detailed exposition, together with compilations of source apportionment results and descriptions of the participants' numerical methods are presented in Ref. [1].

GENERATION OF SIMULATED DATA SETS

Three data sets were constructed as outlined in Table I. Each source profile consisted of 19 elements and the isotope ratio $^{14}C/^{12}C$ (not provided in data set I), with concentrations representative of a fine particle aerosol component. Using contrived source geographies (see Figures 1 and 2) and the RAM Gaussian-plume-dispersion model Version 80336 [2], ambient concentrations of each species were generated for each of forty 12-hour averaging periods. All simulations were run with real meteorological data obtained from the National Climatic Center (Asheville, North Carolina). Starting with Julian day 249 of 1976, hourly surface data from St. Louis/Lambert, Missouri were combined with twice-daily mixing height information from Salem, Illinois in the meteorological preprocessing program RAMMET [3] to obtain hourly wind speeds, wind directions, stability classes, mixing heights, and air temperatures. These output quantities were then employed in the generation of ambient concentrations by the dispersion model. The original unprocessed meteorological information was provided to the participants. An additional component of source variation in the model, not provided to participants, resulted from the use of hourly source strength fluctuations due to daily operating schedules created for each classification of sources. For example, type A and B steel plants followed the same diurnal pattern based on a weekly schedule.

Expressing the generating equation from Table I in matrix form,

$$C = AS + e \qquad (1)$$

the vector of "observed" concentrations (ng/m^3) is represented by C, the matrix of source profiles (mg/g) by A, and the source contribution vector (μg/m^3) by S. As Table I indicates, each data set included normally-distributed measurement errors for the individual observations (e_i, random) and for the source profiles (e_m, systematic). Set III also included log-normal profile errors (e_H, random) for each of the sources, *i.e.*, the profiles themselves changed from one period to the next. In all cases participants were informed of the distributions and standard deviations of the error components. A "world list" of 13 source profiles also was provided. All sources were active in Sets II and III, but only 8 were active in Set I. In addition, Set I included one source type with a sandblast profile, of which the participants were not informed. The simulation data sets thus had a well-defined structure in terms of the linear model and error terms. The sets were realistic because the transport and mixing of emissions (from source to receptor) was accomplished via dispersion, and random and systematic errors were added in accordance with typical field values. (e_H errors, however, were chosen somewhat smaller than what realism would dictate.) The data and model did assume that all species were conserved from source to receptor.

The difficulty of the data sets increased in the order I → II → III, because of the larger number of active components in the latter two data sets and especially because of the considerable log-normal variability in source profiles in Set III. In many respects, however, Set I was the most interesting because participants were not told how many sources from the world list were active, and a totally unknown source was also present. In summary, the simulation data sets were believed to comprise fairly realistic but well-controlled sets of test data for reliable evaluation of the precision and accuracy of the final data evaluation step of the source apportionment process.

To illustrate the general nature and level of difficulty of the exercise, a few highlights for Set I are given in Figure 3. Here, the 13 source types given to the participants are indicated in the first column, together with two sources unknown to the participants. The first unknown type, "road," was simply a linear combination of "auto" (75%) and "soil" (25%), used to represent the reentrainment of road dust by motor vehicles, for all three data sets. While soil was also used as an independent source in the data set generation, auto was not. Thus, the auto source intensity was simply 75% of the road intensity, whereas the total soil intensity was the sum of 25% of the road intensity and a quantity independent of the road source. "Sandblast" was a source totally unknown to the participants. Average concentrations of the nine active sources at the (single) receptor, given in the second column of Fig. 3, were adjusted in the data generation step so that a range of difficulty (or detectability) would be represented, as indicated by the S_j/σ_{Sj} bar graph. The theoretical detection limit of 3.29 [4] is indicated by the arrow on the horizontal axis. The weakest source, steel-A, was well below the detection limit, whereas several sources were easily detectable — including the unknown sandblast — since the true average concentrations exceeded the standard errors (which would result from weighted least-squares deconvolution) by many-fold.

Figure 4 depicts the range of variation over the 40 sampling periods for the intensity of the incinerator source at the receptor site in Set I. For this typical source, a large fraction of the intensities are seen to exceed the (approximate, weighted least-squares) detection limit.

RESULTS OF THE INTERCOMPARISON

Tables II and III give the average source strengths determined by the Workshop participants from their analyses of simulated data Sets I and II. No results for Set III were reported by the participants, due partly to the exceedingly short time ($\leq$ 2 weeks) between the receipt of the data and the beginning of the Workshop. Table IV identifies the individual participants and the codes for their numerical methods.

In presenting the results in an understandable and fair manner, two complications must be addressed. The first relates to the difference between the Chemical Mass Balance (CMB) and factor analysis approaches to source apportionment. The former uses a supplied set of source profiles, while the latter generates its own set from the variability of the ambient data. In general, the two sets will not agree exactly and judgment is needed to make a correspondence between the two. In the case of Set I, factor analysis was part of the approach used by every participant, and there was fairly good agreement as to the number and nature of the sources found. Accordingly, for Set I the supplied sources were grouped in a manner to best correspond to the source categories found from the factor analyses. The groupings are indicated in Table II. The problem does not arise with Set II, since only CMB analyses are given for that Set. Even with CMB analysis, however, the question of the number of distinguishable sources is pertinent, and the use of the Singular Value Decomposition (Ref. 5) should be quite relevant.

The second complication arises from the use of the "road" source in the generation of all three data sets, with this source being a fixed linear combination of the "auto" and "soil" sources. A factor analysis approach is by nature sensitive to the combined source rather than its components, whereas a CMB approach separates the components, and the combination will go unrecognized. For this reason it is natural to express the results from factor analyses in terms of the "road" source (Table II), but the results from CMB analyses in terms of the "auto" source (Table III).

Set I Results

Examining the Set I results in Table II, we conclude the following:

- No false positives (absent sources reported as present) occurred.

- False negatives (sources missed) were primarily restricted to those that were "weak" (large relative standard deviation [RSD]) or highly "collinear" (correlated) with other sources. This result was especially evident in the approaches utilizing factor analysis to ascertain source categories and number of sources. For example, of the nine sources present, TTFA found only six initially and seven in the post-Workshop solution.

- The unknown source, sandblast, was purposely made large (signal/noise >>1; see Figure 3). Three of the four participants responding reported values for an extra component, all of which were consistent with the truth.

- Arsenic was seen as a lone component in all factor analyses, so arsenic data were discarded before fitting Set I.

- Two of the methods, MLR(T)/FA and TTFA, provided independent estimates of the source profiles which were reasonably consistent with the true profiles [1]. For the unknown profile, for example (sandblast), the normalized residuals ranged from -3.2 to +2.0 for TTFA (post-Workshop solution), and from -3.6 to +5.6 for MLR(T).

- The quality of fit for the source impact estimates, as reflected in the normalized residuals, for all methods was roughly comparable. Excluding the Road estimates, the maximum absolute normalized residuals ranged from 4.5 to 16. Methods which inferred their own profiles [MLR(T) and TTFA] showed large negative residuals for Road estimates [-10. and -34., resp.].

Set II Results

Referring next to Set II (Table III) where all 13 sources were operating, results for two CMB methods had been reported (for the average source contributions) by the end of the Workshop. Despite some collinearity among the sources, both methods yielded results that were roughly consistent with the true values. Estimated standard errors however, differed; those for CMB(RR) being larger than those given by CMB(EV). One would conclude that eight of the sources would not have been "detected" (RSD > 60%) by this (RR) technique. The second set of results for CMB(RR), submitted after the Workshop, differed in how the data were weighted in the fitting process. In the first set, the weight for an average elemental concentration took into account both its measurement standard error and its day-to-day variability. In the second, the weight was based on the quoted concentration (measurement) error only. Surprisingly, the criteria used to choose the ridge regression K-value led to selecting K = 0 for the second CMB(RR) solution, corresponding to ordinary weighted least-squares CMB.

A graphical presentation of the Set II results, given in Figure 5A, shows clearly that most of the results lie well within a factor of two of the truth. Striking exceptions are those most affected by collinearity: Aggregate and Oil-B. Not shown is a TTFA four-source solution submitted prior to the Workshop: auto = 6.2 ± 0.2 $\mu g/m^3$, incinerator = 3.2 ± 0.1 $\mu g/m^3$, coal = 5.3 ± 0.3 $\mu g/m^3$, soil = 19.6 ± 0.3 $\mu g/m^3$. This approach thus severely underestimated the number of sources present in Set II, presumably due to the fact that so many sources were collocated.

Set III Observations

No results for Set III were submitted, but it was evident (because of the additional log-normal error component) that deconvolution would have been still more difficult than for Set II. The importance of additional physicochemical characterization of source and ambient samples for such difficult but realistic deconvolution situations is evident from the demonstrated effects of collinearity among source profiles. An explicit illustration of this was given by the National Bureau of Standards where the effect of ^{14}C characterization was investigated. For Set III, without ^{14}C measurements, some 16 very strong pairwise correlations were found among the 13 source estimates; the addition of this isotopic characterization reduced the number to simply one severe interference (between steel-A and steel-B). (The interference questions were studied numerically by examining the variance-covariance matrix resulting from weighted least-squares analysis.)

One obvious case where ^{14}C was crucial was when the interference was reduced between wood and automotive sources. Clearly, the addition of other correlation-breaking observations, such as other isotopic analyses and individual particle and crystalline phase analyses, may be the key for resolving a host of sources with adequate precision for control purposes.

UNCERTAINTIES

One topic that deserves further investigation in future intercomparisons and method development is the matter of uncertainties. It is encouraging (and represents real progress) that all reported results included approximate standard error estimates. Some obvious differences in treating different error components, degrees of freedom, and confidence intervals were seen, but we attempted to convert all results to an equivalent basis. Limitations in current treatments — especially with respect to systematic error and the treatment of negative estimates — are summarized below:

- None of the reported results included a completely adequate statement of uncertainties.
- For methods using the profiles as submitted (some form of CMB):
  - only the effective variance CMBs included the effect of source profile (σ_m) uncertainties, but these were treated as random rather than systematic error components.
  - other CMBs gave uncertainties derived from observation measurement error (σ_i) only. This form of uncertainty was generally by choice, however, and not indicative of an inherent limitation in the method (e.g., ridge regression).
- Multivariate and factor analysis results (Set I) quoted uncertainties that were to be taken as "no greater than the standard error"; no information on bias bounds was provided.
- Ridge regression (Set II) provided no quantification of bounds for the bias that is introduced by this mathematical method. However, according to Williamson, the absence of significant negatives or unreasonably large positive values precludes the possibility of extreme biases.
- Uncertainty about such items as the appropriate number of degrees of freedom prevented the construction of confidence intervals based upon Student's-t.
- Deviations ($|$error$|$) compared to stated uncertainties (standard errors), as shown in Fig. 5B, make it quite clear that the uncertainties ranged from somewhat under-estimated to somewhat over-estimated.
- For Set I (fewer than 13 sources operating), the questions of uncertainty intervals for the number of sources or alternative acceptable solutions were generally not addressed.
- Quality of fit measures should accompany all reported results. In at least some cases discussed at the Workshop, it was apparent that such measures (e.g., chi-square) did not always remain within statistically acceptable bounds.

CONCLUSIONS

Despite the short time (~2 weeks) prior to the Workshop for data analysis, deconvolution results were generally satisfactory, providing adequate estimates for nearly all sources present. In quantitative terms, the best solutions for the two data sets that were analyzed gave source intensity estimates that on average were within about 30% of true values. No one method was "best" in all circumstances. The multivariate methods were successful in treating Set I which contained a relatively small number of sources, one of whose profiles was unknown. The least-squares (CMB) methods were clearly the better choice for Set II which contained a large number of active sources with no unknown source profiles. For real data sets containing many sources some of which are probably unknown, it seems apparent that both CMB and factor analysis approaches should be applied and some degree of consistency between them should be required before results are regarded as valid.

Limitations in the analysis of the data sets related primarily to limitations in resolving power and differing approaches to the statement of uncertainty — especially regarding systematic error and distributional effects. The latter limitation has already been discussed. The former is intrinsic in the data (not the deconvolution methods) and is expected to be alleviated by (1) adding correlation-breaking data (e.g., isotopic and microscopic) which tend to carry unique information concerning different sources, (2) adopting sampling strategies that permit the application of such techniques as wind trajectory analysis, and (3) incorporating the singular value decomposition to objectively assess resolving power. Strengths and weaknesses in the simulation data, *per se*, were evident in both the generation process and the results of the intercomparison. The greatest strength was the fact that rigorously controlled models, with known error structure and source concentrations, could be applied. The ranges of concentrations and source types were realistic and interesting in that participants could readily quantify some components but scarcely detect others. Also important, for validating anticipated difficulties in the deconvolution process, were: the unknown number of components, the extra component, the linear combination component, and the collinearity among source profiles. Suggestions for improved future simulation exercises are noted in Table V.

ACKNOWLEDGMENT

Special credit is due the following colleagues who provided important guidance and/or served as intercomparison participants: David Balfour, John Cooper, Stuart Dattner, Richard DeCesar, Glen Gordon, Steven Heisler, Philip Hopke, Jitu Shah, George Thurston, and Hugh Williamson. One of us (RWG) gratefully acknowledges support from a Fellowship awarded by the National Research Council.

REFERENCES

1. R. Stevens, T. Dzubay, C. Lewis, L. Currie, D. Johnson, R. Henry, G. Gordon, B. Davis, "Mathematical and Empirical Receptor Models: Quail Roost II", U.S. Environmental Protection Agency, ms (1982).

2. J. H. Novak, D. B. Turner, "An efficient Gaussian-Plume multiple-source air quality algorithm," J. Air Pollut. Contr. Assoc., 26(6) 570 (1976).

3. D. B. Turner, J. H. Novak, Users Guide for RAM, EPA-600/8-78-016a, U.S. Environmental Protection Agency, Research Triangle Park, NC (1978).

4. L. A. Currie, "Detection and quantitation in x-ray fluorescence spectrometry," X-Ray Fluorescence Analysis of Environmental Samples, T. G. Dzubay, ed., Ann Arbor Science, Inc., Ann Arbor, Michigan, (p 289) (1977).

5. R. C. Henry, "Stability Analysis of Receptor Models that use Least Squares Fitting," Proc. APCA Specialty Conference on Receptor Models Applied to Contemporary Pollution Problems (1983).

TABLE I. STRUCTURE OF THE SIMULATED DATA SETS

Generating Equation

$$\tilde{C}_i^{(k)} = \sum^{P} [\tilde{A} - e_m - e_H^{(k)}]_{ij}\, S_j^{(k)} + e_i^{(k)}$$

where:

k	=	sampling period $[1 \le k \le 40]$
$\tilde{C}_i^{(k)}$	=	"observed" concentration of species-i, period-k $[1 \le i \le N,\ N \le 20]$
$S_j^{(k)}$	=	true intensity (at receptor) of source-j $[1 \le j \le P,\ P \le 13]$
$\tilde{A}_{ij}$	=	"observed" source profile matrix (element-i,j)
e_i	=	random measurement errors, independent and normally distributed
e_m	=	systematic source profile errors, independent and normally distributed (systematic, because fixed over the 40 sampling periods)
e_H	=	random source profile variation errors, independent and log-normally distributed

Data Set Characteristics

[P = No. of active sources; P_o = No. in the world list = 13]

Set I : P = 9 (including one unknown source)*; errors = e_i, e_m; City Plan -1 (Fig. 1)

Set II : P = P_o (all known); errors = e_i, e_m; City Plan -2 (Fig. 2)

Set III : P = P_o (all known); errors = e_i, e_m, e_H; City Plan -2

*For data Set-I, participants were told only that $P \le P_o$

TABLE II. ESTIMATED SOURCE IMPACTS ($\mu g/m^3$) COMPARED WITH TRUE VALUES FOR SET I[a]

Source	CMB(EV)/FA[b]	CMB(W)/FA[c]	MLR(T)/FA[d]	TTFA	TTFA[e]	Truth
Steel-A Steel-B	1.0 ± 0.3	-- --	-- --	-- --	-- --	0.05 --
Oil-A Oil-B	6.1 ± 1.0	8.7 ± 0.9	7.9 ± 0.8	2.3 ± 0.4	2.1 ± 1.0	-- 2.0
Incinerator	1.5 ± 0.2	0.7 ± 0.3	2.4 ± 0.4	1.8 ± 0.1	1.9 ± 0.14	1.3
Glass	--	--	--	--	--	--
Coal-B	--	--	--	--	2.2 ± 0.73	2.4
Coal-A Aggregate Basalt Soil[f] } [g]	15.9 ± 1.1	15.9 ± 0.2	10.6 ± 0.9	12.7 ± 0.5	12.5 ± 0.75	 -- 4.7 8.0
Road[f]	5.3 ± 0.4	8.3 ± 0.2	5.0 ± 0.2	3.0 ± 0.1	4.0 ± 0.09	7.1
Wood	--	--	--	7.4 ± 0.7	4.3 ± 0.52	3.3
Sandblast[h]	--	1.9 ± 1.4	4.0 ± 0.3	5.2 ± 0.4	4.1 ± 0.20	4.2
Total:	29.8	35.5	29.9	32.4	31.1	33.05

[a] Uncertainties represent one standard deviation.
[b] CMB result for the average of the observation periods.
[c] Average of CMB results for each observation period.
[d] Fitted intercept = 3.3 ± 1.5 $\mu g/m^3$.
[e] Results submitted after completion of Workshop.
[f] Data were generated using road as a source component, where road = 0.25 soil + 0.75 auto.
[g] A (composite) "crustal" component.
[h] Sandblast was an extraneous source, not included among the 13 profiles furnished to the participants.

TABLE III. ESTIMATED SOURCE IMPACTS ($\mu g/m^3$) COMPARED WITH TRUE VALUES FOR SET II[a]

Steel-A	N	CMB(EV-1)[b]	CMB(EV-1)[c]	CMB(RR)[c]	CMB(RR)[c,d]	CMB(EV-2)[b,d,e]	Truth
Steel-A	(19)	0.58 ± 0.033	0.97 ± 0.14	0.29 ± 0.73	0.66 ± 2.4	0.55 ± 0.11	0.42
Steel-B	(11)	0.24 ± 0.017	--	0.31 ± 0.49	0.26 ± 1.6	0.22 ± 0.08	0.29
Oil-A	(17)	0.44 ± 0.033	0.47 ± 0.12	0.11 ± 0.66	0.45 ± 0.56	0.47 ± 0.10	0.56
Oil-B	(11)	1.2 ± 0.13	1.7 ± 1.1	2.3 ± 3.3	0.11 ± 2.7	0.72 ± 0.25	0.41
Incinerator	(27)	1.5 ± 0.05	1.5 ± 0.1	1.3 ± 0.84	1.5 ± 0.29	1.5 ± 0.43	1.81
Glass	(3)	0.16 ± 0.03	--	0.15 ± 1.3	0.31 ± 1.1	0.28 ± 0.11	0.46
Coal-B	(21)	3.7 ± 0.17	3.8 ± 1.1	4.3 ± 3.1	4.9 ± 2.3	3.8 ± 0.91	3.3
Coal-A	(11)	4.3 ± 0.3	6.2 ± 1.3	3.0 ± 4.9	4.6 ± 4.7	5.3 ± 1.40	3.7
Aggregate	(11)	2.1 ± 0.3	--	7.3 ± 4.0	2.5 ± 6.9	2.1 ± 0.57	0.46
Basalt	(24)	8.2 ± 0.5	11.2 ± 1.7	9.0 ± 3.4	8.2 ± 6.1	9.5 ± 1.28	11.3
Soil	(33)	9.5 ± 0.4	7.8 ± 1.4	6.3 ± 2.1	8.6 ± 3.5	8.4 ± 1.26	7.9
Auto	(40)	2.5 ± 0.05	2.5 ± 0.1	2.3 ± 0.78	2.5 ± 0.2	2.5 ± 0.31	2.5
Wood	(11)	0.5 ± 0.1	0.6 ± 0.5	1.1 ± 0.38	1.0 ± 0.54	0.91 ± 0.16	0.94
Total:		34.92	36.74	37.76	35.59	36.25	34.05

[a]Uncertainties represent approximately one standard deviation.
[b]Average of CMB results for each observation period (40 total).
N is the number of periods for which source was used in CMB(EV-1).
[c]CMB result for the average of the observation periods.
[d]Results submitted after completion of Workshop.
[e]Average intercept = 0.52 $\mu g/m^3$.

TABLE IV. CODE FOR NUMERICAL METHODS — APPLIED [Ref. 1]

DATA SET I

Effective Variance Chemical Mass Balance/Factor Analysis [CMB(EV)/FA], Charles W. Lewis, MD-47, U.S. Environmental Protection Agency, Environmental Science Research Laboratory, Research Triangle Park, NC 27711

Weighted CMB/Factor Analysis [CMB(W)/FA], George D. Thurston, Department of Environmental Health Sciences, Harvard School of Public Health, 665 Huntington Avenue, Boston, MA 02115

Multiple Linear Regression by Tracers/Factor Analysis [MLR(T)/FA, Stuart L. Dattner, Texas Air Control Board, 6330 Highway 290 East, Austin, TX 78723

Target Transformation Factor Analysis [TTFA], Philip K. Hopke, Environmental Research Laboratory, University of Illinois, 1005 W. Western Avenue, Urbana, IL 61801

DATA SET II

Effective Variance CMB [CMB(EV-1)], Steven L. Heisler, Environmental Research and Technology, Inc., 696 Virginia Road, Concord, MA 01742

Weighted Ridge Regression CMB [CMB(RR)], Hugh J. Williamson, Radian Corporation, P.O. Box 9948, Austin, TX 78766

Effective Variance CMB Regression [CMB(EV-2)], John A. Cooper, NEA, Inc., 10950 SW 5th St., Beaverton, OR 97005

TABLE V. RECOMMENDATIONS FOR FUTURE SIMULATION EXERCISES

- Generate replicate observations to provide direct measures of imprecision and bias.
- Overcome plume narrowness (RAM model) by initial dispersion; include a proper subroutine for line sources (highways).
- Provide participants with geographical information (for both sources and receptors).
- Use multiple ($\sim$3) receptors to evaluate ability to pinpoint unknown sources and estimate upper emissions limits for specified sources.
- Treat σ_H as discrete (feedstock) change, rather than random with each sampling period.
- Change sampling times from 1200, 2400 to 0700, 1900 to better expose diurnal variations.
- Introduce unknown sources to test for false positives and negatives (and profile estimation) by including $\geq$2 unknown sources and increasing size of world list.
- Use both fine and coarse fractions.
- Include morphological data.
- Include sensitivity studies of source intensity detection threshold and source resolution of similar profiles.

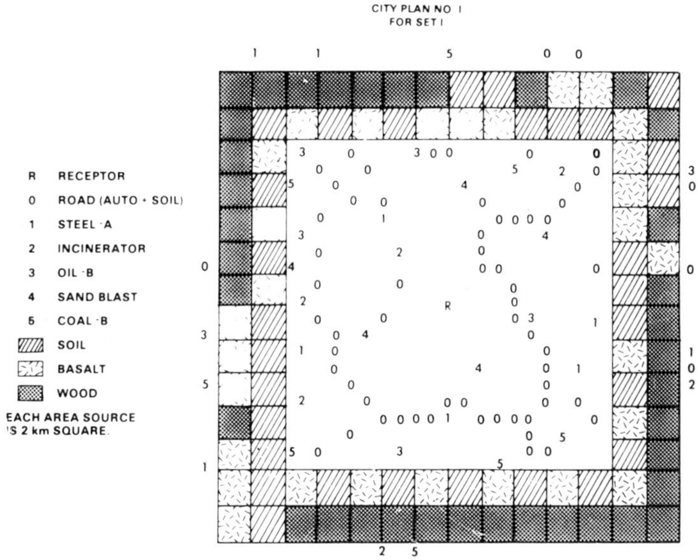

Figure 1. City Plan No. 1 (for Data Set I).

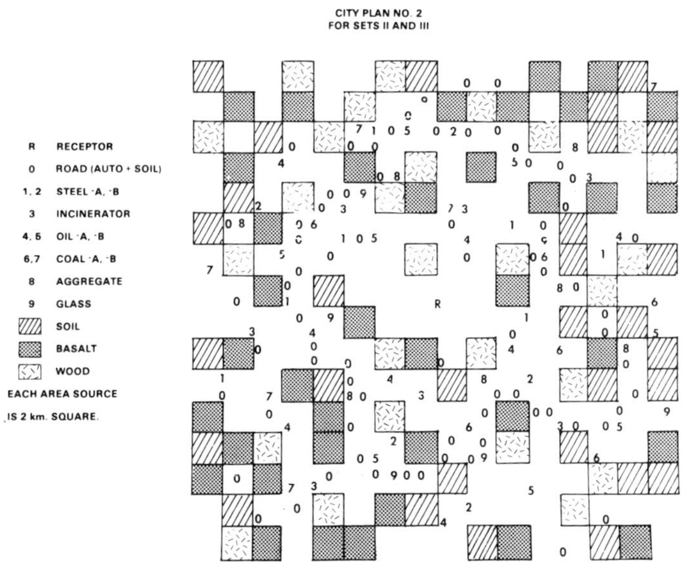

Figure 2. City Plan No. 2 (for Data Sets II and III).

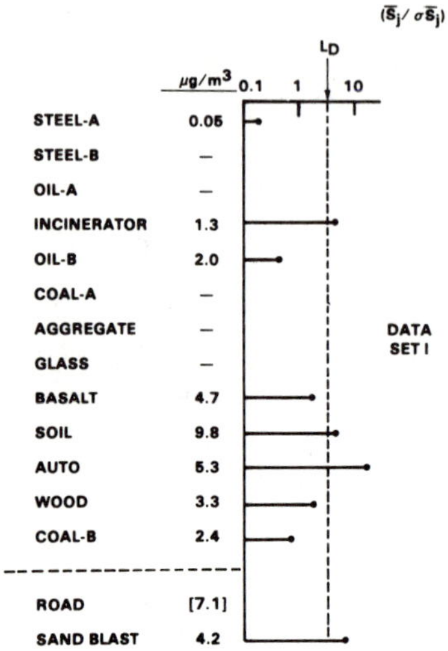

Figure 3. Characteristics for Data Set I. First two columns, above the dashed line, list the 13 sources in the world list and their true concentrations (S_j) at the receptor. Sources below the dashed line were the unknown linear combination (road) and the unknown source (sandblast). The bar graph is a log display of signal/noise (i.e., 1/relative standard error) for each of the active sources. L_D is the approximate detection limit.

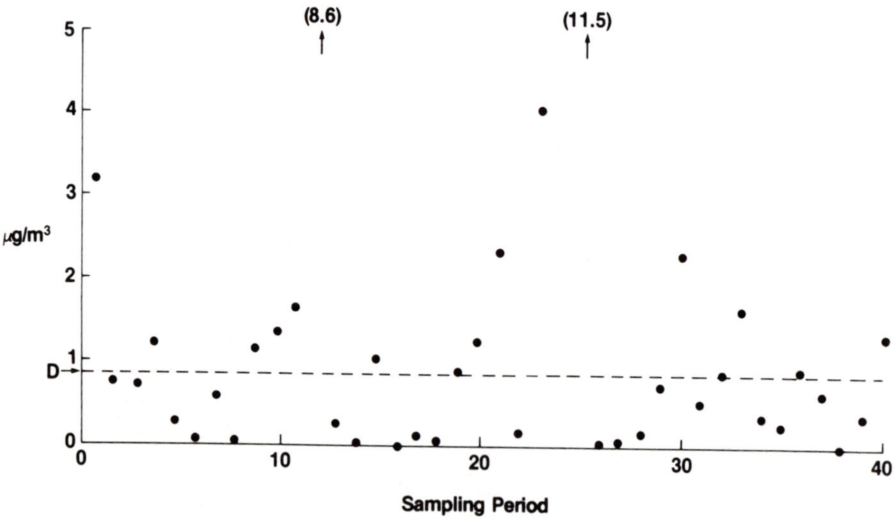

[Data Set I - Incinerator, Source-10]

Figure 4. Incinerator impact (S_j) at receptor for Data Set I.

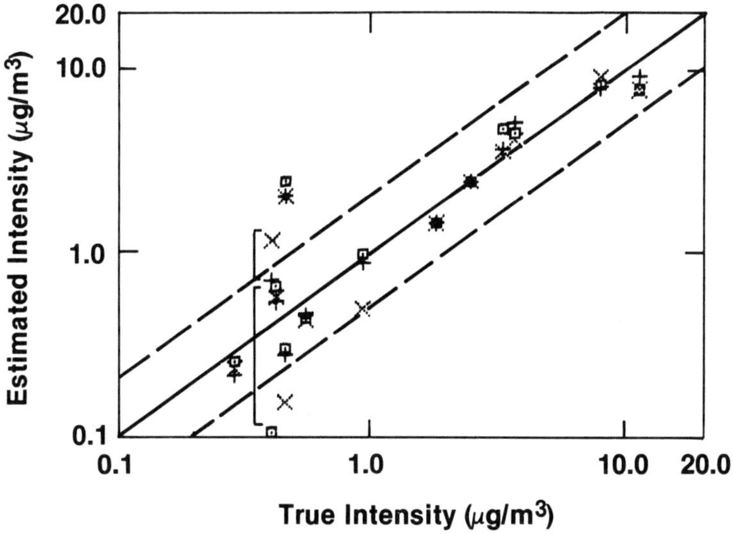

Figure 5A. Plot of estimated average source strengths for each of the 13 sources in Set II using chemical mass balance (EV-1, EV-2) and Ridge Regression. Dashed lines indicate region within a factor of two of the truth. [X=CMB(EV-1), +=CMB(EV-2), ▣=CMB(RR)].

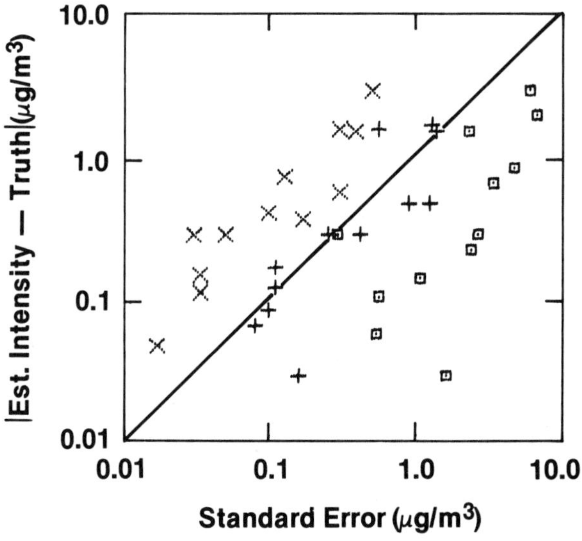

Figure 5B. Actual (absolute) errors for each of the 13 x 3 estimates given in Fig. 5A and Table 3, plotted as a function of the stated standard errors. (On the average, one would expect about two-thirds of the points for any correct method to lie below the diagonal.)

APPLICATION AND VERIFICATION STUDIES OF TARGET TRANSFORMATION FACTOR ANALYSIS AS AN AEROSOL RECEPTOR MODEL

Philip K. Hopke* →
Kathryn G. Severin and
Shau-Nong Chang
Institute for Environmental Studies
University of Illinois
1005 W. Western Avenue
Urbana, Illinois

For several years, target transformation factor analysis (TTFA) has been under development for use in aerosol mass apportionment studies. There are several approaches to applying TTFA depending on the nature of the correlation matrix used. The correlation between the measured variables (R-mode) or between the particle samples (Q-mode) can be employed as the basis for the analysis. A detailed description of these two approaches is presented. These methods are then applied to a created data set prepared by the National Bureau of Standards from known source profiles and a dispersion model. The ability of both modes to resolve these data sets has been tested with these data with the result that for this data set, the R-mode appears to give superior results.

* Author to whom correspondences should be addressed.

INTRODUCTION

Over the last several years, there have been extensive efforts to develop retrospective methods of quantitatively assessing the impact of particulate emission sources on the air quality at a specific site. These receptor models have relied on the elemental composition patterns of the sources to permit various statistical techniques to apportion the measured aerosol concentrations. These models have been reviewed by Gordon [1], Cooper and Watson [2] and Cooper [3].

The basic premise of receptor models is that the amount of an element observed in a sample is a linearly independent sum of contributions from the various particle sources in the airshed. For example, the volume concentration of lead should be equal to the amount of lead from motor vehicles, from incinerators, from non-ferrous metal smelters, etc.

$$\mathrm{Pb(T)} = \mathrm{Pb(Auto)} + \mathrm{Pb(Incinerator)} + \mathrm{Pb(Smelter)} + \ldots \qquad (1)$$

However, the particles emitted by automobiles do not consist entirely of lead. The lead contribution from automobiles can then be considered as the product of the elemental concentration of lead in motor vehicle exhaust particles, $a_{Pb,\ auto}$, times the mass concentration of motor vehicle particles, f_{auto}

$$\mathrm{Pb\ (Auto)} = a_{Pb,\ auto}\ f_{auto} \qquad (2)$$

This mass balance approach can be extended to all of the m elements measured in the n samples where p sources contribute particles to the airshed.

$$x_{ij} = \sum_{k=1}^{p} a_{ik}\ f_{kj} \qquad (3)$$

where x_{ij} is the concentration of the ith element in the jth sample, a_{ik} is the average concentration of the ith element in particles from the kth source, and f_{kj} is the mass of particles from the kth source contributing to the jth sample. If the number of sources and the elemental composition are known _a priori_, then equation (3) can be solved for the mass contributions by a multiple regression analysis. This approach, generally called the Chemical Element Balance (CEB) method, has been widely employed [4-7].

In order to avoid the need for such extensive initial information on the number and nature of the sources for a region, an alternative approach has been proposed [8-13] based on a modified form of factor analysis, called target transformation factor analysis (TTFA). Several improvements have been recently reported [14,15]. These changes and others described below improve the quality of resolution that can be obtained for airborne particle data.

TARGET TRANSFORMATION FACTOR ANALYSIS

The objectives of the factor analysis are 1) to determine p, the number of independent sources that contribute to the system, 2) to identify the elemental profiles of particles emitted by those sources, and 3) to determine the mass contribution of each source to each sample of collected particulate matter.

To accomplish these objectives, an eigenvalue analysis is made of a correlation matrix. In the factor analysis model previously employed [8-13], the interrelations between samples have been utilized as a starting point for the analysis. Following its introduction by Imbrie and Van Andel [16], this approach known as Q-mode factor analysis has been widely employed in geological studies [17-18]. In a study of mass spectrometry data, Rozett and Petersen [19] found that the Q-mode analysis gave a more sensitive response to the proper number of factors and was preferred over the R-mode that examined the interrelationships between the variables. In response to Alpert and Hopke [9], Redman and Zinsmeister [20] point out that the two approaches should provide the same result but that the Q-mode should be biased because of the inherent correlation between elements. In the response to these comments, Alpert and Hopke [21] point out the R-mode is biased by the inherent serial correlations between samples in a set of airborne compositional data so that neither approach fulfills the assumption of independent samples drawn from a multivariate normal population. It, therefore, is of value to compare R-mode and Q-mode analysis.

The fundamental model of factor analysis is given by equation 3 that can be rewritten in matrix form as

$$X = AF \tag{4}$$

In the Q-mode analysis an nxn correlation matrix is formed by

$$C_Q = (XV^{-1})^t(XV^{-1}) \tag{5}$$

where V is a diagonal matrix whose elements are defined by

$$v_{ii} = \left[\sum_{j=1}^{n} x_{ij}^2\right]^{1/2} \tag{6}$$

where n is the number of samples, m is the number of elements, and t denotes the transpose matrix. For the R-mode, the correlation matrix would be formed by

$$C_R = (XV^{-1})(XV^{-1})^t \tag{7}$$

yielding an m x m matrix. From these two matrices, the analyses proceeds to yield the A and F matrices as outlined in figure 1. In the Q-mode analysis, the source profile matrix, A, is calculated from the data and the eigenvalue matrix while the mass contribution matrix, F, is calculated from the data and the eigenvectors in the R-mode analysis.

Q-mode R-mode

Correlation Coefficient:

$$C_Q = (XV^{-1})^t (XV^{-1}) \qquad C_R = (XV^{-1})(XV^{-1})^t$$

$$\text{where } v_{ii} = (\sum_{j=1}^{n} x_{ij}^2)^{\frac{1}{2}} \quad ; \; i = 1, m$$

$$v_{i \neq j} = 0$$

Diagonalization:

$$Q_Q^{-1} C_Q Q_Q = \Lambda_Q \qquad Q_R^{-1} C_R Q_R = \Lambda_R$$

$$Q_Q^t = Q_Q^{-1} \qquad Q_R^t = Q_R^{-1}$$

$$Q_Q^t (XV^{-1})^t (XV^{-1}) Q_Q = \Lambda_Q \qquad Q_R^t (xv^{-1})(xv^{-1})^t Q_R = \Lambda_R$$

$$[(xv^{-1})^t Q_Q]^t [(xv^{-1}) Q_Q] = \Lambda_Q \qquad [(xv^{-1})^t Q_R]^t [(xv^{-1})^t Q_R] = \Lambda_R$$

If $A \equiv (XV^{-1})Q_Q$, If $F \equiv [(XV^{-1})^t Q_R]^t = Q_R^t XV^{-1} = Q_R^t V^{-1} X$,

then $F = VQ_Q^t$ then $A = Q_R V$

Figure 1. Outline of the diagonalization process for Q- and R-mode factor analysis.

However, there are an infinity of possible solutions and that obtained by the diagonalization may not be representative of physical reality. The first problem is to select the number of factors to be retained. At this point the data is reconstructed using one eigenvector, then two eigenvectors and so forth with several measures of the quality of fit being calculated at each step. These measures include chi-squared

$$\chi^2 = \sum_{i=j} \frac{(x_{ij} - \bar{x}_{ij})^2}{\sigma^2_{ij}} \qquad (8)$$

where x_{ij} is the reconstructed data point using p factors and σ_{ij} is the uncertainty in the value of x_{ij}. The Exner function [22] is a similar measure and is calculated by

$$E^p = \left[\frac{\sum_{i=j} (x_{ij} - \bar{x}_{ij})^2}{\sum_{i=j} (x^o - \bar{x}_{ij})^2} \right]^{1/2} \qquad (9)$$

where x^o is a grand ensemble average value. The empirical indicator function suggested by Malinowski [23] is also calculated.

$$RSD = \left[\frac{\sum_{j=p+1}^{n} \lambda j}{n(m-p)} \right]^{1/2} \qquad (10)$$

$$IND = \frac{RSD}{(n-p)^2} \qquad (11)$$

where λ_j are the eigenvalues from the diagonalization. This function has not proven as useful with particle composition data as Malinowski has found it to be with spectroscopy results [23]. The root-mean-square error and the arithmetic average of the absolute values of the point-by-point errors are also calculated. All of these parameters can prove to be useful indicators of the dimensionality of the problem and permit the determination of the number of factors to be used in the subsequent portions of the analysis.

Once A and F have been truncated to their first p columns and rows, respectively, then the problem arises as the identification of what sources they represent. The use of a target transformation rotation has been shown to be a useful way of identifying the source profiles. The question arises as to how much information about the composition of the particles is required for the rotation. The rotation is an attempt to force an overlap between the input test profile and a factor axis by a weighted least-squares fit [14].

$$r = (A^t\ WA)^{-1}\ A^t\ Wb \tag{12}$$

where r is one of the columns in a rotation matrix, b is the input source profile, and W is a diagonal weighting matrix whose elements are typically the inverse of the variances of the elements in the data set. These rotation vectors are then applied to equation 9 as

$$X = (AR)(R^{-1}F) = A'F' \tag{13}$$

Roscoe and Hopke [14] have shown that for a simple data set, these profiles can be developed by an interative process beginning with very simple test profiles. These profiles consist of 1.0 for one element and 0.0 for all of the others. These profiles are called uniqueness vectors. From the rotation vector and b, a predicted profile b' can be calculated. The error between the predicted and original values gives an indication if b' is a reasonable representation of a source. In the iterative process, b' becomes the suggested source profile for the next rotation to produce a b''. This process continues until there is on the average a change of less than 10^{-4} in any of the values in the profile. The iterated profiles obtained from all of the uniqueness vectors can then be examined and combined to form a matrix of source profiles. In the earlier reports of TTFA [8-12], F' was calculated as $R^{-1}\ VQ^t$. In an effort to improve the fit, an alternative procedure has been developed to calculate the mass contributions from the observed elemental concentrations [13] and the developed source profiles. The complete matrix of refined source profiles, B, can be used to develop a column of F by a least-squares fit to a column of X using the same weight matrix previously employed.

$$f' = (B^t\ WB)^{-1}\ B^t\ Wx \tag{14}$$

The calculated source profiles and mass contributions may still need to be scaled as described in earlier reports [8,10,11].

After making these determinations, it is of value to be able to estimate the uncertainty in each of the calculated parameters. A procedure has been developed by Roscoe and Hopke [15] from the concepts presented by Clifford [24]. In this process, the matrix of differences between the original and calculated data are formed a row or a column at a time

$$E = X(Obs) - X(Calc) \tag{15}$$

An average variance of the data points in that row or column is then calculated by

$$\sigma^2 = \frac{E^t\ WE}{m - p} \tag{16}$$

where W, the weight matrix is that used in the previously described analysis, m is the number of elements and p is the number of sources included. To apportion the total variance in the data, the matrix

$$P = (A^t WA)^{-1} \tag{17}$$

is formed. The errors in the calculated results may then be estimated by

$$\sigma(Y_i) = \sigma[P_{ii}]^{1/2} \tag{18}$$

where $\sigma(Y_i)$ is the estimated error in parameter Y_i. Complete details are presented in the earlier report [15]. These procedures have been incorporated in a series of programs: FANTASIA, the program to perform the TTFA, and OERROR, the program to perform the error analysis, and other ancillary programs. This series of programs have been described by Hopke, Alpert, and Roscoe [25].

RESULTS

Artificial data were provided by the National Bureau of Standards [26]. The set used in this study was the first of three and consists of 40 samples for which 19 elemental concentrations were provided. The data set was created using a contrived source geometry and the RAM Gaussian-plume dispersion model. The details of the contruction of these data are presented in the report in these proceedings by Currie [26]. The composition of the nine sources included in the data set are given in Table 1. In the data set, arsenic was found to be below detection limits for the vast majority of samples. In a classical factor analysis used to screen the data [27], arsenic appears as a unique factor. It was, therefore, eliminated from the data set. A classical factor analysis showed a very stable solution for 7 factors using the criterion of variance greater than 1.0 after an orthogonal rotation [27-29]. This approach provides a useful rule of thumb for the maximum number of factors to retain.

The results of the Q-mode TTFA analysis are presented in Tables 2 to 4, and the R-mode values are given in Tables 5-7. In Tables 2, and 5, the results of the diagonalization and reproduction are presented. For the Q-mode analysis, examination of the chi-squared and the average % error (Table 2) suggests the presence of 6 factors. After careful study of the 6 and 7 factors solutions, the six factor Q-mode solution seemed to be more satisfactory. Table 3 gives the refined source profiles developed from the set of unique vectors by the iterative process previously described. The average elemental concentration contributions and the fit to the average composition are given in Table 4. The average mass contributions are found to be wood smoke 7.4 $\mu g/m^3$ (23%), soil 12.7 $\mu g/m^3$ (39%), refuse incineration 1.8 $\mu g/m^3$ (5.5%), oil-fired boiler 2.3 $\mu g/m^3$ (7.2%), sandblasting 5.2 $\mu g/m^3$ (16%), and motor vehicles 3.0 $\mu g/m^3$ (9.2%).

Table 1. Actual Compositions of Sources Used to Create N.B.S. Data Set No. 1

Element	Oil-B	Incinerator	Soil	Road	Wood	Sandblast	Coal-B	Steel	Basalt
C	200	40	66.4	429	600	5.0	8.0	150	15
Na	22	120	4.0	52.5	3.0	8.0	4.0	9.3	20.2
Al	5.4	4.2	222	62	7.0	7.5	140	13.2	86
Si	70	7.0	543	147	8.9	400	200	9.0	166
S	110	130	0.26	6.06	1.00	8.8	180	25	0.23
Cl	12	270	0.13	8.06	10.0	28	17	33	0.20
K	1.40	200	12.0	5.0	8.0	7.0	8.0	10.2	8.6
Ca	8.6	11.0	20.5	13.9	9.0	8.2	40	63	46
Ti	1.10	0.90	12.7	3.17	0.020	4.0	9.0	1.5	7.9
V	0.67	0.027	0.25	0.143	0	0.031	0.8	0.19	0.22
Cr	0.42	0.85	0.31	0.141	0.002	5.8	0.3	2.8	0.12
Mn	0.140	0.33	2.0	0.94	0.020	1.4	0.2	15	1.6
Fe	13.0	6.1	97	29.5	1.0	63	70	123	82
Ni	0.06	0.100	0.090	0.070	0.005	3.2	0.6	3.1	0.085
Cu	0.75	3.6	0.20	0.050	0	6.9	0.4	2.3	0.10
Zn	1.30	26	0.41	2.51	0.70	4.7	0.9	2.8	0.08
As	0.25	0.150	0.010	0.010	0	0.099	0.2	0.08	0.012
Br	0.39	0.82	0.010	31.3	0.080	0.047	0.01	0.09	0
Pb	3.8	17.0	0.26	90.1	0.100	5.8	1.2	8.0	0

Table 2. Diagonalization Results and Reproduction Summary for Q-mode Analysis NBS Data Set No. 1

FACTOR	EIGENVALUE	RMS ERROR	CHI-SQUARE	EXNER	AVERAGE % INDICATOR	AVERAGE % ERROR
1	3.60E+01	4.48E+01	3.20E+02	.3469	4.75E-05	126.61
2	2.97E+00	1.48E+01	8.11E+01	.1525	2.23E-05	70.95
3	3.26E-01	1.09E+01	6.30E+01	.1123	1.75E-05	58.58
4	2.37E-01	7.30E+00	6.94E+01	.0710	1.18E-05	51.51
5	8.10E-02	4.65E+00	8.18E+01	.0487	8.75E-06	48.64
6	3.19E-02	3.06E+00	1.78E+01	.0366	7.06E-06	37.27
7	2.18E-02	2.14E+00	1.55E+01	.0253	5.26E-06	34.07
8	1.25E-02	1.59E+00	9.28E+00	.0156	3.52E-06	25.21
9	5.65E-03	7.19E-01	1.04E+01	.0078	1.90E-06	23.31
10	1.42E-03	4.57E-01	5.62E+00	.0038	1.02E-06	14.34
11	2.66E-04	3.38E-01	2.80E+00	.0025	7.35E-07	10.55
12	9.80E-05	2.55E-01	5.51E+00	.0018	5.72E-07	10.20
13	6.92E-05	1.17E-01	1.70E+00	.0010	3.50E-07	6.46
14	1.50E-05	6.51E-02	1.62E+00	.0007	2.75E-07	5.10
15	9.50E-06	4.14E-02	1.43E+00	.0004	1.95E-07	3.70
16	2.95E-06	3.14E-02	1.51E+00	.0003	1.63E-07	2.99
17	2.29E-06	1.94E-02	1.68E+00	.0002	1.17E-07	1.90

Table 3. Refined Emission Source Profiles (mg g^{-1}) Q-Mode Analysis of NBS Data Set No. 1

ELEMENT	WOOD	SOIL	REFUSE	OIL	SANDBLAST	MOTOR VEHICLE
C	464. ± 24.	34. ± 10.	33. ± 40.	140. ± 40.	3.6 ± 12.8	642. ± 30.
NA	8.3 ± 9.0	4.3 ± 3.8	115. ± 16.	12. ± 12.	13. ± 5.	84.7 ± 11.2
AL	0. ± 50.	246. ± 21.	76. ± 81.	8.0 ± 63.1	83. ± 27.	9.1 ± 59.6
SI	15. ± 80.	518. ± 34.	.001 ± .646	37. ± 105.	375. ± 44.	11. ± 99.
S	.040 ± .123	12. ± 6.	149. ± 23.	87. ± 18.	11. ± 8.	47. ± 17.
CL	3.7 ± 5.1	4.2 ± 2.2	245. ± 9.	6.2 ± 6.7	2.2 ± 2.8	16. ± 7.
K	.26 ± 7.96	9.6 ± 3.4	213. ± 14.	7.9 ± 10.6	17. ± 5.	.51 ± 9.94
CA	0. ± 18.	7.3 ± 7.6	12. ± 31.	41. ± 24.	49. ± 10.	54. ± 23.
TI	.41 ± 2.43	11.3 ± 1.1	.85 ± 4.11	4.6 ± 3.3	3.7 ± 1.4	8.5 ± 3.1
V	.074 ± .216	.30 ± .10	.052 ± .364	1.94 ± .29	.14 ± .12	.007 ± .688
CR	.25 ± .49	.78 ± .21	.088 ± .821	.11 ± .65	3.58 ± .27	.78 ± .61
MN	.42 ± .33	2.20 ± .14	.22 ± .55	.36 ± .43	.83 ± .18	1.1 ± .5
FE	.69 ± 3.38	85.0 ± 5.7	36. ± 23.	4.4 ± 17.7	113. ± 8.	6.9 ± 16.8
NI	.080 ± .330	.29 ± .14	.50 ± .56	.93 ± .44	2.09 ± .18	.71 ± .42
CU	.001 ± .398	.18 ± .23	6.12 ± .92	.63 ± .72	4.18 ± .30	.98 ± .68
ZN	.42 ± 1.66	1.2 ± .7	12. ± 3.	.52 ± 2.19	3.1 ± 1.0	5.4 ± 2.1
BR	.014 ± .952	.009 ± .400	.021 ± .609	.037 ± .257	.010 ± .519	73.3 ± 1.2
PB	.27 ± 3.04	1.5 ± 1.3	13. ± 6.	3.4 ± 4.1	.35 ± 1.66	223. ± 4.

Table 4. Average Elemental Contributions (ng/m^3) for Each Source for NBS Data Set No. 1 (Q-Mode Analyis)

ELEMENT	WOOD	SOIL	REFUSE	OIL	SANDBLAST	MOTOR VEHICLE	AVG PRED CONTRIB	AVG OBS CONTRIB	AVG % ERROR
C	.342E+04	.437E+03	.585E+02	.334E+03	.188E+02	.194E+04	.621E+04	.621E+04	9.0
NA	.611E+02	.549E+02	.204E+03	.274E+02	.681E+02	.256E+03	.672E+03	.716E+03	54.5
AL	.254E-05	.313E+04	.135E+03	.186E+02	.435E+03	.275E+02	.375E+04	.298E+04	30.0
SI	.110E+03	.660E+04	.203E-02	.854E+02	.196E+04	.336E+02	.879E+04	.875E+04	17.7
S	.293E+00	.154E+03	.264E+03	.203E+03	.562E+02	.143E+03	.821E+03	.890E+03	26.5
CL	.270E+02	.530E+02	.435E+03	.144E+02	.116E+02	.498E+02	.591E+03	.653E+03	18.4
K	.194E+01	.123E+03	.379E+03	.185E+02	.878E+02	.154E+01	.611E+03	.506E+03	30.1
CA	.386E-04	.929E+02	.206E+02	.945E+02	.258E+03	.163E+03	.629E+03	.803E+03	71.8
TI	.300E+01	.144E+03	.151E+01	.108E+02	.193E+02	.256E+02	.204E+03	.212E+03	22.8
V	.542E+00	.376E+01	.926E-01	.452E+01	.749E+00	.210E-02	.967E+01	.778E+01	59.3
CR	.186E+01	.999E+01	.157E+00	.246E+00	.187E+02	.235E+01	.333E+02	.302E+02	66.2
MN	.307E+01	.280E+02	.397E+00	.842E+00	.434E+01	.331E+01	.400E+02	.393E+02	20.1
FE	.511E+01	.108E+04	.647E+02	.103E+02	.593E+03	.209E+02	.178E+04	.186E+04	14.6
NI	.593E+00	.367E+01	.886E+00	.216E+01	.109E+02	.215E+01	.204E+02	.174E+02	82.2
CU	.859E-03	.232E+01	.109E+02	.147E+01	.218E+02	.296E+01	.395E+02	.380E+02	32.4
ZN	.307E+01	.148E+02	.214E+02	.121E+01	.160E+02	.162E+02	.726E+02	.870E+02	29.1
BR	.104E+00	.114E+00	.378E-01	.855E-01	.521E-01	.222E+03	.222E+03	.224E+03	63.3
PB	.200E+01	.194E+02	.232E+02	.783E+01	.181E+01	.675E+03	.729E+03	.701E+03	14.6

Table 5. Diagonalization Results and Reproduction Summary for R-mode Analysis NBS Data Set No. 1

FACTOR	EIGENVALUE	RMS ERROR	CHI-SQUARE	EXNER	AVERAGE % INDICATOR	AVERAGE % ERROR
1	1.3520E+01	4.9878E+01	9.5050E+02	.751410	2.8086E-04	180.74
2	1.9239E+00	3.4644E+01	4.0590E+02	.567590	2.4687E-04	115.38
3	1.1266E+00	2.5718E+01	4.0293E+02	.424468	2.1695E-04	97.62
4	1.0134E+00	1.8928E+01	1.3423E+01	.229053	1.3911E-04	29.20
5	2.4871E-01	1.6514E+01	1.1906E+01	.145333	1.0623E-04	23.11
6	4.4662E-02	1.3255E+01	1.3375E+01	.124472	1.1114E-04	23.70
7	3.1766E-02	1.1959E+01	3.1198E+00	.107191	1.1897E-04	14.54
8	2.4694E-02	8.3365E+00	2.2045E+00	.091531	1.2892E-04	12.78
9	1.6720E-02	5.8237E+00	2.7253E+00	.079188	1.4514E-04	11.67
10	1.0803E-02	3.5099E+00	3.3557E+00	.070066	1.7239E-04	11.56
11	9.5315E-03	3.2162E+00	3.4877E+00	.060893	2.0920E-04	10.53
12	9.2421E-03	2.5435E+00	4.3095E+00	.050431	2.5472E-04	8.91
13	8.4261E-03	1.5050E+00	3.9300E+00	.038489	3.0665E-04	6.69
14	5.1508E-03	1.2622E+00	3.4780E+00	.028849	4.0153E-04	4.75
15	3.1809E-03	9.5868E-01	4.2558E+00	.020770	5.9343E-04	3.58
16	1.4431E-03	7.4068E-01	7.7023E+00	.015796	1.2437E-03	3.38
17	1.3160E-03	3.8839E-01	1.7242E+01	.009146	4.0738E-03	3.05

Table 6. Refined Emission Source Profiles (mg g^{-1})
R-Mode Analysis of NBS Data Set No. 1

	WOOD	OIL-B	INCINERATOR	COAL	SOIL	SANDBLAST	MOTOR VEHICLES
C	591. ± 42.	185. ± 26.	35. ± 37.	15. ± 51.	68. ± 12.	16. ± 25.	573. ± 17.
NA	4.0 ± 20.0	14. ± 13.	64. ± 18.	35. ± 25.	0061 ± 4095	8.5 ± 11.8	54.8 ± 8.0
AL	10. ± 31.	11. ± 20.	0022 ± 5139	210. ± 40.	166. ± 9.	0099 ± 5516	44. ± 13.
SI	12. ± 92.	44. ± 57.	.53 ± 9.45	230. ± 120.	466. ± 25.	508. ± 54.	73. ± 37.
S	.25 ± 6.51	120. ± 17.	64. ± 24.	180. ± 40.	5.7 ± 7.2	.025 ± .579	10. ± 11.
CL	16. ± 11.	5.4 ± 6.7	196. ± 10.	23. ± 14.	.35 ± 2.90	37. ± 7.	16. ± 5.
K	13. ± 5.	2.9 ± 2.8	122. ± 4.	17. ± 6.	7.5 ± 1.3	8.5 ± 2.7	0044 ± 6499
CA	14. ± 14.	26. ± 9.	1.6 ± 11.4	81. ± 16.	41.4 ± 3.6	1.5 ± 7.7	13. ± 6.
TI	0. ± 3.	2.7 ± 1.4	5.4 ± 2.0	16. ± 3.	12.9 ± .7	2.7 ± 1.4	.77 ± .90
V	0.00 ± .11	.660 ± .068	.25 ± .10	.41 ± .14	.315 ± .030	.15 ± .07	0029 ± 4322
CR	0.0 ± .4	.026 ± .203	1.2 ± .3	.35 ± .40	.54 ± .09	5.49 ± .20	.027 ± .130
MN	0.0 ± .4	.13 ± .20	1.3 ± .3	1.9 ± .4	1.97 ± .09	1.1 ± .2	.61 ± .13
FE	.31 ± 9.32	35. ± 6.	2.4 ± 8.2	143. ± 12.	96.4 ± 2.6	45.6 ± 5.5	9.2 ± 3.8
NI	0.00 ± .11	.094 ± .066	.12 ± .10	.50 ± .13	.196 ± .029	3.06 ± .07	0037 ± 0423
CU	0.0 ± .7	.40 ± .42	4.13 ± .59	.014 ± .820	.52 ± .19	6.63 ± .40	.060 ± .269
ZN	.45 ± 2.43	.25 ± 1.52	10. ± 3.	.53 ± 2.97	.28 ± .66	6.0 ± 1.5	4.2 ± 1.0
BR	.12 ± 1.13	.48 ± .71	.37 ± .98	0.0 ± 1.4	.89 ± .31	.70 ± .67	52.4 ± .5
PB	.014 ± .614	3.4 ± 2.9	9.2 ± 4.1	1.6 ± 5.7	5.0 ± 1.3	5.9 ± 2.8	152. ± 2.

Table 7. Average Elemental Contributions (ng/m^3) for Each Source for NBS Data Set No. 1 (R-Mode Analyis)

ELEMENT	WOOD	OIL-B	REFUSE	COAL B	SOIL	SANDBLAST	MOTOR VEHICLE	AVG PRED CONTRIB	AVG OBS CONTRIB	AVG % ERROR
C	.251E+04	.384E+03	.666E+02	.345E+02	.846E+03	.667E+02	.227E+04	.618E+04	.621E+04	2.8
NA	.170E+02	.291E+02	.121E+03	.778E+02	.768E-01	.349E+02	.217E+03	.497E+03	.716E+03	102.6
AL	.433E+02	.223E+02	.411E-03	.468E+03	.208E+04	.405E-02	.175E+03	.279E+04	.298E+04	14.8
SI	.523E+02	.910E+02	.100E+01	.522E+03	.583E+04	.208E+04	.289E+03	.887E+04	.875E+04	11.5
S	.105E+01	.247E+03	.121E+03	.407E+03	.712E+02	.103E+00	.415E+02	.889E+03	.890E+03	5.2
CL	.659E+02	.112E+02	.372E+03	.519E+02	.432E+01	.151E+03	.624E+02	.719E+03	.653E+03	13.5
K	.563E+02	.607E+01	.232E+03	.381E+02	.936E+02	.346E+02	.175E-03	.461E+03	.506E+03	13.9
CA	.581E+02	.540E+02	.298E+01	.182E+03	.518E+03	.632E+01	.510E+02	.872E+03	.803E+03	19.6
TI	0.	.562E+01	.103E+02	.367E+02	.162E+03	.110E+02	.305E+01	.228E+03	.212E+03	11.9
V	0.	.136E+01	.473E+00	.919E+00	.394E+01	.598E+00	.115E-02	.730E+01	.778E+01	42.5
CR	0.	.534E-01	.233E+01	.785E+00	.672E+01	.225E+02	.108E+00	.324E+02	.302E+02	34.5
MN	0.	.278E+00	.255E+01	.430E+01	.247E+02	.432E+01	.243E+01	.386E+02	.393E+02	9.2
FE	.134E+01	.716E+02	.447E+01	.320E+03	.121E+04	.187E+03	.366E+02	.183E+04	.186E+04	3.3
NI	0.	.194E+00	.237E+00	.113E+01	.246E+01	.125E+02	.145E-01	.166E+02	.174E+02	20.4
CU	0.	.824E+00	.784E+01	.319E-01	.646E+01	.271E+02	.238E+00	.425E+02	.380E+02	31.7
ZN	.191E+01	.515E+00	.190E+02	.119E+01	.354E+01	.244E+02	.165E+02	.671E+02	.870E+02	38.0
BR	.500E+00	.989E+00	.701E+00	0.	.111E+02	.286E+01	.207E+03	.224E+03	.224E+03	77.0
PB	.597E-01	.713E+01	.175E+02	.366E+01	.625E+02	.243E+02	.603E+03	.718E+03	.701E+03	16.3

Table 5 presents the R-mode diagonalization and reproduction results. There are indications of 6 to 8 factors, and the study of these possible solutions lead to 7 factor solution as the best fit to the data in agreement with the classical factor analysis. The refined source compositional profiles are given in Table 6. The average elemental concentration contributions and the fit to the average composition are given in Table 7.

DISCUSSION AND CONCLUSIONS

This data set was created using nine source types. Neither Q- or R-mode analysis correctly determined the presence of 9 sources nor did the classical factor analysis. There are several reasons for this result. The soil and basalt sources were co-located around the perimeter of the imaginary area around the receptor site. The variation at the receptor site is then entirely due to meteorological conditions and they co-vary in the same manner. Thus, a factor analysis that operates on the variation will not be able to separate them. Since the variation in concentrations of Soil and Basalt is due entirely to the meteorology, the factor analysis would not be able separate these sources even if there were substantial differences in their concentration profiles. The other source type missed in the analysis is the steel mill. It was included as a very weak source. It does not represent a unique source of any element and it does not contribute sufficient mass for its variation to be observed above the random variation built into the data set. The R-mode analysis was able to obtain a 7 factor solution while it was not possible to do so with the Q-mode results. We can force fit the known source profiles using the Q-mode eigenvectors, but we cannot iteratively obtain good source profiles as we could with the R-mode results. Both approaches did give a reasonable description of the sandblast source composition. This source was not given to the analysts with the data as were the other profiles. The other unusual feature of the data set was the coupling of a motor vehicle composition with soil to give "road" material. Since the two sources were coupled the factor analysis could not separate them and the analysis, therefore, deduced the composition of the road source rather than motor vehicles. For real data, a similar result might provide a better estimate of the true impact of traffic on airborne particulate loadings than trying to separate vehicle exhaust from reintrained soil.

Although the factor analysis did not fully resolve this data set, the R-mode analysis did provide reasonably good estimates of the impacts of the various sources. A comparison of the average results of the two forms of analysis with the actual values are made in Table 8. In the R-mode analysis, a coal source was resolved permitting much better predictions of the average mass values. In order to compare the ability to resolve individual samples, linear regressions between the mass values calculated by the factor analysis and the actual values were performed. Table 9 presents these results. For well-resolved sources like sandblasting and soil/basalt, excellent agreement was obtained as

Table 8. Estimated Source Impacts ($\mu g/m^3$) Compared with True Values for NBS Data Set

Source	Q-Mode	R-Mode	Truth
Steel-A	--	--	0.05
Oil-B	2.3 ± 0.4	2.1 ± 1.0	2.0
Incinerator	1.8 ± 0.1	1.9 ± 0.1	1.3
Coal-B	--	2.2 ± 0.7	2.4
Basalt Soil	12.7 ± 0.5	12.5 ± 0.8	4.7 8.0
Road	3.0 ± 0.1	4.0 ± 0.1	7.1
Wood	7.4 ± 0.7	4.3 ± 0.5	3.3
Sandblast	5.2 ± 0.4	4.1 ± 0.2	4.2
Total:	34.2	31.1	33.05

Table 9. Comparison of R-Mode Calculated Mass Contributions with the Actual Mass Values for each Sample

SOURCE	CORRELATION	SLOPE	INTERCEPT
Wood	.800	0.873 ± 0.113	642 ± 367
Oil-B	.867	1.208 ± 0.209	454 ± 584
Incinerator	.968	1.502 ± 0.077	20 ± 84
Coal-B	.689	0.656 ± .191	383 ± 450
Soil/Basalt	.949	0.957 ± .051	294 ± 572
Sandblast	.990	1.029 ± 0.026	-195 ± 124
Road	.992	0.588 ± 0.006	- 67 ± 49

reflected by high correlation coefficients, a slope near 1.0 and a zero intercept. Coal burning, although resolved by the R-mode analysis, was not fit very well. these results indicate that the factor analysis obtained the major features of the data but did not always resolve the detailed structure. It must be remembered that part of the problem may be in the data set. The co-location and mulitple location of sources may not be entirely realistic and may cause some of the difficulties encountered in the analysis. On the other hand, these data have been extremely valuble in testing our methodology and would seem to indicate that more intensive study of the R-mode approach is warranted.

ACKNOWLEDGEMENT

This work was supported in part by the U.S. Envirmental Protection Agency under contract 68-02-3449 and by the Department of Energy under contract DE AC02-80EV10403. We would like to thank Drs. Lloyd Currie and Robert Gerlach of the National Bureau of Standards for creation of data set and for helpful discussions of the results.

REFERENCES

1. G.E. Gordon, "Receptor models," Environ. Sci. Technol., 14:792 (1980).

2. J.A. Cooper and J.G. Watson, Jr., "Receptor oriented methods of air particulate source apportionment," J. Air Poll. Control Assoc., 30:1116 (1980).

3. J.A. Cooper, "Review of the chemical receptor model of aerosol source apportionment," in Atmospheric Aerosol: Source/Air Quality Relationships (E.S. Macias and P.K. Hopke, Eds.) American Chemical Society, Washington, Symposium Series No. 167, p. 75.

4. M.S. Miller, S.K. Friedlander, and G.M. Hidy, "A chemical element balance for the Pasadena aerosol," J. Colloid Interface Sci., 39:165 (1972).

5. D.F. Gatz, "relative contributions of different sources of urban aerosols: Application of a new estimation method to multiple sites in Chicago," Atmospheric Environ., 9:1 (1975).

6. G.S. Kowalczyk, C.E. Choquette, and G.E. Gordon, "Chemical element balances and identification of air pollution sources in Washington, D.C.," Atmospheric Environ., 12:1143 (1978).

7. J.E. Core, P.L. Hanrahan, and J.A. Cooper, "Air particulate control strategy development: A new approach using chemical mass balance methods," in Atmospheric Aerosol: Source/Air Quality Relationships (E.S. Macias and P.K. Hopke, Eds.)

American Chemical Society, Washington, Symposium Series No. 167, p. 107.

8. P.K. Hopke, R.E. Lamb, and D.F.S. Natusch, "Multielemental characterization of urban roadway dust" Environ. Sci. Technol., 14:164 (1980).

9. D.J. Alpert and P.K. Hopke, "A quantitative determination of the sources in the Boston urban aerosol," Atmospheric Environ., 14:1137 (1980).

10. D.J. Alpert and P.K. Hopke, "A determination of the sources of airborne particles collected during the Regional Air Pollution Study," Atmospheric Environ., 15:675 (1981).

11. P.K. Hopke, "The application of factor analysis to urban aerosol source resolution," in Atmospheric Aerosol: Source/Air Quality Relationships (E.S. Macias and P.K. Hopke, Eds.) American Chemical Society, Washington, Symposium Series No. 167, p. 21.

12. C.K. Liu, B.A. Roscoe, K.G. Severin and P.K. Hopke, "The application of factor analysis to source apportionment of aerosol mass," Amer. Ind. Hygiene Assoc. J., 43:314 (1982).

13. K.G. Severin, B.A. Roscoe, and P.K. Hopke, "The use of factor analysis in source determination of particulate emissions," Particulate Science and Technology, (In Press, 1983).

14. B.A. Roscoe and P.K. Hopke, "Comparisons of weighted and unweighted target transformation rotations in factor analysis," Computers and Chemistry, 5:1 (1981).

15. B.A. Roscoe and P.K. Hopke, "Error estimation of factor loadings and sources obtained with target transformation factor analysis," Anal. Chimica Acta, 132:89 (1981).

16. J. Imbrie and T.H. Van Andel, "Vector analysis of heavy mineral data," Bull. Geol. Soc. Am., 75:1131 (1964).

17. A.T. Miesch, "Q-mode factor analysis of geochemical and petrologic data matrices with constant row-sums," U.S. Geological Survey Professional Paper 574-G (1976).

18. K.G. Joreskog, J.E. Klovan, and R.A. Reyment, Geological Factor Analysis, Elsevier Scientific Publishing Company, Amsterdam. 1976, 178 p.

19. R.W. Rozett and E.M. Petersen, "Methods of factor analysis of mass spectra," Anal. Chem., 47:1301 (1975).

20. T.C. Redman and A.R. Zinsmeister, "Comments on 'A quantitative determination of sources in the Boston Urban Aerosol'," Atmos. Environ., 16:1567 (1982).

21. D.J. Alpert and P.K. Hopke, "Response to comments by Redman and Zinsmeister," Atmospheric Environ., 16:1567 (1982).

22. O.G. Exner, "Additive physical properties," Collect. Czech. Chem. Commun., 31:3222 (1966).

23. E.R. Malinowski, "Theory of errors in factor analysis," Anal. Chem., 49:606 (1977).

24. A.A. Clifford, Multivariate Error Analysis, Applied Science Publishers, London. 1973, pp 70-82.

25. P.K. Hopke, D.J. Alpert, and B.A. Roscoe, "FANTASIA - a program for factor analysis to apportion sources in aerosols," submitted to Computers and Chemistry, March 1982.

26. L. Currie, "Model validation using simulated data sets," Proceedings of the APCA Specialty Conference on Receptor Models Applied to Contemporary Pollution Problems, Danvers, MA, October 1982.

27. B.A. Roscoe, P.K. Hopke, S.L. Dattner, and J.M. Jenks, "The use of principal components factor analysis to interpret particulate compositional data sets," J. Air Pollut. Contr. Assoc., 32:637 (1982).

28. P.K. Hopke, "Comment on trace element concentrations in summer aerosols at rural sites in New York State and their possible sources," by P.P. Parekh and L. Hussain, Atmospheric Environ., 16:1279 (1982).

29. P.K. Hopke, "An Introduction to Multivariate Analysis of Environmental Data", in Analytical Aspects of Environmental Chemistry, D.F.S. Natusch and P.K. Hopke, eds., J. Wiley and Sons, Inc., New York. 1983.

EVALUATION OF MULTIVARIATE AND CHEMICAL MASS BALANCE APPROACHES TO AEROSOL SOURCE APPORTIONMENT, USING SYNTHETIC DATA AND AN EXPANDED PACS DATA SET*

Richard T. DeCesar
Oregon Graduate Center
19600 N. W. Walker Road
Beaverton, Oregon

and

John A. Cooper
NEA, Inc.
7700 S. W. Nimbus Avenue
Beaverton, Oregon

The relative merits and limitations of chemical mass balance (CMB) and target transformation factor analysis (TTFA) have been evaluated using an expanded Portland Aerosol Characterization Study (PACS) data set and a synthetic data set. The original PACS data set contains chemical data for 27 species measured on 1300 size-segregated ambient filters collected over a one-year period and selected for analysis on the basis of frequency-of-occurance to represent the airshed's typical ventilation and activity conditions. The chemical data was expanded to 40 species for 37 sources and over 500 of the archived ambient filters including a continuous ten day period when samples were collected at 4 hour intervals. The synthetic ambient data set was constructed by the National Bureau of Standards by using a dispersion model, chemical profiles of source emissions and a perturbation routine to simulate measurement and source uncertainties.

The CMB and TTFA methods were evaluated on the basis of their ability to resolve and quantify the impact of contributing sources, amount and type of input data required, reproducibility, effectiveness at identifying unknown sources and ability to generate appropriate confidence limits for estimated source contributions.

The two methods produced comparable quantitative impact estimates for major sources which were suspected *a priori* to influence the aerosol. The CMB approach demonstrated superior source resolution for estimating the impact of sources which contribute either occasionally or at low levels.

*This research was sponsored in part by funds provided by a grant from the U.S. Environmental Protection Agency, No. R-807375-01-1.

INTRODUCTION

Over the past fifteen years, receptor oriented approaches have been examined as possible source impact assessment tools.[1] It has only been recently, however, that they have evolved to the point where they now form a distinct discipline and clear alternative to source oriented dispersion models. The number and variations of these receptor approaches have grown rapidly over the past few years without adequate evaluation or definition of their boundary conditions, strengths and weaknesses. The validation of these methods and definition of their boundary conditions are crucial as they begin to be used by non-specialists for development of control strategies. A major limiting factor to validating these procedures and defining their strengths and weaknesses has been the lack of an adequate data set for their application. Most past applications have either been to data sets designed for a purpose other than receptor models, been limited by analytical methods, and/or did not determine all of the major chemical species. In addition, most of the previous studies did not adequately characterize all of the major regional sources or include source matrix uncertainties when estimating predicted source strength errors.

A recently completed study, the Portland Aerosol Characterization Study (PACS),[2] was the first large urban area study designed from the beginning to provide all of the essential data required by one of the receptor models, the chemical mass balance method (CMB).[3] In this study, both the fine and coarse particle fractions from 37 sources representing 95% of the emission inventory in the Portland AQMA were chemically characterized for 27 chemical species. About 1300 individual fine and total ambient air particulate filter samples collected over a one-year period were chemically characterized for the same 27 chemical species from which 90% of the ambient aerosol mass could be accounted. Over 1700 CMB calculations were determined for 32 days selected out of 94 days sampled, to represent typical annual meteorological conditions. The PACS data and sample set represent the most complete, high quality data set currently available for testing and evaluating receptor models.

The primary objectives of this study were to expand the PACS data set for the purpose of evaluating the source resolving power of selected receptor models and the influence of adding additional chemical features to the data set and develop a data set that can be used by others to evaluate new models.

BACKGROUND

There is currently a great deal of confusion in the published literature as to what constitutes a receptor model. Such terms as chemical mass balance, factor analysis, multivariate statistical methods, principal component analysis, x-ray diffraction, automated particle analysis, regression analysis, etc. are casually used to describe different receptor models. Little distinction is made between a chemical or physical feature that is measured, the method used to measure the feature and the interpretive approach. There is, in fact, only one basic receptor model which is in turn a subset of a more fundamental model on which both dispersion and receptor models are based, i.e., the conservation of mass model. This model is a mathematical description of a perceived physical world; one in which it is assumed that mass arriving at a receptor (sampling site) from source j was transported with conservation of mass by atmospheric dispersion. Although this assumption may be a valid approximation for some applications, it is frequently not valid and the methodology of each approach to the fundamental model is built around methods of mitigating the effect of deviations from the basic model.

Although receptor models might be subdivided according to the feature measured (chemical, particle type, etc.), the measurement method (microscopy, x-ray diffraction, etc.) or by the interpretive approach,[1] the latter is considered to be most appropriate. In this case, there are two fundamental approaches to the receptor model; either a statistical analysis of the variability or a regression analysis of the mass. Variability analysis methods such as factor analysis, cluster analysis, principal component analysis, etc., provide valuable qualitative and semiquantitative insight into the nature of the sources and their contribution to the variability but are not able to provide quantitative information on source contributions to the particulate mass. The regression analysis approach, on the other hand, is quantitative; both over the total mass on a series of filters and over the chemical distribution on a single filter as with the chemical mass balance method. Hybrid methods such as target transformation factor analysis are combinations of the two basic approaches which assist with the quantification of the mass. There is another category which includes other miscellaneous qualitative approaches such as enrichment factors, trajectory analysis, etc.

The receptor model can be applied to any data set of conserving features of particles and gases such as elemental, chemical or partical type concentrations as measured by any appropriate quantitative analysis technique. Thus, chemical data developed by a valid x-ray diffraction analysis or particle type data generated by microscopic methods can be interpreted by any of the mass regression or variability analysis approaches. Data generated by these analytical techniques do not represent different receptor models, but means of generating useful analytical data.

MODEL EVALUATION CRITERIA

Although a large number of different model evaluation criteria could be listed depending on the user's objectives, the criteria felt to be most useful are similar to those used to evaluate analytical spectroscopic methods. These criteria include selectivity, resolution, signal to noise ratio, detection limits, reproducibility, quantitation and uncertainties, ease of use, costs, dynamic range of applicability, etc. These same evaluation criteria are also applicable to receptor source apportionment approaches.

The first requirement of a receptor model analysis is to generate a signal which exceeds the noise that can be used to identify a source (qualitative analysis) and quantify its impact. If a variability analysis approach is being used, variation in the concentration of measured features becomes the gross signal and the objective becomes one of separating the influence of the source on the variability from background influences such as meteorology. To maximize the signal in this case, maximum feature variability is required which suggests short sampling times; there is a point, however, where analytical sensitivity or uncertainty becomes the major source of noise and consequently the limiting factor.

Each of the above criteria have been used to evaluate CMB and TTFA approaches to source apportionment in a manner similar to that indicated above. The main focus of this paper, however, is the ability of an approach to resolve and quantify an impact as well as such criteria as input data requirements, reproducibility, effectiveness at identifying unknown sources and a method's ability to generate appropriate confidence limits for estimated source contributions.

TEST DATA SETS

Four selected data sets were used to evaluate and illustrate the relative effectiveness of the CMB and TTFA approaches to receptor modeling: two data sets from the PACS and two NBS simulated data sets. The NBS data sets were prepared as part of the Quail Roost II Workshop in Receptor Modeling.[4] The simulated ambient data sets (Data Sets 1 and 2) were constructed using a short term Gaussian plume dispersion model, hourly meteorological data, chemical profiles of source emissions and a mass balance equation. The data sets were made more realistic by propagating throughout their formation, estimates of sampling and analysis errors and source variability.

The real data sets used for evaluation consisted of two subsets of the expanded PACS data set (PACS II). The first PACS II data set used (data set 3 in this paper) is composed of coarse fraction values developed from 7 consecutive days (1/19/78 - 1/25/78) of 4 hour ambient sampling at the Portland Central Air Monitoring Station (CAMS).

Data set 4, also a part of the PACS II data set, consists of chemical measurements made on the fine fraction ambient samples collected at 8 hour intervals from 3/14/78 through 3/23/78 at the Portland Industrial Air Products site.

DATA INTERPRETATION AND RESULTS

Tables I-IV contain the results of applying the CMB and TTFA approach to data sets 1 through 4. Columns CMB-1 and CMB-2 were generated at the Oregon Graduate Center (OGC) using a CMB analysis procedure to quantitatively apportion the aerosol sources. The heart of this procedure is an effective variance least-square multiple regression algorithm.[3] This algorithm is contained within an interactive computer code which allows operator selection of which source profiles and elements to include in the regression. Column CMB-1 contains the mean source impacts calculated from 40 individual CMB analyses. The errors associated with the impacts in column CMB-1 are the standard deviations of the mean computed by,

$$\sigma_m = \sqrt{\frac{\Sigma(\overline{X} - X_i)^2}{n(n-1)}}$$

where $\overline{X}$ is the mean impact, X_i is an individual impact and n is the number of samples. To calculate the means and standard deviations in column CMB-1, a value of 0.0 $\mu g/m^3$ was used as the impact of a source not identified as contributing to a particular sample. Column CMB-2 contains the results of a single CMB analysis on the average chemistry of data set 1. The results in columns CMB-1 and CMB-2 were completed with the additional knowledge that a Sandblaster source, not in the original 13 source world list, was impacting data set 1. However, columns CMB-1 and CMB-2 were completed before an unknown source (Sandblaster) profile was made available. After completing a few initial fits, however, it was apparent that an unknown source was a major contributor to data set 1 and would therefore have a significant influence on the results. This unknown source appeared to contain a large amount of silicon and also large percentages of Cr, Ni and Cu compared to the other source profiles based on unexplained mass for these elements. If such a situation would arise in a real data set, the analysis would be discontinued until a probable identification of the source could be found. However, since the data set was to be analyzed with the information provided, the analysis was continued and an attempt to minimize its influence on the other source contributions was made by removing the elements Cr, Cu and Ni from the majority of the fits. The unknown source probably had its greatest effect on the contribution attributed to Coal Plant A which had a source profile dominated by Si. With Cr, Cu and Ni removed from the fit, the source profile of Coal Plant A in effect, tried to fill in the gap left by the unknown source. Because of this, the contribution attributed to Coal Plant A must be considered highly tentative. The analysis suggests that the unknown source affected the contributions calculated for the other sources to varying, but by lesser degrees.

Columns TTFA-1 and TTFA-2 are from results by Hopke and coworkers.[5] Column TTFA-1 was completed prior to the workshop with only the original information. Column TTFA-2 was completed with the additional information on the Sandblaster source. Column TTFA-3 was developed at OGC with the additional Sandblaster information using the computer program FANTASIA.[6]

In Table II, columns CMB-1 and CMB-2 were developed by Heisler of Environmental Research and Technology, Inc.[4] prior to the workshop, using an effective variance least-squares fitting procedure. Column CMB-1 is the average of 40 individual fits and column CMB-2 is the result of a single fit to the average chemistry of data set 2.

Columns CMB-3 and CMB-4 of Table 2 were completed at OGC using the same procedures as in columns CMB-1 and CMB-2 of Table 1. These two columns, CMB-3 and CMB-4 were generated after the workshop at a time when the average source impacts were available.

Column TTFA-1 in Table II was completed by Hopke and coworkers[6] with only the original information on data set 2. Column TTFA-2 of Table II was developed at OGC after the workshop with the same procedure used for column TTFA-3 of Table 1.

The data in Tables III and IV were developed at OGC. The CMB columns were generated using the same procedure as column CMB-1 of Table I. The TTFA columns were generated using the same procedure as column TTFA-3 of Table I.

METHOD EVALUATION

Tables 1-4 contain the estimated source impacts for 4 sets of ambient data developed using two of the most extensively applied interpretive approaches of receptor modeling, i.e. effective variance Chemical Mass Balance and Target Transformation Factor Analysis. The characteristics of the data sets analyzed to produce Tables I-IV permit, for the first time, an objective evaluation of the relative performance of the CMB and TTFA approaches.

Resolution

When applied to data sets I-IV the CMB method is consistently able to resolve and identify a greater number of the contributing sources than the TTFA approach. The number of extra sources resolved ranged from 1 for data set 1, to 9 additional sources resolved by the CMB approach for data set 2. Generally, the estimated impacts show the additional sources to be small contributors to the average aerosol. However, their impacts on certain sampling days can reach very significant levels. A clear example of the superior source resolving ability of the CMB approach compared to the TTFA approach is illustrated in Table II. The two sources which make the largest contribution to data set 2 are Soil and Basalt. Two independent applications of TTFA to these data result in the resolution of a single geological type source. On the other hand, all four applications (2 each, by 2 independent groups) of CMB methodology effectively resolved and quantitated the contributions of the Soil and Basalt sources. The general trend of superior source resolution ability of the CMB vs. the TTFA approach holds true for simulated data sets 1 and 2 and real data sets 3 and 4.

Quantitation

For the major sources contributing to data sets 1-4, the CMB and TTFA approaches generate comparable quantitative estimates of the source impacts. In addition, the contributions estimated for the major sources by both approaches agree reasonably well with the true values given in Tables I and II. However, Tables I-IV illustrate a number of notable differences in the ability of the two methods to accurately quantify the impact of less significant sources. As stated above, the CMB approach consistently resolved a greater number of sources and usually the additional sources are the minor contributors. In Table I, for example, the CMB derived results include impact estimates for Steel-A, Basalt and Coal-B, while none of the three sets of TTFA results contain estimates for Steel-A or Basalt and only one of the three contains an estimate for Coal-B. The source contributions listed in Table I for the CMB methodology for Coal-B and Basalt are reasonably consistent with the true values. The only application of TTFA which was able to quantify the impact of Coal-B produced an estimate which agrees well with the true value and the CMB calculated contribution.

Table II clearly shows a substantive difference in the effectiveness of the CMB and TTFA approaches to quantify the average impacts of sources which are known *a priori* to influence the aerosol composition. Two independent applications of CMB methodology (CMB-1 and CMB-3) were able to produce quantitative estimates for all 13 contributing sources which are generally in excellent agreement with the true values. This is in sharp contrast to the four out of thirteen sources which were quantified by the two independent applications of TTFA. The contributions estimated by TTFA-1 for the four sources agree reasonably well with the true values if the impact attributed to Soil is taken to be a sum of the Basalt and Soil contributions.

The source apportionment results for real data sets 3 and 4 follow the same pattern as those for simulated data sets 1 and 2. The CMB and TTFA approaches generate comparable estimated impacts for the major sources that are suspected *a priori* to influence the aerosol. For data sets 3 and 4 the CMB approach estimated the contributions of more sources than the TTFA approach. Since data sets 3 and 4 are composed of real data, there are no true values for comparison. However, the impacts estimated for the additional sources by the CMB approach appear reasonable. Taken as a whole, Tables I-IV strongly suggest that the CMB approach is the superior tool for estimating the impacts of sources which are known or suspected of contributing to the particles in an airshed, but are minor contributors.

Uncertainty Estimates

Table I provides a comparison between the error estimating schemes used with the CMB and TTFA approaches. Of the eight correctly identified sources listed under CMB-1, only one, the low level Steel-A source, has a corresponding standard deviation that places the estimated impact more than

3 σ's from the true value as illustrated in Figure 1. Two of the applications of the TTFA approach to data set 1 have corresponding uncertainties attached to the estimated source impacts. In column TTFA-1, four of five, and in column TTFA-2, three of seven correctly identified sources have corresponding standard deviations that place the estimated impacts more than 3 σ's from the true value. This comparison strongly indicates that for data set 1, the CMB approach generates source impacts with realistic uncertainties while the TTFA approach often severely underestimates the uncertainty in the calculated source contributions.

Identification of Unknown Sources

The comparison study indicated that the TTFA approach is superior to the CMB approach as a tool for indicating the nature of contributing sources which are not suspected *a priori* of influencing the aerosol at a receptor. Although the TTFA approach proved capable of indicating the nature of unsuspected sources, neither TTFA nor CMB was effective at identifying unsuspected sources which contributed to the aerosol. The best example of this phenomenon obtained as part of this study is found in the application of TTFA and CMB to data set 1. The chemical profile of a Sandblaster was one of the sources used by NBS in generating data set 1. However, this information was not originally furnished to the participating investigators who were only told of the possibility of one or more unknown sources contributing to data set 1. All three TTFA generated profiles for the unknown sandblaster source are in good agreement with the true Sandblaster composition. Thus, TTFA in this case, properly indicated the nature of the unknown source but this indication of the source's nature was not sufficient to identify the source.

Reproducibility

There has been some concern about the reproducibility of results developed by the TTFA and CMB procedures due to the number of subjective decisions required by both approaches. The results reported here suggest that the CMB procedure can be very reproducible while verifying the concern over the subjective nature of TTFA. In Table II, for example, the columns labeled "CMB-1 - CMB-3" and "CMB-2 - CMB-4" form two sets of CMB results where identical input data was used within the set and independent investigators developed the results within each set. The source by source agreement within the two sets of columns is excellent. This is a strong indication that state-of-the-art CMB analysis has developed to a point where it is systematic and effective in resolving, identifying and estimating the impacts of contributing sources.

Table I indicates that concern over the subjective nature of TTFA is valid. The overall agreement between the three columns of TTFA data in Table I is much less than the agreement between the two sets of CMB results in Table II. The most obvious difference between the TTFA results in Table I is the additional source listed under TTFA-2. Another conspicuous discrepancy is the large variation in estimated impacts for the Wood source, which range from 3.1 $\mu g/m^3$ to 7.4 $\mu g/m^3$. Another indication that the TTFA approach is deficient in terms of reproducibility is given by the wide range in percent Pb (4.07-22.03%) and Br (1.7-7.3%) values in the Motor Vehicle chemical profiles developed by three applications of TTFA to data set 1 and two applications of TTFA to data set 2. This is in spite of the fact that data sets 1 and 2 were generated using the identical chemical profile for Motor Vehicle Exhaust.

Cost

Once the source and ambient data has been entered in a computer, for data sets of the size reported on here, there is very little difference between the TTFA and CMB procedures in terms of cost per sample. Since the two approaches require the same input data for maximum effectiveness, neither approach has an economic advantage in a detailed airshed study. Columns CMB-2 and CMB-4 indicate, however, that the CMB approach has a decisive economic advantage as a tool for ongoing airshed monitoring. Columns CMB-2 and CMB-4 highlight the ability of the CMB approach to identify and accurately estimate the impact of major sources contributing to a single long term sample. Although it has never been done, it seems probable that the sources of a long term (one month, for example) real ambient sample could be apportioned in a similar manner. The TTFA approach, on the other hand, requires a relatively large number of samples to resolve even the major sources and is therefore not appropriate for long term or periodic samples.

CONCLUSIONS

The results of applying both the CMB and TTFA methods to four data sets lead to the following conclusions:

- The CMB method consistently identified and apportioned the impact of a greater number of contributing sources than the TTFA approach.
- For the major sources which were suspected *a priori* to influence the aerosol, the two methods produced comparable quantitative impact estimates.
- The CMB approach is a superior tool for estimating the impacts of sources which are suspected of contributing to the particles in an airshed, but do so either occasionally or at a low level.
- Of the two approaches, the CMB method generated more realistic uncertainties for the corresponding impact estimates.
- The TTFA approach was more effective at indicating the nature of sources which are not suspected *a priori* of influencing the aerosol at a receptor.
- For maximum effectiveness both approaches require accurate and precise chemical characterization of ambient and source samples.
- For maximum effectiveness both approaches require ambient samples collected over short time intervals.
- The CMB results were more reproducible by independent investigators.
- The CMB approach was suggested as a superior tool for cost effective, long-term airshed monitoring.
- Maximum confidence in a studies conclusions will be attained when both CMB and multivariate methods, are used.
- Most of the currently employed methods of receptor modeling can apportion the impact of major sources quite easily. There is, however, a progression of mathematical sophistication and source resolvability within groups of methods. For example, a substantive improvement in

source resolution occurs when switching from the tracer method to the effective variance least squares fitting approach to Chemical Mass Balance.

- Separating ambient samples into fine and coarse size fractions results in the physical separation of source impacts which, in turn, provides improved source resolution and impact assessment.

REFERENCES

1. See extensive references listed in the following review: J. A. Cooper, J. G. Watson, Jr., "Receptor Oriented Methods of Air Particulate Source Apportionment," J. Air Pollut. Control Assoc.,30: 1116 (1980).

2. J. A. Cooper, J. G. Watson, "Portland Aerosol Characterization Study," final report to the Oregon State Department of Environment Quality,(July 1979).

3. J. G. Watson, "Chemical Element Balance Receptor Model Methodology for Assessing the Source of Fine and Total Suspended Particulate Matter in Portland, Oregon," Ph.D. Dissertation, Oregon Graduate Center, (February 1979).

4. R. Stevens, T. Dzubay, C. Lewis, "Proceedings of Quail Roost II Workshop on Receptor Modeling, Section 5 Source Apportionment Applied to Simulated Data Sets," from preliminary copy transmitted by C. Lewis, (June, 1982).

5. P. K. Hopke, D. J. Alpert, B. A. Roscoe, "FANTASIA - A Program for Factor Analysis to Apportion Sources in Aerosols," submitted to Computers and Chemistry (1982).

6. P. K. Hopke, K. G. Severin, S. Chang, "Application of Target Transformation Factor Analysis for Apportionment of Atmospheric Aerosol Mass," prepared for Quail Roost II Workshop on Receptor Modeling, (March, 1982).

7. R. T. DeCesar, J. A. Cooper and S. Fields, "Source Apportionment of Simulated Ambient Data Sets Using Chemical Mass Balance Methodology," prepared for Quail Roost II Workshop on Receptor Modeling, (April, 1982)

TABLE I. Estimated Source Impacts (μg/m³) Compared with True Values for NBS Simulated Data Set #1

Source	CMB-1[a]	CMB-2[b]	TTFA-1[c]	TTFA-2[d]	TTFA-3[e]	True Value
Steel-A	0.34 ± 0.07	0.4 ± 0.3	———	———	———	0.05
Steel-B	———	———	———	———	———	———
Oil-A	———	———	2.3 ± 0.4	2.1 ± 1.0	———	———
Oil-B	2.0 ± 0.5	———		———	2.8	2.0
Incinerator	1.0 ± 0.3	0.9 ± 0.3	1.8 ± 0.1	1.9 ± 0.1	3.0	1.3
Glass	———	———	———	———	———	———
Coal-B	2.3 ± 0.5	3. ± 1.	———	2.2 ± 0.7	———	2.4
Coal-A	(8. ± 1.)	(12. ± 4.)	———	———	———	———
Aggregate	———	———	5.2 ± 0.4	———	———	———
Basalt	2.9 ± 0.8	———	———	———	———	4.7
Soil	11. ± 1.	12. ± 1.	12.7 ± 0.5	12.5 ± 0.8	18.	9.8
Auto	5. ± 1.	5.2 ± 0.9	3.0 ± 0.1	4.0 ± 0.09	3.1	5.3
Wood	3.2 ± 0.5	4. ± 2.	7.4 ± 0.7	4.3 ± 0.5	1.2	3.3
Sandblast	———	———	———	4.1 ± 0.2	4.0	4.2

[a] Ref. 7 completed at OGC after Quail Roose II Workshop, average of 40 individual CMB fits.
[b] Completed at OGC after Workshop, a single CMB fit to the average ambient chemistry.
[c] Refs. 4 and 6 completed by Hopke and co-workers prior to the workshop.
[d] Ref. 4 completed by Hopke and co-workers after the workshop.
[e] Completed at OGC after the workshop.

TABLE II. Estimated Source Impacts ($\mu g/m^3$) Compared with True Values for NBS Simulated Data Set #2

Source	CMB-1[a]	CMB-2[b]	CMB-3[c]	CMB-4[d]	TTFA-1[e]	TTFA-2[f]	True Value
Steel-A	0.58 ± 0.03	0.97 ± 0.14	0.6 ± 0.2	0.7 ± 0.2	------	------	0.42
Steel-B	0.24 ± 0.02	-----	0.22 ± 0.05	-----	------	------	0.29
Oil-A	0.44 ± 0.03	0.5 ± 0.1	0.5 ± 0.02	0.4 ± 0.2	------	------	0.56
Oil-B	1.2 ± 0.1	2. ± 1.	0.7 ± 0.3	-----	------	------	0.41
Incinerator	1.5 ± 0.1	1.5 ± 0.1	1.5 ± 0.2	1.3 ± 0.3	r	2.8	1.81
Glass	0.16 ± 0.03	-----	0.3 ± 0.1	0.4 ± 0.3	------	------	0.46
Coal-B	3.7 ± 0.2	4. ± 1.	3.8 ± 0.9	5. ± 1.	r	4.7	3.3
Coal-A	4.3 ± 0.3	6. ± 1.	5. ± 2.	7. ± 2.	------	------	3.7
Aggregate	2.1 ± 0.3	-----	2. ± 1.	-----	------	------	0.46
Basalt	8.2 ± 0.5	11. ± 2.	10. ± 3.	12. ± 3.	------	------	11.3
Soil	9.5 ± 0.4	8. ± 1.	8. ± 2.	7. ± 2.	r	22.	7.9
Auto	2.5 ± 0.1	2.5 ± 0.1	2.5 ± 0.2	2.5 ± 0.3	r	5.4	2.5
Wood	0.5 ± 0.1	0.6 ± 0.5	0.9 ± 0.3	1.1 ± 0.3	------	------	0.94

[a]Ref. 4 completed by Heisler of Environmental Research and Technology, Inc. prior to the workshop; average of 40 individual CMB fits.
[b]Ref. 4 completed by Heisler of ER&T, Inc. prior to the workshop, a single CMB fit to average ambient chemistry.
[c]Ref. 4 and 7 completed at OGC after workshop; average of 40 individual CMB fits.
[d]Completed at OGC after the workshop, a single CMB fit to the average ambient chemistry.
[e]Ref. 4 completed by Hopke and Co-workers prior to the workshop.
[f]Completed at OGC after the workshop.
r - Source has been resolved and identified but an estimate of the inpact is not currently available.

TABLE III. Source Impacts Estimated by CMB and TTFA for Coarse Fraction Samples Collected at Portland CAMS Site, January, 1978 ($\mu g/m^3$)

Source	CMB[a]	TTFA[a]
Marine	0.06 ± 0.04	1.8
Geological	23.	20.
Residual Oil	0.63 ± 0.07	----
Vegetative Burning	0.13 ± 0.09	----
Steel	0.2 ± 0.1	----
Secondary Sulfate	0.5 ± 0.2	----
Transportation	3.8 ± 0.5	1.1
Ti-Rich	----	15.2
Unexplained	11. ± 3.	2.1

[a]Completed at OGC.

TABLE IV. Source Impacts Estimated by CMB and TTFA for Fine Fraction Samples Collected at Portland Industrial Air Products Site, March, 1978 ($\mu g/m^3$)

Source	CMB[a]	TTFA[a]
Marine	0.5 ± 0.3	----
Geological	4.6 ± 0.7	7.1
Residual Oil	0.56 ± 0.08	----
Vegetative Burning	3. ± 1.	----
Kraft Recovery Boiler	7. ± 2.	3.2
Hogged-Fuel Boiler	0.11 ± 0.09	----
Aluminum Production	0.2 ± 0.1	----
Steel-Furnace	0.8 ± 0.2	----
Ferromanganese Production	0.2 ± 0.1	----
Excess Volatile Carbon	6. ± 2.	----
Excess Non-Volatile Carbon	0.8 ± 0.3	----
Secondary Sulfate	3.6 ± 0.5	2.3
Transportation	2.9 ± 0.6	0.6
Zinc Source	----	4.3

[a]Completed at OGC.

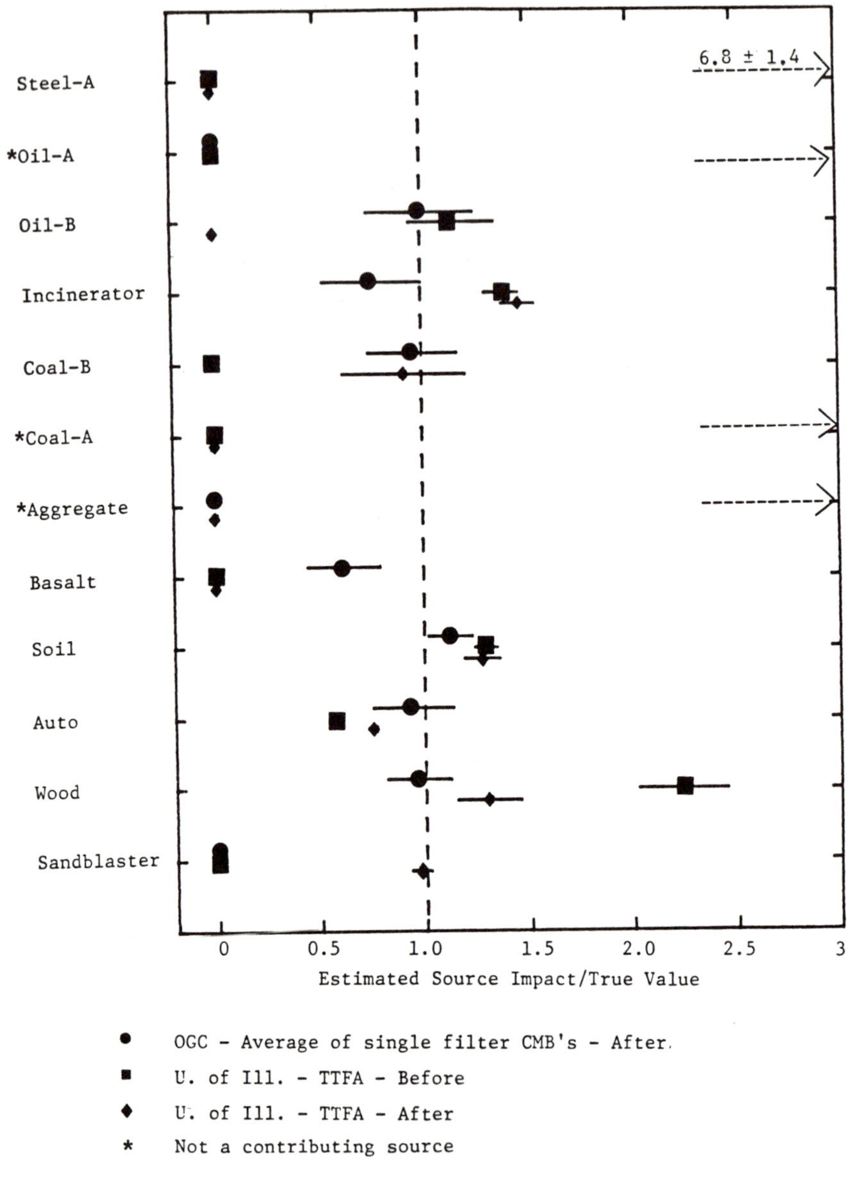

Figure 1 CMB and TTFA estimated source impacts/true value for NBS simulated data set #1.

STABILITY ANALYSIS OF RECEPTOR MODELS
THAT USE LEAST SQUARES FITTING

Ronald C. Henry,
Manager, Visibility Studies
ENVIRONMENTAL RESEARCH & TECHNOLOGY, INC.
2625 Townsgate Road
Westlake Village, CA

This paper presents an analysis of the fundamental causes of instability and large uncertainties in the results of Chemical Mass Balance (CMB) receptor models that depend on least squares resolutions. The mathematical technique of Singular Value Decomposition (SVD) is applied to the weighted source composition matrix to determine which sources are significantly affected by multicollinearities and which are not. Using the SVD, a new method is presented for determining estimable linear combinations of otherwise inestimable sources. Examples are given and a set of rules for the procedure presented. Finally, a new analysis of ridge regression solutions of CMB equations is given which shows that the ridge regression solution may often be equivalent to a CMB problem with a physically impossible source contribution matrix.

STABILITY ANALYSIS OF RECEPTOR MODELS
THAT USE LEAST SQUARES FITTING

Introduction

Most practical applications of receptor modeling have relied on the Chemical Mass Balance (CMB) model because of its inherent simplicity and credibility. Typically, CMB models use weighted least squares procedures to determine the source contributions which best fit the overdetermined system of chemical element mass balance equations. An important topic is calculating the uncertainty in these source contributions caused by errors in the receptor measurements and the assumed source compositions. If an approximate linear relationship exists between two or more source compositions, then errors can be greatly magnified giving rise to possibly negative source contributions and very large estimated uncertainties in the calculated source contributions. This problem is referred to as the multicollinearity or instability problem and is a major cause of uncertainty in CMB model results.

The most widely used method of dealing with multicollinearity is to remove one or more sources involved in the multicollinearity from the source composition matrix. The estimated uncertainties are then greatly reduced; but this is only an apparent improvement, since the possibility that the wrong sources were eliminated from the CMB is a source of uncertainty that is not included in the calculation.

Ridge regression has also been presented as a means of reducing uncertainties in calculated source contributions caused by multicollinearities.[1] Ridge regression alters the original problem to obtain a new, more stable, least squares CMB. This paper presents a new result which shows that the CMB source matrix equivalent to the ridge regression solution may not be physically meaningful.

The main purpose of this paper is to present some simple rules that can be used to determine which sources in a CMB can be estimated to a given accuracy and which cannot be so estimated. A new result is given that allows one to find estimable linear combinations of sources which by themselves cannot be accurately estimated. Without further physical evidence, these results define the limits of knowledge obtainable to a given degree of accuracy from a CMB calculation.

The mathematical basis of this paper is the Singular Value Decomposition (SVD) of the source contribution matrix. The connection of SVD to the multicollinearity problem is given below. Following this, a basic set of rules for stability analysis of a CMB problem is given with a simple but nontrivial example. The process for determining estimable linear combinations of sources is presented next, followed by examples. Finally, an example of analysis of the ridge regression approach is given.

Singular Value Decomposition

The fundamental properties of CMB least squares problems are best understood by the mathematical formalism of singular value decomposition of the source contribution matrix. The usual CMB model is

$$C = AS \tag{1}$$

where,

C	=	m x 1 vector of observed elemental concentrations at the receptor in $\mu g/m^3$;
A	=	m x n matrix of elemental source compositions as a fraction of total mass; and
S	=	n x 1 vector of source contributions at the receptor in $\mu g/m^3$.

The weighted least squares solution to equation (1) is:

$$S = (A^tWA)^{-1} A^tWC \tag{2}$$

where superscript "t" denotes matrix transpose and "-1" matrix inversion. W is a matrix of weights, usually equal to the inverse of the variances of the observed elemental concentrations, or the effective variances as given by Watson[2] to include errors in the source composition matrix. For these choice of weights, the error variances in the source contributions can be shown to be[3] the diagonal elements of the matrix $(A^tWA)^{-1}$.

For simplicity of notation in the following, the weighting matrix, W, is suppressed. This is accomplished by transforming the weighted least squares problem into an ordinary least squares problem as follows. Let $W = GG^t$, then rescale the CMB equation (1) by G:

$$GC = GAS \tag{3}$$

Let $C_1 = GC$ and $A_1 = GA$. Then the least squares solution to the rescaled problem is

$$\begin{aligned} S &= (A_1^tA_1)^{-1} A_1^tC_1 \\ &= (A^tG^tGA)^{-1}AG^tGC \\ &= (A^tWA)^{-1}AWC. \end{aligned} \tag{4}$$

This shows that rescaling by G and solving the usual way gives the same result as the weighted least squares equation (2). In practice, all that is usually required is to divide the observed elemental concentrations and the source elemental composition by the standard errors (<u>not</u> variances) of the elemental uncertainties and then proceed with the ordinary least squares solution. <u>In the remainder of this paper it is assumed that this normalization has been done</u>.

The singular value decomposition theorem[4] states that any matrix can be uniquely written as

$$A = UDV^t \tag{5}$$

where,

A	=	m x n matrix;
U	=	m x m orthogonal matrix;
V	=	n x n orthogonal matrix; and
D	=	m x n diagonal matrix, i.e., the only nonzero elements are d_i.

The columns of the U matrix are the eigenvectors of AA^t, the columns of V are eigenvectors of A^tA and the diagonal elements of D are the square roots of the eigenvalues of A^tA. Standard computer packages can be used to perform these calculations.[4]

Using the properties of orthogonal matrices and the singular value decomposition (SVD) of the <u>weighted</u> source composition matrix in equation (1), it can be shown (Appendix A.1) that the matrix of uncertainties is given by $VD^{-2}V^t$ where D^{-2} represents a diagonal matrix whose nonzero elements are d_i^{-2} and the V matrix is derived from the SVD of the weighted source composition matrix. From this expression, it is easy to see the connection between small eigenvalues and large uncertainties in calculated source contributions. The singular values are the square root of the eigenvalues of A^tA, thus d_i^{-2} is the inverse of the ith eigenvalue. If this eigenvalue is very small, then its inverse is very large giving rise to large uncertainties in CMB least squares calculations. An example of SVD of a typical source composition matrix is given next.

Table I gives the composition of seven sources for sixteen elements based on Watson[2] and the assumed uncertainties, used to obtain the weights, for each element. Table II contains the "V" part of the SVD in Equation (5). Also given in Table II are the singular values that appear on the diagonal of the D matrix in Equation (5).

The Estimable Space

Each row of the V matrix corresponding to a given source in Table II can be interpreted as the coordinates of each source in a seven-dimensional space defined by the eigenvectors that make up V. Silvey[5] proves the basic result that accurate least squares calculations are possible only in directions of eigenvectors corresponding to large eigenvalues. These eigenvectors define the "estimable space," and only sources that lie almost entirely in this space will be accurately estimable by CMB least squares methods.

Examination of Table II shows that there are two very small eigenvalues suggesting that the estimable space for this example is spanned by the first five of the seven eigenvectors. If, as in this case, the weights used are the inverse of the elemental errors in $\mu g/m^3$, then the inverse of the singular value is the uncertainty associated with that eigenvector in $\mu g/m^3$. Silvey[5] shows that any source in the estimable space will have an uncertainty equal to or less than the uncertainty associated with the smallest eigenvalue in the estimable space. From Table II it is seen that the maximum standard error associated with the five-dimensional estimable space is 1.03 $\mu g/m^3$. If a maximum standard error of 0.75 $\mu g/m^3$ was required then the estimable space would be three-dimensional in this case.

Taking approximately 1 $\mu g/m^3$ as our acceptable error, the length of each source in the five-dimensional estimable space is calculated as the square root of the sum of squares of the first five coordinates of each source in Table II. The results, Table III, show that four sources lie almost entirely in the estimable space. One source, vegetative burn, lies almost entirely outside this space and the two dust sources lie only partially in the estimable space. This means that there is essentially no possibility of estimating these last three sources to an accuracy of 1 $\mu g/m^3$.

The fact that the estimable space is five-dimensional means that it is possible to estimate at most five independent sources from a CMB of this source matrix. However, only four sources clearly lie in the estimable space. Silvey[5] points out that a linear combination of two inestimable sources may be estimable. In this case, it is easy to guess that the sum of continental and urban dust may be estimable. In the next section, a method for finding the linear combination of these two sources with the smallest error is presented.

Estimable Linear Combinations of Sources

The above discussion shows that it is not possible to estimate the source contribution of urban and continental dust to an accuracy of 1 $\mu g/m^3$ or better. It could be useful if some linear sum of these two could be estimated accurately. If the amount of continental dust could be arrived at by microscopy or estimated from a background station, then the contribution of urban dust could be estimated by difference with a known level of accuracy. Therefore, the next task is to mathematically determine the linear combination of these two sources contributions that is the most estimable, i.e., has the smallest error variance.

Since the source matrix has been weighted by realistic errors in the elements, arguments adapted from Silvey[5] allow one to calculate the error variance of a linear combination of source contributions in terms of the V matrix and its eigenvalues as given in Table II. The required linear combination is written as

$$L = x_1 S_1 + x_2 S_2 \qquad (6)$$

where S_1 and S_2 are the urban and continental dust contributions to the airborne mass. Appendix A.2 derives the general mathematical procedure used to find x_1 and x_2 that minimize the error in L. The result is that the quantity

$$0.7581 \text{ (continental dust)} + 0.6521 \text{ (urban dust)} \qquad (7)$$

can be estimated with a standard error of 0.996 $\mu g/m^3$. This compares to individual standard errors in the continental and urban dust source contributions of 5.9 and 6.8 $\mu g/m^3$, respectively.

It is possible that the minimum variance linear combination determined above is not estimable, i.e., it may not have sufficient length in the estimable space. To check this, the coordinates in the V space of the combination are formed and the length of this vector determined. It should lie almost entirely in the estimable space. If not, this linear combination is not estimable. In the above example, the linear combination lies entirely in the estimable space.

Note that estimable combination is a combination of the source contributions. Ordinary weighted least squares methods can be used to calculate each source contribution which are then used to calculate the estimable combinations.

It can be concluded from the above SVD analysis of the source matrix in Table I that the marine, leaded auto, residual oil, ferromanganese and the linear combination of urban and continental dust given above can be estimated to an accuracy of 1 $\mu g/m^3$ or better. The vegetative burn source is not estimable to this accuracy. Furthermore, these results are the best that can be obtained from the given source matrix and elemental errors without further information.

Summary of Rules

The above procedure for complete analysis of the fundamental limits of knowledge obtainable from a CMB with least square fitting are summarized in the following six steps:

1. Calculate the singular values and associated eigenvectors of the source matrix weighted by realistic errors in the elemental concentrations.
2. Select the maximum allowable error, M_E, in the estimate of the source contribution.
3. Determine the dimension of the estimable space as the number of singular values, d_i, for which $d_i^{-1} \leq M_E$.
4. Calculate the length of the source vectors in the estimable space. Those sources with about 95% or more of their length in the estimable space can be accurately determined by CMB.
5. Choose inestimable sources for which estimable linear combinations are desired. Using methods of Appendix A.2 calculate combinations with minimum variance $\leq M_E$.
6. Check that the length of the estimable linear combinations are ≥ 0.95.

In the six-step procedure given above, there are only two user-supplied parameters, the maximum acceptable error and the criterion of 0.95 for a minimum acceptable length in the estimable space. The only other required input is the inestimable sources for which estimable linear combinations are desired. An interactive computer program can easily be programmed to handle the complex calculations. In fact, the computer could be programmed to calculate all possible estimable linear combinations of inestimable sources.

A Second Example

Consider the 13-source, 20-element source composition matrix used by Stevens et al.[6] in their intercomparison study of receptor models. A SVD analysis was performed on this source matrix using as weights the inverse of the standard errors of the elements calculated from expressions given by Stevens et al. and assuming that the sources were present at their average values.

The first step is to perform a SVD of the weighted source matrix. The second step is to choose a maximum acceptable uncertainty. In this case, a limit of 0.75 $\mu g/m^3$ is selected. Next, the dimension of the estimable space is found by finding how many singular values there are with inverse less than 0.75. From Table IV it is found that the estimable space is seven-dimensional. Table V gives the lengths of the sources in the estimable space. The first four exceed the 0.95 limit, indicating that they can be estimated to the required accuracy. As an example of an estimable combination, the combination (found by methods of Appendix A.2)

$$0.384\ (\text{Oil B}) + 0.7094\ (\text{Glass}) + 0.5911\ (\text{Coal B}) \qquad (8)$$

has a standard error of 0.57 $\mu g/m^3$ and a length in the estimable space of 0.998. Thus, it is an estimable linear combination. Other estimable linear combinations of six or seven inestimable sources can be found, but are not likely to be useful.

Ridge Regression

It has been claimed[1] that ridge regression allows estimation of sources that cannot be separated by ordinary least squares. This section presents a new analysis of ridge regression which shows that this is not likely to be the case.

Ridge regression alters the ordinary weighted least squares solution by adding a smoothing parameter. This parameter is chosen so that the new least squares solution has smaller uncertainty. Of course, the addition of the smoothing parameter causes a systematic error in the answer. Various methods are used to determine a suitable value of the smoothing parameter to give a small uncertainty and a small bias in the answer.

In Appendix A.3 a new result is derived which allows explicit calculation of a perturbed source matrix whose least squares solutions are identical to the ridge solution for any given value of the smoothing parameter. For example, by trial and error, it was found that a smoothing parameter of 0.1 when applied to ridge regression of the seven-source matrix in Table I gives standard errors of about 1.2 $\mu g/m^3$ or less for the source contributins. Using the formulae derived in Appendix A.3, this ridge solution is found to be equivalent to an ordinary weighted least squares solution of the source contribution matrix in Table VI.

Comparison of Tables I and VI shows significant differences between the source compositions. The smallest differences are shown by the estimable sources, while the vegetative burn and dust sources show large discrepancies. The ridge solution also shows unrealistic negative elemental mass fractions. For example, the ridge solution is able to separate continental and urban dust because it implicitly sets the urban dust mass fractions of manganese, potassium, chlorine, aluminum nickel, bromine and nonvolatile carbon to negative values.

The apparent ability of ridge regression to solve the multicollinearity problem is seen to be based on implicitly changing the CMB problem to one which is not physically meaningful. The analysis and rules presented above, therefore, give the fundamental limits of knowledge obtainable from a CMB calculation, ridge regression notwithstanding. The problems with ridge regression presented here are not the only ones. The review article of Draper and Van Nostrand[9] points out many difficulties with the application of ridge regression and questions previous empirical verifications of ridge regression by analysis of simulated data.

Discussion

The above results are only a few of the most significant that can be obtained from the SVD approach to CMB calculations. It is also possible to easily determine if two nearly identical sources can be accurately distinguished or if additional elements or other properties will help separate sources. There is also a close relationship between SVD and factor analysis that can be explored. These and other applications must wait for future publication.

The above analysis based on SVD can and should be applied before any CMB receptor model study is begun. It is a very useful design tool in that it allows calculation of the fundamental limits of the ability of the CMB to accurately estimate sources. By identifying estimable linear combinations of sources, the above methodology makes possible the optimal use of additional information on source contributions.

All of the above SVD calculations were performed on a personal computer. The methodology and theory has been extensively tested on practical examples and cross checked by independent methods where possible, and checked for internal consistency as well. It is suggested that this methodology be added as a preprocessor to existing CMB least squares programs.

Finally, it is pointed out here without proof that the SVD methodology can be extended to the case where there is error in the source compositions by using weights based on the effective variances.

APPENDIX A

Mathematical Developments

Nomenclature

A	=	m x n matrix of source compositions <u>weighted</u> by inverse of the standard errors of the elements;
S	=	n x 1 vector of source contributions;
A	=	UDV^t = the singular value decomposition of A; and
$[d_i]$	=	diagonal matrix with d_i on the diagonal.

Properties of orthogonal matrices used later are that UU^t and U^tU are identity matrices of appropriate dimensions.

A.1 Relationship Between Singular Values and Uncertainties

The covariance matrix of the least squares solution of the CMB problem

$$C = AS \tag{A-1}$$

is $(A^tA)^{-1}$.

Using the SVD and the properties of orthogonal matrices U and V,

$$\begin{aligned} (A^tA)^{-1} &= (VDU^tUDV^t)^{-1} \\ &= V[d_i^{-2}]V^t \end{aligned} \tag{A-2}$$

This shows that small singular values lead to large values in uncertainties.

A.2 Method of Calculating Minimum Variance Linear Combinations of Sources

Let there be P inestimable sources. The linear combination

$$L = \sum_{i=1}^{P} x_i S_i \tag{A-3}$$

with minimum variance is sought.

Let $\vec{V}_i$ be one of the n columns of the V matrix from the SVD of the weighted source composition matrix, then Silvey[5] proves the result that if

$$X = \sum_{i=1}^{n} \alpha_i \vec{V}_i \tag{A-4}$$

then

$$\mathrm{Var}\,(X^t S) = \sum_{j=1}^{n} (\alpha_j/d_j)^2 \quad . \tag{A-5}$$

Here S is taken to be the weighted least squares estimate of the source contributions. As an example of the use of A-5, let $X = (0\ldots0,1,0\ldots,0)^t$ where 1 is in the ith position, then

$$X^t \cdot S = S_i \ ; \text{ and} \tag{A-6}$$

$$\begin{aligned} \alpha_j &= X \cdot \vec{V}_j \\ &= v_{ij} \ , \end{aligned} \tag{A-7}$$

where v_{ij} is the ith component of $\vec{V}_j$.

Therefore, the variance of S_i is expressed in terms from the SVD as

$$\mathrm{Var}\,(S_i) = \sum_{i=1}^{n} (v_{ij}/d_j)^2 \quad . \tag{A-8}$$

Similarly, the variance of the required linear combination in A-3 can be computed in terms of the V matrix and the singular values. Let S be the nx1 vector of the source contributions as before, and X be the nx1 vector to be determined. Then, A-3 is written as:

$$L = X^t S \ , \tag{A-9}$$

if it is required that the first P rows of S are the inestimable sources and all the components of X after the Pth are zero.

Expressing X as a combination of the $\vec{V}_i$,

$$X = \sum_{k=1}^{n} \alpha_k \vec{V}_k \ , \tag{A-10}$$

taking the dot product of both sides

$$\alpha_j = X \cdot \vec{V}_j = \sum_{i=1}^{n} x_i v_{ij} \tag{A-11}$$

Since only the first P of the x_i are nonzero

$$\alpha_j = \sum_{i=1}^{P} x_i v_{ij} \quad . \tag{A-12}$$

From Equation A-5,

$$\mathrm{Var}\ (X^tS) = \sum_{j=1}^{n}\left[\left(\sum_{i=1}^{P} x_i v_{ij}\right)/d_j\right]^2 \tag{A-13}$$

Let

$$Q = V_P\ [d_i^{-1}] \tag{A-14}$$

where V_P is a Pxn matrix made up of the P rows of the V matrix corresponding to the P inestimable sources and the d_i are the singular values. Equation A-13 is written in matrix terms as

$$\mathrm{Var}\ (X^tS) = (Q^tX)^t\ (Q^tX) = X^tQQ^tX \tag{A-15}$$

where the vector X has been truncated to a Px1 vector by dropping the last n-P zero elements.

The expression in A-15 is known as a quadratic form and the value of X which minimizes it is the eigenvector corresponding to the smallest eigenvalue of the symmetrix matrix QQ^t. Also, the minimum value is equal to the smallest eigenvalue. For proof, see Courant and Hilbert[7] or Rao[8]. Thus, finding the appropriate linear combination is accomplished by an eigenvector analysis of QQ^t and selection of the eigenvector associated with the smallest eigenvalue.

A.3 Derivation of Least Squares Problem Equivalent to Ridge Regression

The ridge regression solution is based on the matrix

$$(A^tA + \lambda I)^{-1}A^t \tag{A-16}$$

where λ is the smoothing parameter and I is an identity matrix[3]. It is required to find a matrix, B, such that the ordinary least squares solution for B is the same as the ridge solution. This is true if B satisfies

$$(A^tA + \lambda I)^{-1}A^t = (B^tB)^{-1}B^t \tag{A-17}$$

From the SVD of A,

$$(A^tA + \lambda I)^{-1} = (V[d_i^2]V^t + \lambda VV^t)^{-1} VDU^t \qquad \text{(A-18)}$$

using the fact that $VV^t = I$.

Then, simple matrix algebra shows that

$$(A^tA + \lambda I)^{-1}A^t = V[d_i/(d_i^2 + \lambda)]U^t \qquad \text{(A-19)}$$

If $$B = U[(d_i^2 + \lambda)/d_i]V^t \qquad \text{(A-20)}$$

it is easily shown that

$$(B^tB)^{-1} B^t = V[d_i/(d_i^2 + \lambda)]U^t \qquad \text{(A-21)}$$

This proves that B, as defined in Equation A-20, has the required property.

Using the fact that

$$U = AV[d_i^{-1}] \qquad \text{(A-22)}$$

B can be written as

$$B = AV[(d_i^2 + \lambda)/d_i^2]V^t \qquad \text{(A-23)}$$

Note that if $\lambda = 0$, $B = A$ since $VV^t = I$.

Equation A-23 gives the required relationship between the original source composition matrix, A, and the source matrix equivalent to the ridge solution, B.

References

1. H.J. Williamson, W.D. Balfour and C.E. Schmidt, "Chemical element balance analysis using the weighted ridge regression technique," (1982).

2. J.G. Watson, "Chemical element balance receptor model methodology for assessing the sources of fine and total suspended particulate matter in Portland, Oregon," Ph.D. Dissertation, Oregon Graduate Center, University Microfilms, Ann Arbor, MI, (1979).

3. N.R. Draper and H. Smith, Applied Regression Analysis, 2nd ed., John Wiley & Sons, New York, New York (1981).

4. C.L. Lawson and R.J. Hanson, Solving Least Squares Problems," Prentice-Hall, Englewood Cliffs, NJ, (1974).

5. S.D. Silvey. "Multicollinearity and imprecise estimation," J. of Royal Statistical Soc., Series B, 31: 539-552 (1969).

6. R. Stevens et al., "Mathematical and empirical receptor models: Quail Roost II," Environmental Sciences Research Laboratory, Office of Research and Development, U.S. EPA, Research Triangle Park, NC 27711, (1982).

7. R. Courant and D. Hilbert, Methods of Mathematical Physics, Vol. 1, 1st English Ed., Interscience Publishers, New York. (1953), pp 31-34.

8. C.R. Rao, Linear Statistical Inference and Its Applications, 2nd Ed. John Wiley & Sons, New York (1973), p 62.

9. N.R. Draper and R.C. Van Nostrand, "Ridge Regression and James-Stein Estimation: Review and Comments," Technometrics, 21, 450-465 (1979)

Table I. Source composition matrix for singular value decomposition example. Values are given as the percent mass fraction.

Element/Source	Marine (Sea Spray)	Leaded Auto Exhaust	Residual Oil Combustion	Ferro-manganese Production	Vegetative Burning	Urban Dust	Continental Dust	Assumed Error ($\mu g/m^3$)
Nonvolatile C	0.0	3.8	3.1	1.5	3.5	1.85	0.59	0.5
Mg	4.8	0.0	0.0	0.0	0.0	1.3	1.76	0.05
Al	0.0	1.1	0.53	0.64	1.44	8.84	11.7	0.3
Si	0.0	0.82	0.96	0.99	0.89	22.3	25.4	0.7
Cl	40.0	3.0	0.0	0.42	5.55	0.0	0.0	0.3
K	1.4	0.072	0.28	10.5	0.6	1.03	1.0	0.15
Ca	1.4	1.25	1.56	1.3	1.07	2.44	0.93	0.10
Ti	0.0	0.0	0.11	0.046	0.0	0.66	0.76	0.02
V	0.0	0.0	3.44	0.024	0.0	0.023	0.025	0.02
Cr	0.0	0.0	0.047	0.042	0.0	0.045	0.03	0.01
Mn	0.0	0.0	0.046	17.3	0.12	0.123	0.2	0.20
Fe	0.0	2.1	2.97	2.1	0.19	6.0	6.8	0.25
Ni	0.0	0.018	5.36	0.0	0.0	0.0093	0.0002	0.025
Zn	0.0	0.35	0.4	0.58	0.0	0.11	0.041	0.01
Br	0.2	5.0	0.013	0.16	0.055	0.02	0.0	0.05
Pb	0.0	20.0	0.11	0.045	0.0	0.37	0.006	0.20

Table II. V matrix from singular value decomposition of source composition matrix in Table I.[a]

(values x 10,000)

Source/Eigenvector	1	2	3	4	5	6	7
Marine	105	8937	3634	1066	2210	719	613
Leaded auto	390	3057	-8709	3823	56	216	-104
Residual oil	9973	-394	478	384	145	37	-32
Ferro-manganese	522	1664	-3254	-8861	2779	371	-102
Vegetative burn	44	952	162	-97	174	-8083	-5806
Urban dust	276	1763	-332	-1691	-6081	-4340	6168
Continental dust	210	1964	86	-1654	-7098	3891	-5278
Eigenvalue	7.795	2.977	2.175	1.525	0.941	0.0362	0.00935
Singular value	2.792	1.725	1.475	1.235	0.970	0.190	0.0967
Standard error in $\mu g/m^3$ (inverse of singular value)	0.358	0.580	0.678	0.810	1.031	5.26	10.34

[a]Columns are eigenvectors of the weighted matrix A^tWA, where the weights are the inverse squares of the uncertainties given in Table I.

Table III. Length of source vectors in five-dimensional estimable space.

	Source	Length in Estimable Space
1.	Marine	0.9955
2.	Leaded auto	0.9997
3.	Residual oil	0.99998
4.	Ferro-manganese	0.9992
5.	Vegetative burn	0.0986
6.	Urban dust	0.6567
7.	Continental dust	0.7550

Table IV. Singular value analysis of thirteen-source composition matrix of Stevens *et al.*[6]

	Singular Value	Uncertainty ($\mu g/m^3$)
1.	22.34	0.0448
2.	6.876	0.1454
3.	6.096	0.1641
4.	5.899	0.1695
5.	4.589	0.2179
6.	2.022	0.4944
7.	1.465	0.6825
8.	1.0576	0.9455
9.	0.6369	1.570
10.	0.4675	2.139
11.	0.3332	3.001
12.	0.26305	3.802
13.	0.1484	6.739

Table V. Length of sources in seven-dimensional estimable space of thirteen-source example.

	Source	Length
1.	Auto	1.00
2.	Wood combustion	0.997
3.	Incineration	0.997
4.	Oil A	0.979
5.	Steel B	0.907
6.	Glass manufacture	0.722
7.	Soil	0.631
8.	Coal B	0.630
9.	Steel A	0.591
10.	Basalt	0.525
11.	Oil B	0.413
12	Aggregate	0.329
13.	Coal A	0.109

Table VI. Source composition matrix equivalent to ridge regression solution of seven-source matrix (Table I) with a smoothing parameter of 0.1. (values are given as a percentage of total mass.)

Element/Source	Marine	Leaded Auto	Residual Oil	Ferro-manganese	Vegetative Burning	Urban Dust	Continental Dust
Nonvolatile C	-1.51	3.9	3.13	1.35	18.63	-2.53	4.33
Mg	5.15	-0.01	-0.01	0.0	-2.07	1.93	1.42
Al	-1.35	1.25	0.55	0.62	12.11	-0.16	21.25
Si	-0.75	0.73	0.9	0.61	3.03	23.78	28.47
Cl	40.73	3.03	0.02	0.35	14.09	-3.92	2.4
K	1.25	0.01	0.26	11.19	2.35	-0.64	2.41
Ca	1.43	1.15	1.54	1.15	1.37	7.17	-2.95
Ti	-0.02	-0.01	0.1	0.03	-0.06	0.74	0.81
V	-0.01	-0.01	3.48	0.02	0.03	-0.07	0.09
Cr	0.0	-0.01	0.04	0.04	-0.06	0.13	-0.04
Mn	-0.08	-0.07	0.02	18.51	0.57	-2.48	2.2
Fe	-0.18	2.16	2.98	2.11	0.44	6.2	7.75
Ni	-0.01	0.01	5.43	-0.01	0.03	-0.1	0.07
Zn	0.02	0.35	0.4	0.6	-0.22	0.34	-0.17
Br	0.17	5.24	0.01	0.14	0.26	-0.49	0.4
Pb	0.01	20.96	0.09	-0.08	-0.73	0.14	0.16

Response to

STABILITY ANALYSIS OF RECEPTOR MODELS
THAT USE LEAST SQUARES FITTING

Hugh J. Williamson
Senior Scientist
Radian Corporation
P.O. Box 9948
Austin, Texas 78766

The author has correctly pointed out that multicollinearity is a very important problem in source apportionment and that analysis of eigenvectors can be used to identify multicollinearities and investigate their nature. The variance inflation factor[1] can also be used to quantify the magnitude of the effect of multicollinearity on each estimated source contribution.

The author discusses the use of estimable functions to combine the particulate contributions of collinear sources. He subsequently presents an interpretation of source apportionment through ridge regression analysis. The reviewer's concern falls into two areas: (1) the physical interpretability of the estimable functions and (2) the validity of the doubts raised about the use of ridge regression in CMB analysis on the basis of intermediate results in only one numerical test case for a single value of λ. A discussion follows.

The author discusses the use of normalized minimum-variance linear combinations of source contributions involved in multicollinearities. Consider, for example, the estimable function F given in the author's expression (8):

$$F = 0.384 \text{ (Oil B)} + 0.7094 \text{ (Glass)} + 0.5911 \text{ (Coal B)} \qquad (1)$$

The question is, how should F be interpreted? Suppose F had the value 20 $\mu g/m^3$, and the ambient particulate concentration was 100 $\mu g/m^3$. Since the weights are all less than one, F is less than the combined particulate contribution of the three sources. Thus, F tells us that Oil B, Glass, and Coal B combined contribute somewhat more than 20 $\mu g/m^3$ to the ambient particulate concentration, but F alone does not tell us how much more.

Moreover, the unequal weights in the linear combination, while necessary to achieve the mathematical property of minimum variance, are unrelated to the importance of the sources in any practical sense. For example, the fact that the source Glass is weighted almost twice as heavily as Oil B is unrelated to the importance of the two sources as far as their contributions to pollutant levels are concerned. Thus, it would seem that the unequal weights obscure the interpretability of F.

It is suggested that the following simple function F' is more easily interpreted:

$$F' = \text{Oil B} + \text{Glass} + \text{Coal B} \qquad (2)$$

Suppose, for example, that F' had the value 35. This would reveal a useful piece of information: The three sources combined contributed 35 $\mu g/m^3$ to the ambient concentration of 100 $\mu g/m^3$.

The author also derives a source signature matrix S' whose least squares solution equals the ridge solution corresponding to the original matrix for a given value of λ. He shows that, for a particular numerical test case and a value of $\lambda = 0.1$, S' has several negative values with small magnitudes. The differences between S' and the original matrix are shown in the author's Tables I and VI. The author concludes that "The apparent ability of ridge regression to solve the multicollinearity problem is seen to be based on implicitly changing the CMB problem to one which is not physically meaningful."

It is true that in ridge regression the original problem is perturbed to gain stability. The larger λ is, the larger the perturbation is. It is possible to choose a value of λ too large to allow a reasonable solution of the original problem. However, criteria exist to evaluate the physical reasonableness of a ridge solution. If the solution corresponding to a particular value of λ was unreasonable, a ridge solution for a smaller λ, representing a smaller perturbation of the original problem, could be selected.

In the particular numerical example presented by the author, it should be noted that most of the negative values in S' are perturbations of values which were originally zero. In no instance is the entry in S' negative for a unique tracer for a given source. None of the negatives is extreme, the largest in magnitude being only -3.92 percent.

The basic question which has been raised is whether a calculation can be valid even though some negative estimates are used for quantities which physically cannot be negative. The answer to this question is yes, as is illustrated by the following simple numerical example. Suppose one is estimating the mean value of a nonnegative quantity X, and a random sample of three values is selected whose true values are 15, 7, and 0. Suppose also that, due to measurement error, the values 15.9, 7.1, and -0.5 are recorded. The question is, can a meaningful average value be calculated from these data? Or does the negative value -0.5 invalidate the sample? The average of the true values 15, 7, and 0 is 7.3. The average of the three measured values 15.9, 7.1, and -0.5 is 7.5. The two averages, 7.3 and 7.5, are very close. Notice that if the "impossible" value -0.5 had been replaced by the "possible" value 0, the mean value would have had a larger error. The negative error in the value -0.5 counterbalanced the positive errors in the other two values to an extent. If the one negative value had been thrown out and the mean estimate had been based on only the two positive values, a large positive bias would have resulted.

The simple calculation of a mean value, of course, is very different from a ridge regression analysis. However, both are statistical procedures designed to allow positive and negative errors to balance or "average out" to the greatest extent possible. The presence of one negative value in a sample of size three did not prevent the calculation of an acceptable mean value. By the same token, the presence of some negative values with small magnitudes in the S' matrix does not automatically invalidate the final results of a ridge regression analysis. The positive and negative perturbations of elements of S', as well as positive and negative original source signature errors, counterbalance in the statistical calculations.

Moreover, it is not necessary to speculate about the accuracy of ridge regression results on the basis of intermediate calculations. Published studies of the final results themselves provide a direct basis for evaluating ridge regression. Consider the following statement by Vinod in a survey article on ridge regression (RR):[2]

> In Monte Carlo experiments the true regression coefficients are specified. The data structure is usually chosen from a real life regression problem. Random numbers are used to create hundreds of "typical" regression problems. The estimated regression coefficients using RR and OLS are compared in terms of mean-squared error (MSE), i.e., the average squared Euclidian distance between the estimate and the parameter. In the following independent studies by investigators from diverse fields, the superiority of RR over OLS [ordinary least squares] is almost always noted; although there is wide disagreement about the "optimum" RR method.

In support of the statements above, Vinod subsequently references sixteen published papers and refers to "many unpublished dissertations." In view of the proven success of ridge regression in many independent studies, the reviewer suggests that it should not be discarded on the basis of certain intermediate results for a single value of λ in one numerical test case.

The disagreement about the optimum RR method pertains largely to selection of the parameter λ. However, λ is ordinarily chosen from within a stable region, where moderate changes in λ do not affect the coefficients significantly. Thus, slight deviation from the optimal λ would not cause serious errors.

The quotation above was chosen from the extensive literature on ridge regression because (1) it appears in a survey article and summarizes results from many independent studies, and (2) the results to which Vinod refers are objective in nature, being comparisons of true and estimated regression coefficients. Comments by other writers regarding ridge regression are discussed in a report now being prepared by the reviewer.

The author refers to the negative values in S' as "unrealistic negative elemental mass fractions." If intermediate results with "unrealistic" negative values are a concern, then that concern also applies to estimable functions. The author states, "Ordinary weighted least squares methods can be used to calculate each source contribution which are then used to calculate the estimable combinations." Thus, if the signatures of two sources were collinear, ordinary weighted least squares would first be used to estimate their individual contributions, S_1 and S_2. Then the intermediate results, S_1 and S_2, would be used to calculate an estimable function C

$$C = X_1 S_1 + X_2 S_2 \qquad (3)$$

However, it is well known that when strong multicollinearities are present, conventional least squares frequently produces sign reversals.[1,2] In many cases, either S_1 or S_2 will be negative with a large magnitude, and the other will have a counterbalancing large positive error.

But the point is that even when S_1 and S_2 are not individually physically reasonable, a linear function of them may be. The same principle applies in the case of ridge regression. Uncertainties in individual elements of S' have the opportunity to balance or "average out" to an extent in the calculation of the final results, especially if the number of species used is large. Thus, the reasonableness of either ridge regression or estimable functions should be evaluated on the basis of final results, not on the basis of selected intermediate calculations.

The reviewer agrees with the author that preliminary methods for identifying multicollinearities should be used, and that limits exist in a practical problem in the number of separate sources that can be resolved accurately. However, neither the combination of sources into categories nor the use of estimable functions can eliminate multicollinearity entirely. In many problems, avoidance of moderate multicollinearities would require severely limiting the number of separate source contributions (or linear combination thereof) to be considered. Ridge regression offers the possibility of improving the accuracy of CMB solutions, even when the strongest multicollinearities have been eliminated and only moderate ones remain.

References

1. H. J. Williamson, W. D. Balfour, and C. E. Schmidt, "Source Apportionment through Weighted Ridge Regression," Presented at the EPA 1982 Workshop on Receptor Model Validation: Mathematical and Empirical (1982).

2. Vinod, H. D., "A Survey of Ridge Regression and Related Techniques for Improvements over Ordinary Least Squares," Review Econ. and Statist. 60 (1978).

ELEMENTAL TRACERS OF CANADIAN SMELTER AEROSOL TRANSPORTED INTO THE NORTHEASTERN UNITED STATES

Kenneth A. Rahn →

Noëlle F. Lewis, Douglas H. Lowenthal
Center for Atmospheric Chemistry Studies,
Graduate School of Oceanography, University
of Rhode Island, Narragansett, Rhode Island

During the last several years, we have noticed that air masses arriving in Rhode Island from the northwest behind strong cold fronts often contain higher concentrations of indium and arsenic than expected from their generally clean character. The nature of these elements and their directionality suggested that Canadian smelters in the Sudbury Basin of southern Ontario and Quebec, roughly 1000 km from Rhode Island, might be their source. Systematic study of this phenomenon using three years of continuous aerosol samples from Rhode Island has now provided strong evidence in support of this view. By using the ratio of indium to noncrustal manganese as a tracer, we have found that this smelter aerosol tends to come in pulses of 1-4 semi-weekly samples, that these pulses are much more frequent and stronger in winter than in summer, that this aerosol cannot be explained by typical northeastern or midwestern aerosol, and that it does match aerosol observed directly in southern Ontario. Trajectory analysis for the periods of smelter aerosol confirms that flow to Rhode Island from the northwest is by far the most common occurrence. During winter, this smelter aerosol accounts for less than 1-1.5 μg m^{-3} of the sulfate in Rhode Island. In other words, the smelters do not seem to be strong sources of sulfate and acidity for the Northeast.

ELEMENTAL TRACERS OF CANADIAN SMELTER AEROSOL TRANSPORTED INTO THE NORTHEASTERN UNITED STATES

For the last several years, we have noticed that after strong cold fronts pass over Rhode Island, the local aerosol is frequently enriched in certain tracer elements such as In, As, and possibly also Sn and Cu. Because of the nature of these elements and the fact that the air masses in question come rapidly from the northwest, we wondered whether their source could possibly have been one or more of the large smelters in eastern Canada. Table 1 lists the five major Cu, Ni and Zn smelters in southern Ontario and Quebec, together with their annual SO_2 emissions (For completeness, the two other major industrial SO_2-emitters of this region, which are both plants of Algoma Steel, are included). These smelters' annual emissions of SO_2 range from 10 to 812 thousand tons. Two of the smelters (INCO and Noranda) emit more than 500 thousand tons per hear, one (Falconbridge) emits roughly 100 thousand tons per year, and two (Canadian Electrolytic Zinc and Kidd Creek Mines) emit about 10 thousand tons per year. Thus, the trace-element emissions from these smelters will probably be dominated by the INCO, Noranda and Falconbridge plants. Historically, the INCO smelter has received the most attention, because it was the largest point source of SO_2 in the world; its emissions comprised 2-3% of the North American total. Table 1 shows that this is no longer necessarily the case, however, because in years when INCO is shut down part of the time (1978 for example), the Noranda smelter can emit more total SO_2 than INCO. Figure 1 shows that the largest of these SO_2-emitters lie within a 30° angle, centered on the northwest, relative to Rhode Island (Two other small smelters in eastern Quebec and New Brunswick are not shown here). The distances of the ICNO, Noranda and Falconbridge smelters from Rhode Island average roughly 1000 km.

We focused our attention on In as a possible tracer of the smelters for several reasons: Its chemical and physical properties seemed appropriate. It is found in the lower right-hand corner of the periodic table, where other elements enriched in the aerosol are located. Indium is a Group IIIb element, below B, Al and Ga and above Tl. Its oxide volatilizes at 850° C, which makes it nearly as volatile as As and Pb and more volatile than Zn and Cu, all of which are strongly enriched in smelter effluent. Direct measurements in the INCO plume[1], in plumes from several Cu smelters[2], and near a Pb smelter[3] show indeed that In, As and Cd are among the elements most enriched relative to normal pollution aerosol. At High Point, NJ, we have also observed strongly enriched As, In and Zn from the New Jersey Zinc smelter in Palmerton, PA, 100 km to the SW.

In general though, little is known about sources of In to the aerosol. It is generally enriched relative to the crust, with typical enrichment factors being 10-100. Occasional enrichment factors of up to 500 are observed[4]. A curious feature is that the mass of In is concentrated on medium-sized aerosol particles, rather than in the classical coarse and fine modes. Consequently its mass-median diameter is 1.5 μm[4]. In this sense, it is similar to Zn in the aerosol. Until recently, we had no explanation for this size distribution. Now, however, we are beginning to believe that it may result from rapid condensation onto a coarse-particle precursor, such as might be emitted from a smelter without strong pollution controls.

There is no reason in principle why other elements emitted strongly from smelters, such as As, Se, Ni, Cu, Cr, Mo, etc., should not also be effective tracers. We have found however, and show below, that In seems to be the best.

Aerosol from the Canadian smelters is of interest for several reasons: (1) Because of the distances involved, identification of smelter effluent in our samples would provide a valuable test of calculated air-mass trajectories. While the use of air-mass trajectories is very common today, they are essentially unverified over these long distances. Recall that 1000 km is also the distance from the Ohio River Valley to the East Coast, a prime pathway currently being proposed for transport of acidic aerosol. (2) These smelters may be an important component of transboundary exchange of pollutants between the United States and Canada. Although their combined output of SO_2 exceeds one million tons per year, their effluent has not yet been traced across the border into the United States, to the best of our knowledge. We felt that trace elements might be a more reliable way of doing this than tracing SO_2 or sulfate, each of which has a high background from other sources in each country. (3) This situation offered an ideal opportunity to expand and refine our system of elemental tracers, which to date has been applied only to aerosol from entire areas such as western Europe, the Northeast, and the Midwest[5,6].

Our original observations of high In and As after cold fronts were semi-quantitative at best. During the analysis, it was noticed that photopeaks of In and As stood out with unusual clarity on gamma-ray spectra of irradiated samples from these periods (neutron activation is our primary analytical technique). Whether the peaks were in fact higher, i.e., represented unusually high elemental concentrations, was not clear at the time. An alternative explanation was that the peaks were no larger than normal, but were sensed more easily because the background from marine elements such as Na and Cl was lowest in air masses from the northwest.

The recent analysis of a large suite of aerosol samples from Rhode Island provided an opportunity to examine this phenomenon more systematically. We now believe that we have indeed detected effluent from these smelters in Rhode Island. This paper is a preliminary report on our findings.

Sampling and analysis

From 1977-80, a series of 235 aerosol samples were taken in Narragansett, Rhode Island, on the grounds of the Graduate School of Oceanography. Because they were originally to be used merely to characterize aerosol of the Northeast in a general sense, they were changed only semi-weekly. All samples were taken on 47-mm Whatman No. 43 ashless cellulose paper, using a Gast high-vacuum, low-volume pump at flow rates of 20-30 $l\ min^{-1}$. Each sample was subdivided for analysis. Approximately one-eighth was analyzed for water-soluble sulfate by colorimetry, using a variant of the Tabatabai barium chloride-gelatin technique[5] proposed by Wolfson[6]. Roughly one-quarter of the filter was analyzed by nondestructive neutron activation for elements with short-lived radionuclides. Fifteen elements were sought in this way, among them Al, Mn, V, and In. Subsets of these samples from the first two winters and summers were analyzed for additional elements by long-lived neutron activation. A total of 35 elements was sought by this technique, among them As, Se, and Ni. The results are given elsewhere[6].

Discussion of results

The results showed clearly that In did indeed come to Rhode Island in pulses, which were not necessarily associated with pulses of other pollution elements. The peak concentrations were most commonly about 0.04 $ng\ m^{-3}$ (averaged over 3-4 days), but occasionally reached 0.07 $ng\ m^{-3}$ during the first winter (1977-78). High concentrations occurred both summer and winter, but were much more frequent and more intense during winter (This was similar to pulsed transport of midwestern aerosol to Rhode Island, which we also found to be much more common in winter than in summer[6]). The high concentrations appeared either as individual samples or in as many as 3-4 consecutive samples. The long consecutive cases were limited to cold snaps in the dead of winter.

Statistically, In was grouped by itself or with As. This is shown in Table 2, which gives the results of factor analysis of 16 elements during the first two winter and two summer periods mentioned above[6]. During summer 1978 and winter 1979, As and In were grouped together; during the other two periods, In was in a "group" by itself. These groupings were confirmed by cluster analysis.

To deal more quantitatively with the smelter effluent, we have used the ratio of In to noncrustal Mn as a tracer:

$$\text{In/noncr. Mn} = \text{In}/[\text{Tot.Mn} - (\frac{\text{Mn}}{\text{Al}})_{\text{crust}}\text{Al}], \qquad (1)$$

referred to below as the In/Mn ratio for short.

We chose In over As because it tended to be more independent of the rest of the pollution elements statistically (which we believe to mean that it has less interference from coal-combustion aerosol than As has) and because we had analyzed all the samples for In but less than one-half of them for As. We included a denominator containing noncrustal Mn both to normalize for removal/dilution effects and to aid in discriminating the smelter effluent from midwestern aerosol, which should have higher noncrustal Mn. Thus, the numerator was higher in the smelters and the denominator was higher in the Midwest.

Figure 2 gives frequency distributions of the In/Mn ratio for the Rhode Island samples during summer, winter, and for all samples. During summer, there is only one large mode, centered on a ratio of roughly 2.1×10^{-3}. This distribution is very similar to that for New York City (also determined by us from semi-weekly samples), shown in the lower part of the figure. A similar mean ratio can be derived from data for Washington, D.C.[9] By contrast, In/Mn ratios calculated for Northwest Indiana and Niles, Michigan[4] are approximately four times lower, presumably because of the higher noncrustal Mn there than in the Northeast. Thus, it would seem that the single summer mode in Rhode Island represents local (northeastern) aerosol.

The winter situation is quite different, however. The large central mode appears to be broken up into two narrower modes, with central ratios of 1.8 and 3.6×10^{-3}. At this point, we are not sure whether this is real or an artifact of the small number of samples involved. The mode with center at 3.6×10^{-3} seems suspiciously narrow. But a much better-defined mode is seen at the upper end of the distribution, with center at 7.5×10^{-3}. This mode also appears clearly in the aggregated summer and winter plot. Its range corresponds closely to that reported earlier for aerosol in Algonquin Provincial Park, Ontario, roughly 250-300 km from Noranda and Sudbury[10]. The upper end of this mode approaches the ratio found in Liége, Belgium, a steel manufacturing city[4].

A survey of meteorological maps for the samples with high In/Mn ratios revealed that they were universally marked by at least one day of flow from the northwest. This was confirmed by 850-mb air-mass trajectories provided to us by the Ontario Ministry of the Environment, examples of which are shown in Figure 3. Table 3 shows that of 102 daily trajectories calculated, fully 64 came to Rhode Island from the N-NW sector, with the others partitioned over the remaining seven sectors without any particular pattern. (This directional analysis was based upon our best judgment of where the air in question had been during the previous day. It was not completely objective because the trajectories tended to pass east of the major smelters, and because small-scale wind systems such as sea breezes occasionally deflected the last few hours of the trajectory.) We regard this evidence from trajectories to be even stronger than it first appears, however, because when daily trajectories are compared with semi-weekly samples, not every trajectory will have to come from the sources in order to have an integrated sample show their effects. Every sample for which trajectories were calculated, both summer and winter, had at least one trajectory from the N-NW.

We have considered the possibility that sources between Rhode Island and the smelters might actually be responsible for the high In observed. It seems very unlikely. A similar northwestern directionality for In has been found at Glen, New York (60 km west of Albany), some 300 km closer to the smelters than Rhode Island is (These results will be reported later). As noted above, effluent from the smelters was previously seen 250 km to their southeast. Thus, any intervening source would have to lie between Glen, New York and southern Ontario. No sources rich in In are known there; most of that area is forested or agricultural land. Recently, we have found two cases of abundant smelter effluent at Underhill, VT (Rahn, Voytek, Schweitzer; to be reported later), with WNW winds. Because of the lack of significant enrichment of Zn in these samples, we feel that they cannot have come from the Canadian Electrolytic Zinc smelter in Valleyfield, Quebec, some 110 km to the NW. Rather, the more distant but larger smelters in the Sudbury Basin would seem to be responsible.

One interesting aspect of our data which puzzled us at the beginning was that high In concentrations were found throughout the winter of 1978-79 while INCO was on strike (September 1978-June 1979). Although peak IN concentrations were lower than they had been during the previous winter (0.04 vs. 0.07 ng m^{-3}), and the In/Mn ratios were somewhat lower as well, by no means was the upper (smelter) mode absent. The reason for this is now clear from Table 1: INCO represents half or less of the total SO_2 emissions from the smelters, and an unknown fraction of the trace-element emissions (which have not been surveyed except at INCO).

How much sulfate (and hence, acidity) is associated with the smelter plumes? The answer for Rhode Island seems to be very little, although it would be expected to be higher closer to the Canadian border. Table 4 shows that during both summer and winter, the sulfate associated with periods of smelter aerosol is roughly one-half that for the season as a whole. Average sulfate during smelter periods at Rhode Island is 203 $\mu g\ m^{-3}$ in winter and 3-5 $\mu g\ m^{-3}$ in summer.

But a portion of even these low sulfates must be northeastern in origin --it cannot all come from the smelters. We have tried to estimate the smelters' contribution to the observed sulfate by subtracting out the northeastern component as follows:

$$SO^{=}_{4_{SM}} = SO^{=}_{4_{TOT}} - SO^{=}_{4_{NE}} \qquad (2)$$

$$= SO^{=}_{4_{TOT}} - (\text{noncr.V})(SO^{=}_{4}/\text{noncr.V})_{NE} \qquad (3)$$

Here, the mean ratio of sulfate to noncrustal V for a particular season is multiplied by the concentrations of noncrustal V during a smelter period within that season to estimate the northeastern component of sulfate in that sample. For each of the two winters studied here, we have found that the smelter contribution to sulfate is very small on the mean and is well below the uncertainty of the calculation, which is roughly 1-1.5 g m^{-3} sulfate. (The major source of uncertainty is the variation of the $SO^{=}_{4}$/noncr. V ratio from sample to sample during winter.)

On balance, then, we have concluded that the Canadian smelters are only weak sources of sulfate for the Northeast, at least during winter. This agrees qualitatively with the results of Parekh and Husain[11], who found that maritime air masses from the United States transported 4-5 times more sulfate than continental polar air masses from Canada did.

In closing, one may legitimately enquire why other elements emitted strongly from smelters, such as As, Se, Ni and Cu, are not as effective tracers as In is. For Se and As, there are interferences from coal emissions. For Se, there is some factor, still unknown, which leads to considerably higher concentrations in the Northeast than expected. The background of Ni in the Northeast is very high because of all the oil burned there. For Cu, the reason is less clear. Without trace-element studies in all the major smelter plumes, it is nearly impossible to understand why one element should be a good tracer and the next one shouldn't.

Acknowledgements

This work was facilitated by contributions from T.J. Conway, T.R. Fogg, D.L. Smith and K.A. Scweitzer of the University of Rhode Island. M. Lusis and W. Chan of the Ontario Ministry of the Environment offered help and encouragement throughout the course of the work. Air-mass trajectories were kindly provided by Joel Kurtz of the Ontario Ministry of the Environment. Samples were analyzed using facilities of the Rhode Island Nuclear Science Center and the Lowell (Mass.) Institute of Technology. This research was supported by the Office of Naval Research.

REFERENCES

1. Stevens R.D.S., Hunt J.E., Vet R.J. and Chan W.H. (1980) "An airborne plume chemistry investigation". Presented at the 61st Canadian Chemical Conference and Exhibition, Winnipeg, Manitoba, June 4-7, 1978. (Manuscript from Ontario Ministry of the Environment, 880 Bay Street, Toronto).

2. Small M., Germani M.S., Small A.M., Zoller W.H. and Moyers J.L. (1981) "Airborne plume study of emissions from the processing of copper ores in southeastern Arizona." Environ. Sci. Technol. 15, 293-299.

3. Ragaini R.C., Ralston H.R. and Roberts N. (1977) "Environmental trace metal contamination in Kellogg, Idaho, near a lead smelting complex." Environ. Sci. Technol. 11, 773-781.

4. Rahn K.A. (1976) "The chemical composition of the atmospheric aerosol." Technical report, Graduate School of Oceanography, University of Rhode Island, Narragansett, Rhode Island 02882-1197, 265 pp.

5. Rahn K.A. (1981) "The Mn/V ratio as a tracer of large-scale sources of pollution aerosol for the Arctic." Atmos. Environ. 15, 1457-1464.

6. Rahn K.A., Lowenthal D.J. and Lewis N.F. (1982) "Elemental tracers and source areas of pollution aerosol in Narragansett, Rhode Island." Technical Report, Graduate School of Oceanography, University of Rhode Island, Narragansett, Rhode Island 02882-1197, 1 October 1982, 119 pp.

7. Tabatabai M.A. (1974) "A rapid method for determination of sulfate in water samples." Environ. Lett. 7, 237-243.

8. Wolfson J.M. (1980) "Determination of microgram quantities of inorganic sulfate in atmospheric particlulates." JAPCA 30, 688-690.

9. Kowalczyk G.S., Gordon G.E. and Rheingrover S.W. (1982) "Identification of atmospheric particulate sources in Washington, D.C., using chemical element balances." Environ. Sci. Technol. 16, 79-90.

10. Rahn K.A. (1971) "Sources of trace elements in aerosols--an approach to clean air." Ph.D. thesis, University of Michigan, Ann Arbor, Xerox University Microfilms (Order No. 72-4956), 325 pp.

11. Parekh P.P. and Husain L. (1982) "Ambient sulfate concentrations and windflow patterns at Whiteface Mountain, New York." Geophys. Res. Lett. 9, 79-82.

Figure 1. Map of the eastern U.S. and Canada, showing the relation between Canadian SO_2-emitters and Narragansett, Rhode Island.

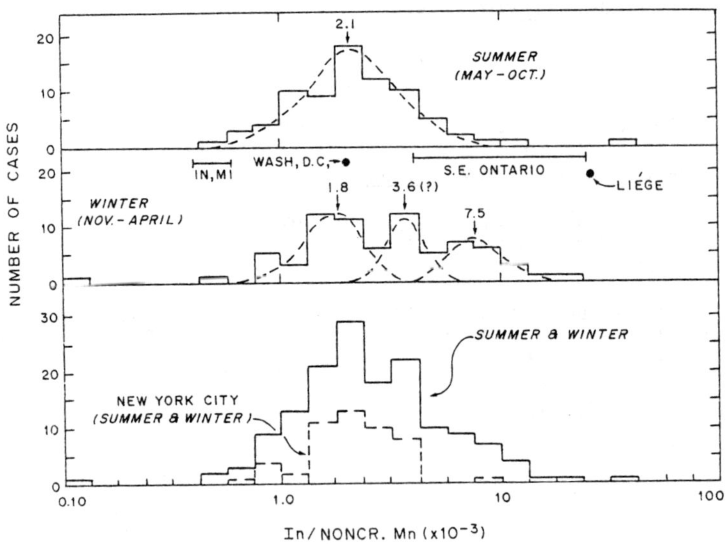

Figure 2. Frequency distributions of the In/Mn ratio for semi-weekly aerosol samples of summer, winter, and the entire year at Narragansett, Rhode Island, 1977-80.

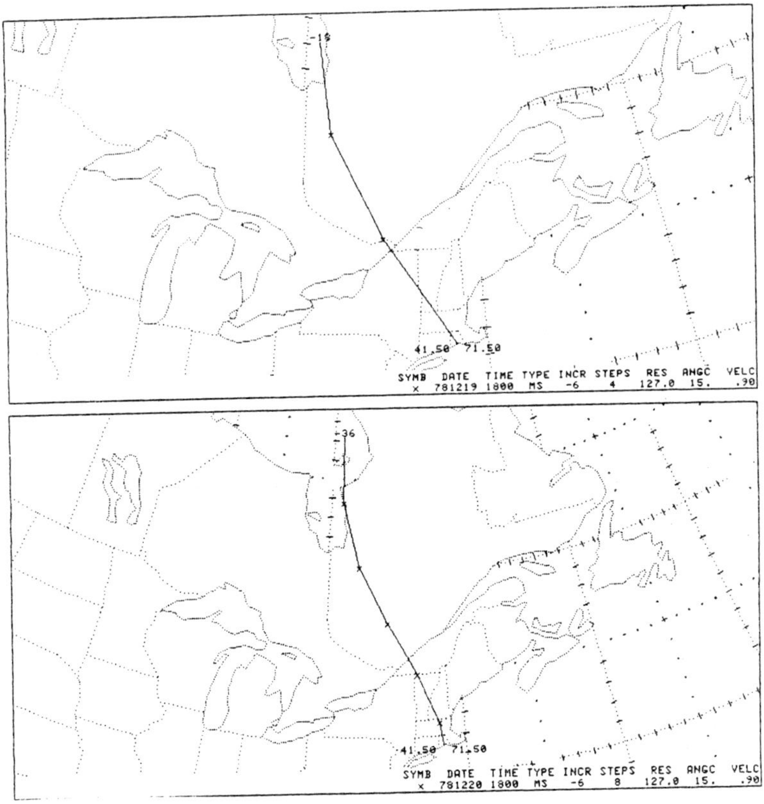

Figure 3. Typical backwards 850-mb air-mass trajectories from Narragansett, Rhode Island. The frictional backing angle was 15 degrees and the celocity was 0.9 times the geostrophic value. Calculations were started at 1800 GMT. Crosses represent 6-hour time steps.

Table 1. Major industrial SO_2-emitters in southern Ontario and Quebec.

Emitter	SO_2 emissions, 10^3 T yr^{-1}			
	1973-81[a]	1978[a]	1979[a]	1980[b]
INCO - Sudbury, Ontario Cu/Ni smelter	940	515		312
Noranda - Noranda, Quebec Cu smelter			566	538
Algoma Steel - Wawa, Ontario Sinter plant		163		
Falconbridge - Sudbury, Ontario Cu/Ni smelter	173	117		122
Algoma Steel - Sault Ste. Marie, Ontario Steelworks		14		
Canadian Electrolytic Zinc - Valleyfield, Quebec Zn smelter				11
Kidd Creek Mines - Timmins, Ontario Cu/Zn smelter				10

a) Source: M. Lusis and D. Yap, Ontario Ministry of the Environment (personal communication).

b) Source: Final Report of Work Group 3B, US-Canada Memorandum of Intent on Transboundary Air Pollution. June 1982.

Table 2. Six-factor loadings for trace elements in winter and summer at Narragansett, Rhode Island (after ref.).

	I	II	III	IV	V	VI
			Winter 1978			
Se	0.80		0.25			
Zn	0.78	0.26	0.34			
$xSO_4^=$	0.75			0.29		
Co	0.61	0.48			0.32	
xMn	0.57	0.41		0.36		
Cr	0.54	0.30	0.46		0.37	
Cl		0.78				
Na	0.36	0.76				
xV	0.42	0.61			0.40	
Fe	0.34		0.81			
Al			0.79	0.33		
Sm	0.26		0.28	0.83		
La	0.26	0.25	0.28	0.80	0.26	
As	0.30	0.32			0.76	
Sb	0.35	0.28		0.30	0.74	
In						0.91

	I	II	III	IV	V	VI
			Summer 1978			
Sm	0.83	0.27	0.30			
Fe	0.80	0.37				
La	0.80	0.35	0.28			
Al	0.66	0.27		0.42	0.30	
Cr	0.58	0.43		0.41		
Zn	0.34	0.78	0.27			
xV	0.27	0.70		0.37		
Sb	0.36	0.66	0.31			
xMn	0.48	0.62	0.35			
Co	0.53	0.59		0.38		
As		0.28	0.74	0.36		
In	0.39	0.29	0.73			
Se	0.39	0.40	0.53	0.44		
$xSO_4^=$	0.26		0.32	0.76		
Na					0.90	
Cl						0.97

	I	II	III	IV	V	VI
			Winter 1979			
Co	0.78	0.36				
Se	0.77	0.26				
Cr	0.76	0.41				
Zn	0.75	0.34			0.28	
Sb	0.73				0.34	
xV	0.71			0.40		
$xSO_4^=$	0.53			0.49	0.39	
Sm	0.26	0.84				
La	0.32	0.82				
Fe	0.36	0.79				
Al	0.35	0.61			0.39	
In		0.28	0.80			
As	0.31	0.37	0.69			
Na				0.73		0.42
xMn	0.34				0.80	
Cl						0.94

	I	II	III	IV	V	VI
			Summer 1979			
Sm	0.84		0.25			
Fe	0.82		0.32			
Al	0.70	0.32		0.37		
La	0.65	0.55				
Se	0.65	0.45				
Cr	0.62	0.26		0.43		
Co	0.60		0.55	0.27		
As	0.56	0.26	0.56			
xMn	0.53		0.42		0.44	
Sb	0.30	0.79	0.26			
xV	0.26	0.66	0.40	0.27		
Zn	0.49	0.63	0.28			
In		0.31	0.79			
Na	0.26			0.79		
$xSO_4^=$					0.88	
Cl						0.98

$xSO_4^=$ = nonmarine

xMn,V = noncrustal

Table 3. Distribution of air-mass trajectories during samples with aerosol from Canadian smelters.

Sector	Number of trajectories
N-NE	9
NE-E	3
E-SE	5
SE-S	2
S-SW	7
SW-W	8
W-NW	4
NW-N	64
TOTAL	102

Table 4. Sulfate associated with aerosol from the Canadian smelters compared to seasonal means.

Season	No. events from Canadian smelters*	Mean $SO_4^=$	Mean $SO_4^=$ for season	Events/Season
Nov. 1977-Apr. 1978	8	3.2	3.6	0.88
Nov. 1978-Apr. 1979	7	2.1	4.0	0.52
May-Oct. 1978	1	2.9	5.8	0.50
May-Oct. 1979	4	5.0	8.1	0.62

*In/Noncr.Mn $\geq 6 \times 10^{-3}$

An Assessment of the Relationship Between Regional Pollution Transport and Trace Elements Using Wind Trajectory Analysis

George D. Thurston, ⟶

John D. Spengler,
Department of Environmental Health Sciences,
Harvard School of Public Health,
Boston, Massachusetts

and Perry J. Samson,
Department of Atmospheric and Oceanic Science,
University of Michigan,
Ann Arbor, Michigan

A quantitative apportionment of the sources of atmospheric sulfate pollution is made difficult by the fact that sulfates are "secondary" pollutants, not usually emitted directly into the atmosphere. Assessments of regional culpability for sulfate episodes may be possible based upon the elemental make-up of aerosols. In this work, the elemental composition of aerosol samples collected between June, 1979 and December, 1980 at a site WNW of Boston, MA is examined as a function of air mass source sector. For each of 241 sampling days, an estimation of air mass source sector(s) was made based upon 72 hour back-trajectories of wind data. Those sampling days influenced by air masses that passed over specific source regions are selected from the data set and characterized as to their elemental composition. In this manner, the data are searched for relationships among fine aerosol sulfur, manganese, selenium, vanadium and mass concentrations and their respective source sectors. During the warm months of the year, air masses passing over the Ohio Valley/Midwest sector are found to result in sulfur concentrations averaging more than 40 percent higher than any other source sector influencing the site. Concentrations of Mn and Se (possible tracers for steel industry and coal combustion emissions, respectively) are higher in air masses which have passed over the Ohio Valley/Midwest sector than for those which passed over the Eastern Seaboard sector. Moreover, the mass fraction of selenium (Se/mass) is also found to be significantly higher for the Ohio Valley/Midwest sector than the Eastern Seaboard sector. These data all indicate that there are compositional differences between aerosols reaching the sampling monitor on Ohio Valley/Midwest versus Eastern Seaboard sector days. However, the ratio of Mn/V is not found to discriminate reliably between air masses transported from the two sectors to this site. It is recommended that multiple element profiles be used in future attempts to apportion transported pollution. Trace element emission inventories should also be developed and applied to calibrate conventional sulfate dispersion models.

Introduction

A major challenge to researchers in the field of receptor modeling is the source apportionment of the sulfate fraction of fine particle mass. Sulfates (usually as sulfuric acid (H_2SO_4), ammonium bisulfate (NH_4HSO_4), or ammonium sulfate ($(NH_4)_2SO_4$)) are primarily "secondary" pollutants, in that they are formed in the atmosphere from SO_2 and are not usually directly emitted by specific sources. As such, it is difficult to associate them with a specific source. Particulate sulfur compounds are therefore often treated as a separate category in source apportionments. A determination of the sources of sulfate pollution has taken on greater importance in the 1980's, as acid precipitation control legislation is being considered by Congress. Elemental tracers have been suggested as a means for ascribing at least regional culpability for transported sulfate pollution in the Northeast U.S.

An ideal tracer for a region's pollutant air mass does not likely exist, but it would be desirable that a tracer have qualities of uniqueness, constant rate of emission, stability, and ease of detection. If a tracer emanates from more than one region, the discrimination of source regions is complicated. If a tracer element is emitted in variable quantities (e.g., higher in winter than summer), the source region may not always be discernible. The tracers should be stable in the atmosphere and during collection, so that assumptions regarding conservation will not be violated. Finally, an ideal tracer should be readily and inexpensively detectable, so that it can be sampled often at brief sampling intervals (e.g., on the order of hours) in order to best discern transport incidents.

Two approaches to identifying tracers of regional air masses are: (1) ambient upwind concentration measurements collected in the "source" region, and (2); ambient downwind "receptor" concentration measurements. In the first, a tracer is identified based upon pollutant measurements made in the pollution "source" region, which indicate a specific tracer to be associated with aerosols from that region. This is analogous to stack sampling of specific sources. For example, Rahn[1,2,3] has noted that regions may differ in the mass concentration ratio of elements: Eastern U.S. aerosols having a characteristic Total Suspended Particle (TSP) ratio of noncrustal Mn to V between 0.1 and 0.2; Midwestern aerosols having a characteristic TSP ratio of noncrustal Mn to V greater than 1.0. Under the assumption of tracer (e.g., elemental ratio) conservation during transport, a discrimination of air mass origin is made based upon these "source" characteristics.

A second approach to characterizing regional contributions to a downwind receptor is employed in this work. It involves identifying associations between specific tracers and pollution transport conditions, based upon samples collected in a downwind region. In a somewhat analogous fashion, Gordon[4] has previously examined short-term (6-hour) samples collected during persistent wind directions in order to discern elemental profiles for nearby upwind sources. Here, those sampling days affected by air masses which have passed over specific pollution source regions are selected out of a large dataset using regional scale wind trajectory analyses. These subsets are then characterized in order to assess the nature of polluted air masses reaching the site from various source regions. Assuming these characteristics are applicable to all air masses from these regions, they can be examined for criteria to discriminate source regions in general.

In particular, this work investigates the behavior of manganese (Mn) and selenium (Se) during sector classified days, in order to search for relationships between these tracers and regional transport. The absolute concentrations of these tracers, as well as their ratios with vanadium and mass, are investigated as possible discriminators of air mass source region. Mn is examined because of its reported differences between the Eastern and Midwestern source regions, especially as a ratio to V[1,2,3]. Moreover, Mn might be expected to be a tracer for midwestern pollution, since a major source of manganese pollution is the iron and steel industry concentrated in the Midwest[5]. Se is examined because it has been found to exhibit a significant correlation with fine particle S at a number of sites[6], and Se is a fine fraction element primarily emitted in the U.S. by coal combustion[7]. The variation of fine particle S concentrations among source sector subsets is also considered.

Data Collection

The work presented in this paper is based upon: (1) extensive particle elemental composition data collected at a site 10 km WNW of Boston, Massachusetts, and; (2) concurrent wind trajectory data indicating air mass origins. The monitoring site is located in suburban Watertown, Massachusetts (42°22'N., 71°11'W.), and has had a Beckman dichotomous virtual impactor particle sampler (Cat. No. 675900) operating since May 1979 as part of Harvard's 6-City Study of the health effects of air pollutants[6,8]. Twenty-four hour size-fractionated Inhalable Particle (IP) samples (coarse mass: $2.5\mu m < d_a < 15\mu m$, fine mass: $d_a < 2.5\ \mu m$) were collected at least every other day at the Watertown site, and have been analyzed for 30 elements, including sulfur, manganese, selenium, and vanadium, using x-ray fluorescence[9].

Of the 272 days on which 24-hour particle samples had been collected in Watertown between June 4, 1979 and December 18, 1980, it was possible to generate air parcel back-trajectories for 241 days, allowing an estimation of parcel origin for these samples. The Air Resources Laboratories' Atmospheric Transport and Dispersion Model(ARL-ATAD) was used to generate the trajectories[10]. The model uses vertically averaged transport winds (from the surface to the top of the mixing layer) as recorded at upper air stations throughout the U.S. to calculate trajectories for four times per day. Trajectories were calculated for periods of 72 hours, moving backward in time from Watertown, Mass. These trajectories estimate the movement of the center of mass of the mixed layer of the atmosphere, so that sources located laterally from a trajectory line may also contribute to a particular sample[11].

Each trajectory generated for Watertown has been classed into a source sector based upon its path. The sectors, defined in order to separate potential source regions, are displayed in Figure 1. The case shown in Figure 1 is indicated to have been influenced by air transport from the Ohio Valley/Midwest sector. The symbols superimposed on trajectories indicate their end hour during the day. The spacing between the symbols indicate the trajectory velocity, with twelve hours travel time between sequential symbols. The example shown in Figure 2 is an exemplary Eastern Seaboard transport day, since the trajectories envelop the Mid-Atlantic states. Many plots were unlike Figures 1 and 2, in that not all four trajectories in a day traveled over a single source region. For this reason, the number of trajectories passing over each sector in each day was recorded. Trajectories passing through (or between) more than one sector were assigned proportionately. For example, one half a trajectory was assigned each sector if a trajectory passed through, or along the border between, two sectors. Since Watertown is less than 10 miles inland of the Atlantic coastline, continuous wind data collected on-site was used to screen from the Ohio Valley/Midwest, Western Canada, and Eastern Canada sector classes any samples possibly influenced by a coastal seabreeze (i.e., from Boston). An air mass was categorized as having passed over a sector if three or more of the four trajectories during the sampling day were estimated as from that sector.

Results

The categorical distribution of trajectories recorded during the study period from June 1979 through December 1980 are displayed in Figure 3 for the Eastern Seaboard (ES), Ohio Valley/Midwest (OV), Western Canada (WC), Eastern Canada (EC), and Atlantic (AT) sector categories. Overall, the Ohio Valley/Midwest and Eastern Canada sectors are the most frequent air mass contributors, with the Eastern Seaboard and Western Canadian sectors third and fourth. The Atlantic sector is indicated as playing a relatively inconsequential role, which is likely due in part to the prevailing westerlies, and in part to the fact that trajectories from over the Atlantic cannot be easily analyzed (i.e., no oceanic upper air soundings). Once the trajectory data are considered in terms of the sulfate concentrations recorded (Figure 3) the distributions shift dramatically. The Ohio Valley/Midwest and Eastern Seaboard trajectories clearly dominate those periods wherein sulfur concentrations above the median are recorded at the Watertown site.

Figure 4 displays the fine sulfur concentration box-plot distributions, by trajectory sector. The sulfur concentrations during the OV and ES sectors are the highest, on average, with the WC and EC sector days recording the lowest sulfur concentrations. The Atlantic sector data are not presented, as these were so few in number ($n = 4$). While the Ohio Valley sector records the highest peak concentrations, the OV and ES means do not differ statistically ($p > .05$) overall.

Figure 5 yields insight into the reason that the overall ES and OV sector fine S means reported in Figure 4 are not statistically different. Most ES trajectories occur in the summer months, peaking at over 35 percent of the trajectories recorded in the month of June. In contrast, the occurrence of OV sector trajectories is more uniform across the year (averaging about 30 percent in every month). The tendency for ES days during the summer months, when SO_2 to sulfate conversion is enhanced by warm, moist conditions, raises its overall mean relative to the OV days, which are more evenly distributed throughout the year. Figure 6 confirms that, once the data are stratified by season (mid-Spring to mid-Fall = Warm Season), the mean S concentration during OV sector days are an average 42 percent higher than during ES sector days ($p < .05$).

Figures 7 and 8 display the sector-specific box plots of the Se and Se/mass distributions occurring at the Watertown site, respectively. The Se concentrations for the OV sector are statistically higher than the other sectors, the mean value being nearly double that in the ES sector. The mean Se concentration for the EC sector falls below twice the mean uncertainty of measurement (as do all EC sector Se concentrations), indicating that the Se concentrations for air masses from Eastern Canada are below detection. For this reason, the EC Se/mass data were excluded from Figure 8. The Se mass fractions presented in that figure indicate that the aerosol compositions differ in this respect between sectors, with the highest mean value recorded for the OV sector.

Figure 9 and 10 present the sector specific box-plot distributions of Mn and the ratio of noncrustal Mn to noncrustal V. The noncrustal concentrations of tracer elements were calculated using Si as a crustal reference element, according to the formula[1]:

$$X_{noncrustal} = X_{aerosol} - (X/Si)_{crustal}\, Si_{aerosol}$$

where X is the concentration of tracer element X and Si is the concentration of Silicon. For crustal concentrations, the data of Mason[12] were used (Mn = 950 ppm, V = 135 ppm, Si = 277,200 ppm). The Mn concentrations are highest, on average, for the OV sector. The WC sector shows wide variation and a mean nearly equal to that for the OV. The ES sector has an Mn concentration significantly less the the OV mean ($p < .05$). Examination of the noncrustal Mn/V distributions in Figure 10 reveals that the mean ratio in every sector is well below the 1.0 minimum value expected, based upon the hypothesis by Rahn[1,2,3], for the OV sector samples. Moreover, the fine particle noncrustal Mn/V ratio does not differ among the regions ($p > .05$), and this ratio does not appear to be a useful discriminator of air mass sector at this site.

Table 1 presents comparison of the OV and ES sector mean concentrations of fine mass, S, V, Mn, and Se, as well as of the mean ratios of fine Se/mass, noncrustal fine Mn/V, and noncrustal IP Mn/V. In light of the skewed nature of some of the distributions presented in Figures 3-10, these comparisons were also conducted on the logs of the values, but no changes in the t-test results were found. It is seen in Table 1 that the fine Mn, fine Se, fine mass, and fine Se mass fraction (Se/mass) differ significantly between these two regions, with OV cases recording a higher mean value overall. The overall mean noncrustal Mn/V ratio did not differ significantly. This appears to be at least in part a result of the fact that, while the Mn concentrations are higher for the OV cases than the ES cases (as expected), the vanadium concentrations are higher for the OV cases than for the ES cases in the warm season, which was not expected.

The fact that the OV sector days record higher V concentrations than the ES days during the warm season is partially attributable to reduced Northeast residual oil emissions in the summer months versus winter months. During these periods of generally reduced ES vanadium emissions, a local pollution source immediately to the west (<10 km) of the monitor apparently becomes a relatively important contributor of fine particle V. A check of SO_2 concentrations during the warm season showed a statistically higher concentration for the OV versus ES days, supporting the contention of a local source influence. It is interesting to note, however, that the correlation of SO_2 with fine particle S at this site is quite weak ($r=.19$), while the correlation of Se with fine particle S is much stronger ($r=.65$).

During the cold season, the inter-regional vanadium concentration pattern shifts, with the ES mean being nearly double the OV mean (though they are not statistically different). The percent Vanadium ((fine V/mass)x100.), however, does differ between the OV and ES cold season days, being significantly higher for the ES sector (.12% ± .016%) than the OV sector (.08% ± .007%). The product of the seasonal shifts in V impact is that, while the Mn/V ratio does not discriminate the source sectors in the summer months, it does so in a statistically significant manner ($p < .05$) during the cold season. It appears that the influence of local V sources, combined with seasonal variability in Eastern Seaboard residual oil burning, confound the use of Mn/V to discriminate air mass sectors at this site.

It is interesting to examine the IP noncrustal Mn/V ratio data presented in Table 1. These IP data (based on the sum of the fine and coarse elemental concentration) are presented to more closely approximate the TSP data employed by Rahn[1,2,3]. These data show the same seasonal relationships found for the fine non-crustal ratio, being statistically different between regions during the cold season only. The ratios are higher, however, due largely to the fact that there is a significant noncrustal Mn component in the coarse fraction, averaging about 40 percent of the IP total. The noncrustal Mn data presented in Figure 1 are fairly typical in that regard, with (4/9.3 x 100=) 43 percent in the coarse fraction. The origin of this coarse fraction noncrustal Mn is not known. Despite this coarse contribution, however, the mean OV sector noncrustal Mn/V ratio is still well below the expected 1.0 ratio.

If, in examining the data in Table 1, one considered the concentration data alone, it might be conjectured that all the aerosol could be of local origin, with the primary difference between the groups being poorer local dispersion on OV versus ES days. Available meteorological data (Boston surface data, Portland, Maine upper air data), however, indicate that there is no important difference between the ventilation rates (=mixing height x layer mean wind speed) of these two groups of days. For example, during the warm season, the mean afternoon ventilation rate was 13,545 (±2,378) m^2/sec for the 20 OV days, versus 10,739 (± 1,372) m^2/sec for the 26 ES cases. This indicates a mean ventilation rate which is higher (though not statistically so) for OV cases. Moreover, examination of the Se/mass data in Table 1 shows that it is not merely the absolute concentrations of Se which average higher on OV days. The mean Se mass fraction is nearly double that occurring on ES days. Thus, ventilation conditions do not appear to explain the differences between sectors. Instead, there is evidence of differences between these sectors in the elemental composition of aerosols reaching the monitor.

Discussion

Examination of an extensive elemental dataset, stratified by air mass sector, has revealed certain similarities and differences among the polluted air masses reaching the Boston area. Consideration of back-trajectory wind data indicates that, for high sulfur aerosol days at the Watertown site, the Ohio Valley/Midwest and Eastern Seaboard sectors are the most frequent contributors. During the warm season of the year, trajectories from these two sectors impact southern New England with approximate equal frequency, but the sulfur concentrations at the Watertown site on Ohio Valley/Midwest sector days average roughly 40 percent higher than during days experiencing air mass transport from the Eastern Seaboard sector. During the cold season, the mean S concentrations do not differ between the two sector groups, but the Ohio Valley/Midwest trajectories are more prevalent. These results are consistent with the assumption that the Ohio Valley/Midwest is a major contributor to sulfate pollution in southern New England, although air masses originating along the Eastern Seaboard are seen to be associated with high sulfates as well.

Comparisons of Ohio Valley/Midwest and Eastern Seaboard sector days' elemental data indicate that the ratio of noncrustal Mn/V is not a useful discriminator of sectors during the (high sulfur) warm season months. The Mn/V ratio differed statistically between these two sectors during the cold season only, when the Eastern ratio was driven to the 0.1-0.2 range by high winter vanadium emissions. The Ohio Valley/Midwest sector's mean noncrustal Mn/V ratio was less than the 1.0 minimum value expected, and the characteristically high Midwest ratios are apparently not sustained during transport to southern New England. Thus, the hypothesis that noncrustal Mn/V can be used as an indicator of pollution source region was not confirmed by the sector analysis of this data.

The inadequacies of elemental ratios for discriminating regions appear to be non-uniqueness, tracer instability, and variable tracer emission. The problem of non-uniqueness is due to the fact that, while elements may be emitted in differing quantities across regions, no single tracer examined is unique to a sector. The addition and loss of tracers during transport also appeared to be a problem. For example, a qualitative examination was made of upwind meteorological data for five relatively low Mn and Se days during which transport was from the Ohio Valley/Midwest sector. This analysis indicated that all of these days experienced precipitation along their path to southern New England. The overall similarity of V concentrations between Ohio Valley/Midwest and Eastern Seaboard sector groups appears to be due in part to the addition of local V emissions, reducing the difference between sector ratios of Mn/V. Moreover, reduced emissions of vanadium in the East during the (non-heating season) warm months causes the Eastern Mn/V ratio to be higher during the summer than during the winter months, further reducing the disparity between Ohio Valley/Midwest and Eastern Seaboard sector ratios.

The mean concentrations of both fine Mn and fine Se were statistically higher for the Ohio Valley/Midwest sector days than for the Eastern Seaboard days. This holds out the prospect that these elements may yet be useful in apportioning regional responsibility for sulfate pollution. Moreover, by examining the elemental data on a mass fraction basis, it has been discerned that the composition of aerosols from the two sectors differ significantly. For example, a higher percent Se is present in aerosols collected during OV sector days versus ES sector days, on average, while a lower percentage of V is present in OV samples than ES samples in the cold season. The further development of multi-element regional aerosol composition profiles may present a promising (and statistically reliable) way to apportion regional responsibility for sulfate pollution.

This paper has indicated that there are meteorological and local source complications involved in analyzing a downwind sample and attempting to infer the source region of an aerosol. In fact, it may not be possible to quantitatively apportion culpability for transported sulfate pollution based simply on downwind trace element measurements. However, the application of receptor model techniques to this problem should be continued in order to determine what can be learned. One application which should be examined further is the use of trace elements to calibrate available sulfate formation, dispersion and transport models now being employed in analyses of the acid rain question (e.g., the complete Heffter model[10]). Primary elemental emissions do not undergo the highly variable transformation processes required for sulfate formation. If emission inventories for these elements are developed, then they can be used to calibrate the dispersion and transport portions of the present sulfate models, without the complicating influence of SO_2 to sulfate transformation equations. Thus, the estimates of the overall sulfate models may be improved, enhancing their usefulness as decision making tools.

ACKNOWLEDGEMENTS

This work was performed under the general support of the Harvard Air Pollution Respiratory Health Study (Six City Study). The health study is funded in part by grants from the National Institute of Environmental Health Sciences (ESO-1180), the Electric Power Research Institute (RP-1001), the U.S. Environmental Protection Agency (68-02-3201 and 3466), and the Wisconsin Power and Light Company. Special thanks to Drs. R.K. Stevens, T.G. Dzubay, and A.P. Altshuller of the E.P.A. Environmental Sciences Research Laboratory for providing the dichotomous samplers, the x-ray fluorescence analysis, and advice. Also, our thanks to all Six City staff who contributed to the collection, processing, and presentation of these data. In particular, to Ms. Allison Maskell for her word processing efforts and to Mr. David Glicksberg for his computer programming work in the preparation of back-trajectory plots, such as those shown in Figures 1 and 2.

REFERENCES

1. K. A. Rahn, "Elemental tracers and sources of atmospheric acidity for the Northeast - a statement of new evidence", Kingston, 1981.

2. K. A. Rahn, N.F. Lewis, "Elemental tracers for source areas of atmospheric acidity in the Northeast". Presented at the conference "Acid Rain Reappraised." John Carroll University, Cleveland, Ohio, April 26, 1982.

3. R.A. Kerr, "Tracing sources of acid rain causes big stir" *Science*, **12**, 881 (1982).

4. G.E. Gordon, "Receptor Models" *Environ. Sci. and Technol.*, **14**, 792 (1980).

5. National Academy of Sciences. *Manganese* , Washington, D.C., 1973.

6. J.D. Spengler, G.D. Thurston, W.A. Turner, R.E. Letz, "Mass and elemental composition of fine and coarse particles in six U.S. cities" *J. of the Air Poll. Control Assoc.* (accepted for publication).

7. National Academy of Sciences. *Selenium* , Washington, D.C., 1976.

8. S.L.K. Briggs, G.D. Thurston, T.S. Dumyahn, J.S. Evans, J.D. Spengler, W.A. Turner, "The effectiveness of dichotomous sampler quality assurance procedures",APCA Paper 82-23.9,Presented at the 75th Annual Meeting of the Air Pollution Control Association, New Orleans, LA. June 20-25, 1982.

9. T.G. Dzubay, R.K. Stevens, "Ambient air analysis with dichotomous sampler and x-ray fluorescence spectrometer". *Environ. Sci. and Technol.*, **9** :663-668.

10. Heffter, J.L. "Air Resources Laboratories atmospheric transport and dispersion model (ARL-ATAD)", ERL ARL-81, Air Resources Labs, Silver Spring, MD, 1980.

11. P.J. Samson, "Trajectory analysis of summertime sulfate concentrations in the northeastern United States", *Journal of Applied Meteorology*, **19**, 1382 (1980).

12. B. Mason, *Principles of Geochemistry*, (John Wiley, New York, 1966).

FIGURE 1.
HEFFTER MODEL BACK-TRAJECTORIES
FOR OCTOBER 22, 1979

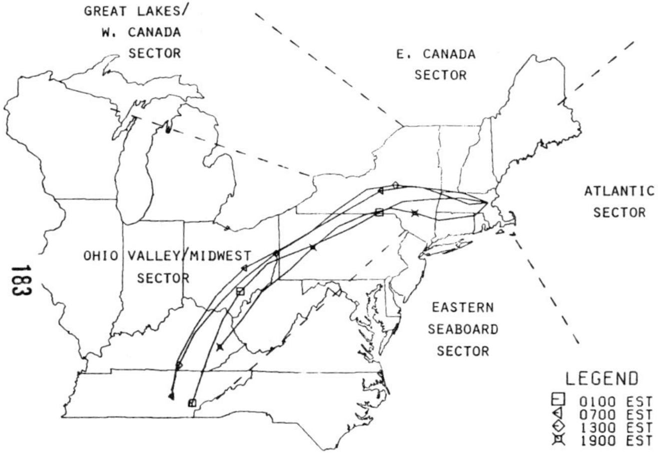

CONCENTRATIONS REPORTED ON THIS DAY WERE:

FINE S	= 4.4 (±0.3) $\mu g/m^3$	NONCRUSTAL IP Mn	= 9.3 ng/m^3
FINE Mn	= 5.7 (±1.3) ng/m^3	NONCRUSTAL FINE Mn	= 5.3 ng/m^3
FINE Se	= 3.5 (±0.5) ng/m^3	NONCRUSTAL FINE Mn/V	= 0.15
FINE V	= 36.5 (±3.6) ng/m^3	NONCRUSTAL IP Mn/V	= 0.23
FINE MASS	= 37.2 (±1.4) $\mu g/m^3$	FINE Se/MASS	= 0.94×10^{-4}

FIGURE 2.
HEFFTER MODEL BACK-TRAJECTORIES
FOR JUNE 10, 1979

CONCENTRATIONS REPORTED ON THIS DAY WERE:

FINE S	= 4.2 (±0.2) $\mu g/m^3$	NONCRUSTAL IP Mn	= 6.7 ng/m^3
FINE Mn	= 2.6 (±1.1) ng/m^3	NONCRUSTAL FINE Mn	= 2.3 ng/m^3
FINE Se	= 0.9 (±0.3) ng/m^3	NONCRUSTAL FINE Mn/V	= 0.28
FINE V	= 8.2 (±2.4) ng/m^3	NONCRUSTAL IP Mn/V	= 0.91
FINE MASS	= 24.4 (±1.4) $\mu g/m^3$	FINE Se/MASS	= 0.37×10^{-4}

FIGURE 3.
NUMBER OF TRAJECTORIES RECORDED, BY SOURCE SECTOR, DURING PERIOD OF STUDY (JUNE 1979 - DECEMBER 1980)

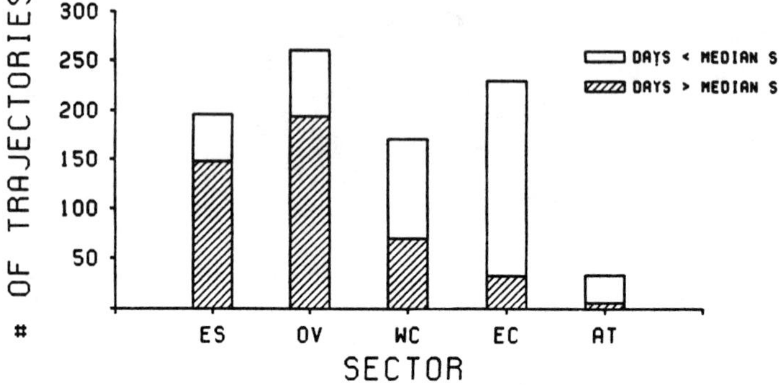

FIGURE 4.
FINE PARTICLE SULFUR DISTRIBUTIONS BY SOURCE SECTOR

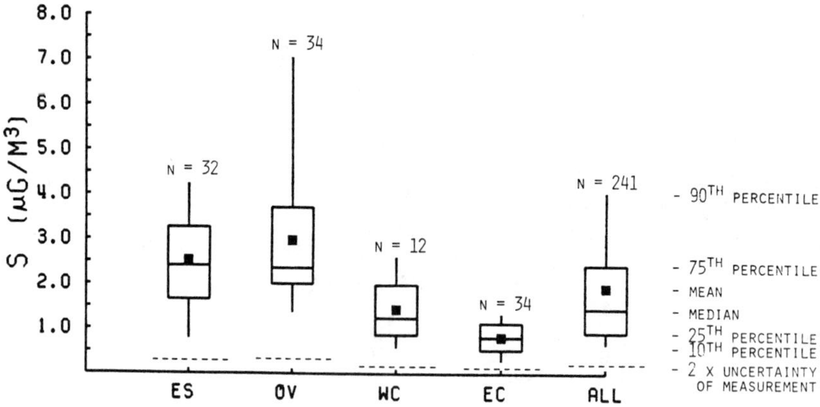

FIGURE 5.
OCCURRENCE OF EASTERN SEABOARD TRAJECTORIES AS
A PERCENT OF ALL TRAJECTORIES, BY MONTH

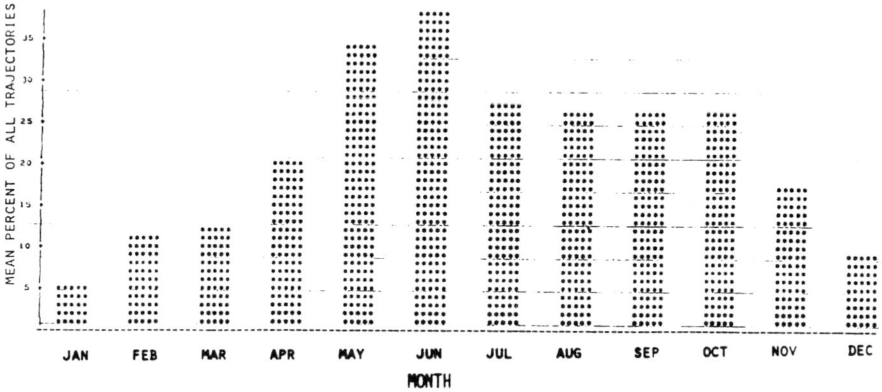

FIGURE 6.
WARM SEASON (MAY 6 - NOV 5) SULFUR DISTRIBUTIONS BY SOURCE SECTOR

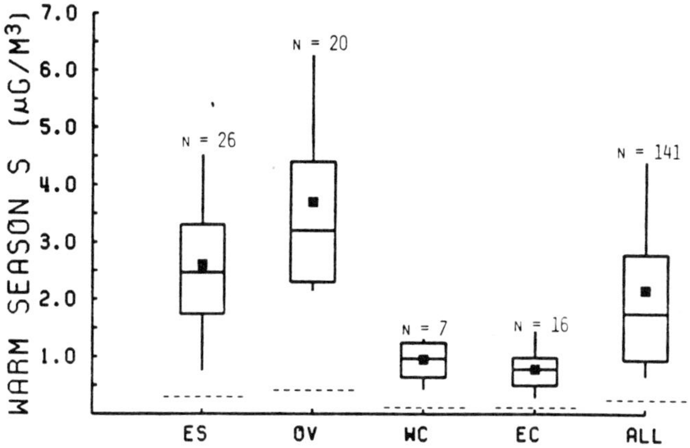

FIGURE 7.
FINE PARTICLE SELENIUM DISTRIBUTIONS BY SOURCE SECTOR

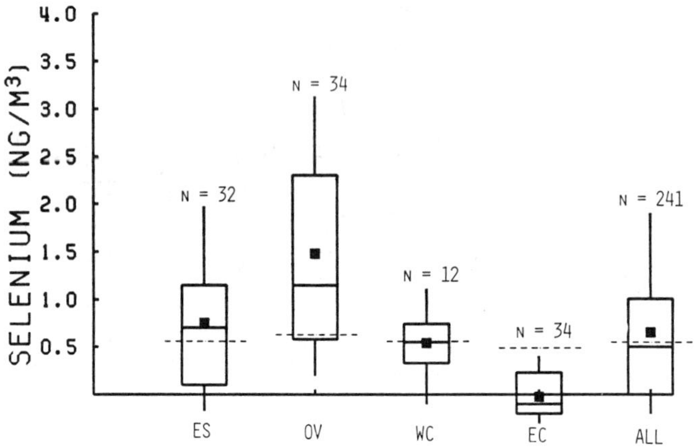

FIGURE 8.
FINE PARTICLE SE/MASS DISTRIBUTIONS BY SOURCE SECTOR

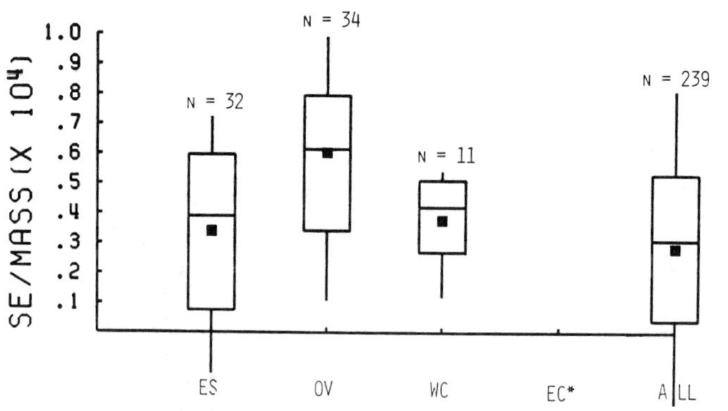

*EC SECTOR NOT PLOTTED (SEE TEXT).

FIGURE 9.
FINE PARTICLE MANGANESE DISTRIBUTIONS BY SOURCE SECTOR

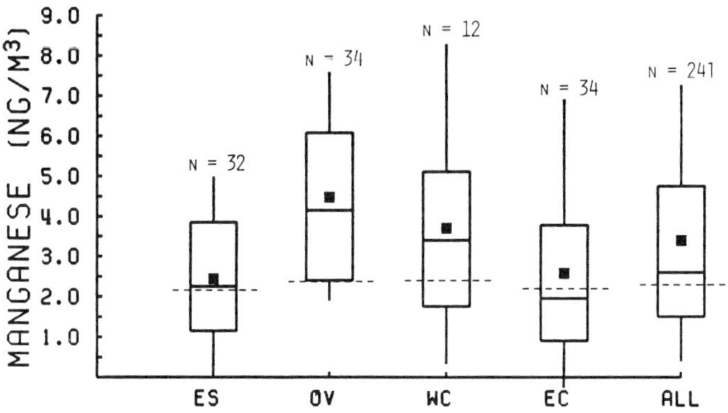

FIGURE 10.
FINE PARTICLE NONCRUSTAL MN/V DISTRIBUTIONS BY SOURCE SECTOR

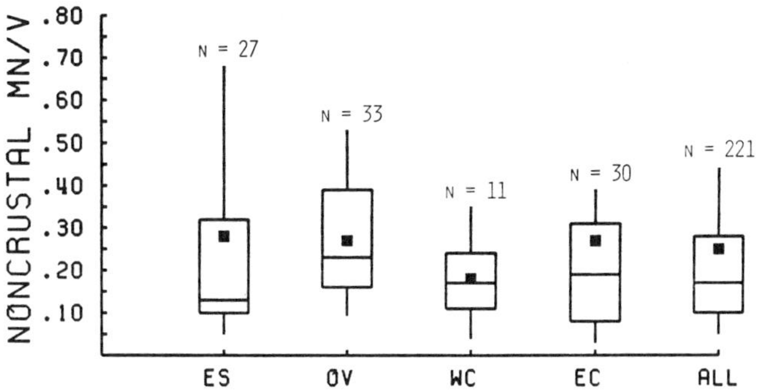

Table 1.
Overall and Seasonal Tests of Difference
Between Ohio Valley/Midwest and Eastern Seaboard Sectors

	OVERALL			WARM SEASON			COLD SEASON		
	ES MEAN (n=32)	OV MEAN (n=34)	p*	ES MEAN (n=26)	OV MEAN (n=20)	p*	ES MEAN (n=6)	OV MEAN (n=14)	p*
Fine S (ng/m^3)	2528. (±232)**	2979. (±295)	0.24	2602. (±263)	3695. (±419)	0.03	2205. (±500)	1956. (±183)	0.56
Fine V (ng/m^3)	15.5 (±2.36)	18.2 (±1.57)	0.34	12.9 (±2.10)	19.7 (±2.17)	0.03	26.7 (±7.58)	16.0 (±2.20)	0.23
Fine Mn (ng/m^3)	2.44 (±0.299)	4.48 (±0.503)	<.01	2.21 (±0.300)	4.24 (±0.672)	0.01	3.42 (±0.884)	4.84 (±0.775)	0.30
Fine Se (ng/m^3)	0.747 (±0.130)	1.47 (±0.183)	<.01	0.727 (±0.137)	1.70 (±0.226)	<.01	0.833 (±0.386)	1.16 (±0.294)	0.54
Fine Mass ($\mu g/m^3$)	18.6 (±1.44)	24.0 (±1.75)	0.02	18.0 (±1.46)	27.8 (±2.29)	<.01	21.1 (±4.51)	18.6 (±2.01)	0.56
Fine Se/Mass ($x10^4$)	0.340 (±0.0638)	0.604 (±0.0662)	< .01	0.344 (±0.0724)	0.600 (±0.0708)	0.02	0.323 (±0.144)	0.609 (±0.128)	0.21
Noncrustal Fine Mn/V	0.281 (±0.0751)	0.272 (±0.0273)	0.91	0.325 (±0.944)	0.222 (±0.0289)	0.31	0.127 (±0.033)	0.348 (±0.047)	0.01
Noncrustal IP Mn/V	0.560 (±0.131)	0.493 (±0.070)	0.65	0.641 (±0.154)	0.477 (±0.0888)	0.36	0.188 (±0.037)	0.520 (±0.119)	0.02

*p value < .05 indicates a statistical difference between sectors.
**Standard error of the mean.

A RECEPTOR-ORIENTED MODEL FOR DIAGNOSING SOURCE CONTRIBUTIONS TO AMBIENT SULFATE CONCENTRATIONS AND SULFUR WET DEPOSITION ON A REGIONAL-SCALE

Perry J. Samson

Department of Atmospheric and Oceanic Science
University of Michigan, Ann Arbor, MI

A receptor-oriented model has been developed as a tool for estimating the contribution of source regions to observed ambient sulfate concentrations and sulfur wet deposition. The model diagnoses both the "natural" potential of a source region to contribute to a receptor, based solely on the physical and chemical properties assumed for the pollutant, and the probability of actual contribution, based on a linear combination of the source region's "natural" potential and its actual emissions strength.

The model has been applied to several receptor locations in the eastern United States and Canada. The results show that the interstate transport of sulfur compounds is a major contributor to the annual average sulfate concentrations and annual sulfur wet deposition at these sites. The model has been compared with observed data and found to predict ambient sulfate concentrations better than sulfur wet deposition. This is not unexpected since the parameterization of the scavenging process is presently greatly oversimplified.

Introduction

Receptor modeling is based upon the analysis of trace element data in an attempt to distinguish signatures in the composition of the observed particle mass which would infer the influence of a particular type of source. When this analysis is applied to transport over regional-scales (> 100 km) these inferences are questionable because of the possibility of:

1. selective scavenging of different trace elements en route,
2. chemical modification of some trace materials over long periods, and
3. the mixing of source regions due to dispersion over long travel times.

Some analyses have been done for trace elements using trajectory analysis in concert with the statistical analysis of elemental composition[1,2]. This work has been useful in delineating the importance of elemental signatures for large, poorly-defined yet distinct source regions. Unfortunately, at present these attempts have been only qualitative and more work is needed to establish the reliability of receptor modeling on a regional-scale.

The purpose of this article is to discuss methodologies available in receptor-oriented dispersion modeling which can be applied to receptor modeling to help clarify ambiguities in discerning source signatures over regional-scales. A receptor-oriented dispersion model has been developed which includes the physical processes of scavenging and dispersion over long time periods and the chemical processes of simple chemical transformation. At present these processes have been parameterized using oversimplified approximations, but the modular form of the model lends itself directly to the incorporation of more complete descriptions of process mechanisms as they become available and tested.

Methods

The mean transport of material to a receptor can be estimated from the computation of upwind (backward) trajectories[3]. The axis of the computed trajectory is assumed to represent the highest probability at a given time upwind of contributing to the receptor. The distribution of potential contribution from regions located laterally from the trajectory will depend upon the degree of vertical mixing coupled with vertical wind velocity shear. As a first approximation it has been assumed that the potential contribution is normally distributed about the trajectory and the distribution increases linearly with time upwind of the receptor[4,5,6].

This model has been specifically applied to the transport, transformation, and deposition of sulfur species. Consequently, the chemical transformation of sulfur dioxide (SO_2) to sulfate ($SO_4^=$) species has been included as a linear transformation process. This is accomplished by assuming linear transformation rates which vary both diurnally and annually from maximum rates in June during daylight hours (when photochemistry is greatest) to minimum rates during the nighttime period. This approximation is a gross simplification of a very complex system which involves both gas-phase transformation processes and in-cloud transformation. The justification for such a parameterization is based on field studies conducted in episodes of high ambient $SO_4^=$ concentrations[7]. Its extension to all meteorological conditions is open to debate, but it is utilized in this model for lack of more information about the variations in this process.

The dry deposition of SO_2 and $SO_4^=$ assumes linear rate coefficients over three hour periods.[4] The dry deposition velocities vary diurnally and seasonally, having minima of 0.15 cm/sec for SO_2 and 0.05 cm/sec for $SO_4^=$ at nighttime regardless of season. The highest values are assumed to occur in the afternoon with 0.8 cm/sec and 0.3 cm/sec assumed for wintertime (January) SO_2 and $SO_4^=$ dry deposition velocities, respectively, increasing to 1.3 cm/sec and 0.45 cm/sec for summertime (July) SO_2 and $SO_4^=$ dry deposition velocities. Rates for other months are linearly interpolated between these values.

The wet scavenging of the sulfur compounds along the trajectories and at the receptor is also assumed to be a linear process. The rate coefficients for wet deposition of SO_2, k_w, and ambient sulfate, κ_w, are determined by

$$k_w = \theta(SO_2)R/h$$

$$\kappa_w = \theta(SO_4^=)R/h$$

where $\theta(SO_2)$ is the washout ratio for SO_2, $\theta(SO_4^=)$ is the washout ratio for sulfate, R is the precipitation rate (mm/hr), and h is the mixing height (mm). The washout ratio is defined as the ratio of the aqueous concentration of pollutant to the ambient concentration in the atmosphere. The value of $\theta(SO_2)$ was chosen as a median value from the range described by Barrie[8]:

$$\theta(SO_2) = 5 \times 10^4$$

and the value of $\theta(SO_4^=)$ was determined from a formulation by Scott[9] for clouds containing ice nuclei:

$$\theta(SO_4^=) = 23.3 \times 10^4 R^{-0.375}$$

It is further assumed that all scavenged sulfur dioxide arrives at the surface as $SO_4^=$ through in-cloud and/or in-precipitation conversion. This removal mechanism does not distinguish between scavenging within the cloud (rainout) and scavenging beneath the cloud (washout) but rather lumps the two together as net removers of the sulfur mass in the transporting layer.

The model has been applied to the transport of sulfur species in the eastern United States and Canada. In order to estimate the wet deposition at the receptor and along the trajectory it was necessary to produce estimates of observed precipitation within the domain of the model grid with a resolution equal to the resolution used in the model (3 h). Figure 1 shows the domain of the model used in this study, each cell roughly 80 km on a side. Hourly precipitation data was gridded for 1978 using available precipitation records from the National Climatic Center (United States) and Environment Canada (Canada).

Emissions data for 1978 were obtained from the United States/Canada Memorandum of Intent Working Group on atmospheric transport modeling[10]. Figure 2 shows the spatial distribution of SO_2 emissions for the year. Four relatively large regions of SO_2 emissions exist:

1. Southern Indiana, southern Illinois, western Kentucky and eastern Missouri;

2. Eastern Ohio, western Pennsylvania, and western West Virginia;

3. New York Metropolitan region; and

4. Central Ontario and western Quebec.

Results

The receptor-oriented approach creates a framework in which the culpability of source regions can be estimated for a given set of assumptions about the nature of the chemical and physical processes involved. One product of the model is a grid of the "natural potential" for contribution to the receptor. This output predicts the spatial contribution over the period of interest assuming unit emissions per unit area over the model domain. This information could be useful for planning purposes because it addresses the "natural potential" of various source regions irrespective of their current emissions strength. An estimation of the "actual" contribution of a source region to the receptor during the period of interest is obtained by multiplying the "natural potential" of the source region by the emissions strength of the region.

Figures 3 and 4 show the spatial distribution of "natu

ral potential" for contribution to ambient $SO_4^=$ and sulfur wet deposition, respectively, at Whiteface Mountain, New York. Figure 3 shows that the estimated "natural potential" for contribution to ambient annual $SO_4^=$ is smoothly distributed about an axis to the west of the receptor. Since the rate of chemical conversion and dry deposition are assumed to be highest in the summertime, this result infers that either the frequency of trajectories from the northwest is roughly equal to the frequency from the southwest in both winter and summer, or any directional bias is removed by a complimentary bias in wet deposition. Figure 4 supports the latter argument because it infers that the "natural potential" for wet deposition is highest with winds from the southwest of the receptor.

It is not surprising that the regions with the highest "natural potential" are located just upwind of the receptor. While the conversion of SO_2 to $SO_4^=$ is a relatively slow process in the atmosphere, the dispersion of culpability produced by the spread of trajectory paths over the period of interest dominates the spatial distribution of "natural potential". Thus, the "natural potential" described in this work is a function of the rate of chemical transformation assumed, the persistence of the trajectories over the period of interest, and the bias of wet deposition towards certain directions.

Combining the "natural potential" with the emissions estimates for the period we obtain the actual potential for source contribution to the receptor for the period of interest. Figures 5 and 6 show the spatial distribution of estimated contribution to annual ambient $SO_4^=$ and sulfur wet deposition, respectively, for 1978. Figure 5 indicates that the sources of the observed $SO_4^=$ was distributed between sources in the eastern Ohio River Basin and central Ontario and Quebec. Figure 6, on the other hand, shows a bias in the sulfur wet deposition contributions towards sources to the southwest of the receptor. These results follow from the previous discussion of differences in the spatial distribution of "natural potential".

The distributions of potential have little meaning if the model is incapable of reproducing observed concentrations. As a test of the predictive capability of the model, daily $SO_4^=$ concentrations were predicted for a six month period from June, 1978 to December, 1978. These predictions were compared with observed $SO_4^=$ concentrations as reported by the New York State Department of Health[11]. Figure 7 shows a scatter diagram of predicted versus observed for the model period. While there is considerable scatter in the results (r=0.44, N=170), the predictions do not seem to be unrelated to the observations. Similar comparisons of the storm-by-storm sulfur wet deposition have not been accomplished.

Conclusions

A framework has been presented in which the upwind meteorology and chemical processes can be introduced in modular form. This tool has been applied to the transport of sulfur species over regional-scales, but could also be applied directly to the regional-scale transport of trace elements. Such application would offer insight into the differential scavenging of elements en route to a receptor and may offer guidance on the role of dispersion in mixing source signatures over long travel times.

It must be stressed that while receptor-oriented modeling shows promise for application as a decision-making tool in the future when realistic chemical and physical mechanisms can be incorporated and tested, the value of the present results are limited. The source-receptor relationships estimated are only valid if the a priori assumptions concerning linear chemical transformation rate and deposition are applicable.

REFERENCES

1. L. Husain and P. J. Samson, "Long-range transport of trace elements," **J. geophys. Res.**, 84: 1237-1240 (1979).

2. G. D. Thurston, P. J. Samson, and J. D. Spengler, "Acid precipitation and long-range transport," Receptor Models Applied to Contemporary Pollution Problems, Air Pollution Control Association, Pittsburgh, PA 15213. 1982 (pp. xx-yy).

3. J. L. Heffter, "Air Resources Laboratories atmospheric transport and dispersion model (ARL-ATAD)," NOAA Tech. Memo. ERL ARL-81, Air Resources Laboratories, Silver Springs, MD, 17 pp. (1980).

4. P. J. Samson, "Trajectory analysis of summertime sulfate concentrations in the Northeastern United States," **J. Appl. Meteor.**, 19:1382-1394 (1980).

5. P. J. Samson and J. L. Moody, "Trajectories and two-dimensional probability fields," **Air Pollution Modeling and its Application I**, C. D. Wispelaere, Ed., Plenum Press, New York (1981).

6. R. R. Draxler and A. D. Taylor, "Horizontal dispersion parameters for long-range transport modeling," **J. Appl. Meteor.**, 21: 367-372 (1982).

7. R. B. Husar, D. E. Patterson, J. D. Husar, N. V. Gillani, and W. E. Wilson, Jr., "Sulfur budget of a power plant plume," **Atmospheric Environment**, 12: 549-568 (1978).

8. L.A. Barrie, "The prediction of rain acidity and SO_2 scavenging in eastern North America," **Atmospheric Environment**, 15: 31-41 (1981).

9. B.C. Scott, "Parameterization of sulfate removal by precipitation," **J. Appl. Meteor.**, 17: 1375-1389 (1978).

10. U.S./Canada Memorandum of Intent, "U.S./Canadian sulfur dioxide emissions for 1978," (Emissions estimates provided by subgroup on atmospheric transport modeling), U.S. Environmental Protection Agency, Washington, D.C. (1982).

11. L. Husain, J. Halstead, P. P. Parekh, and V. A. Dutkiewicz, "A critical investigation of transport of sulfate to the Adirondacks," Second Symp. on the Composition of the Nonurban Troposphere, Amer. Meteor. Soc., Boston, MA, 218-221.

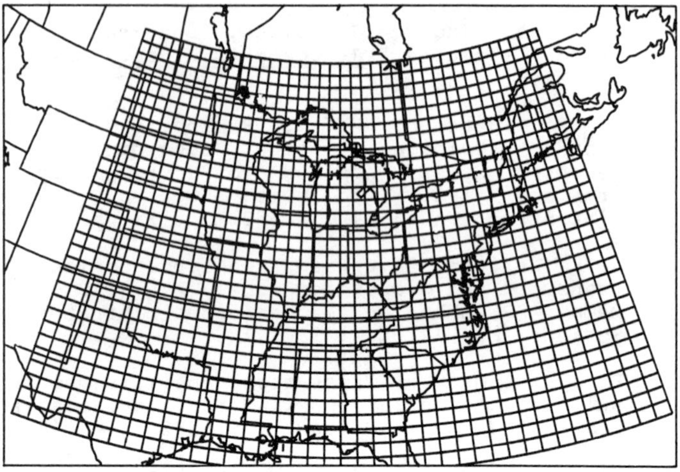

Figure 1. Model grid used for receptor-oriented modeling. Grid cells are approximately 80 km on a side at 42° N. latitude.

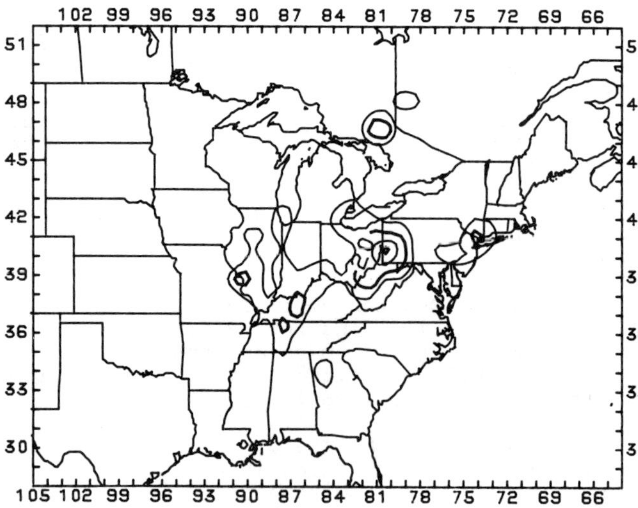

Figure 2. Annual emissions of sulfur for 1978 in units of 10^4 kg km^{-2}. Isopleths are at intervals of 10 kg km^{-2}.

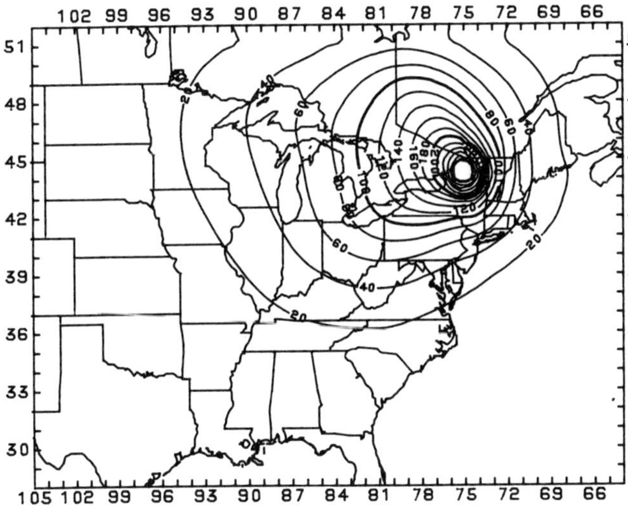

Figure 3. Spatial distribution of the "natural potential" for contribution to the 1978 annual ambient $SO_4^=$ concentration at Whiteface Mountain, New York. Units are $\mu g m^{-3}$ per 100 Tg of sulfur emissions.

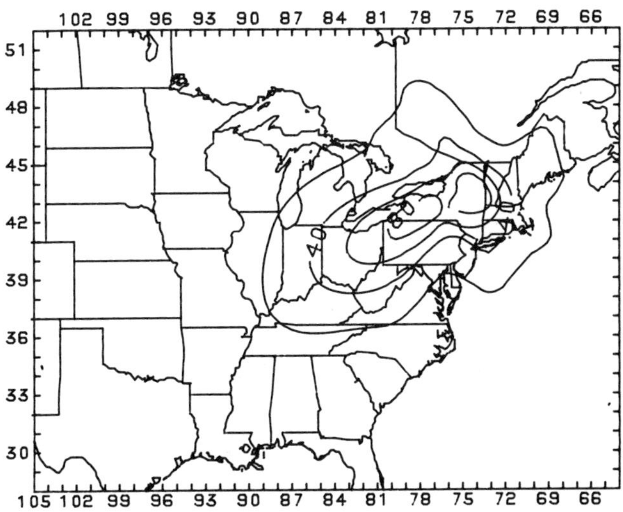

Figure 4. Spatial distribution of the "natural potential" for contribution to the 1978 sulfur wet deposition at Whiteface Mountain, New York. Units are kg ha^{-1} of sulfur deposition per 100 Tg of emissions.

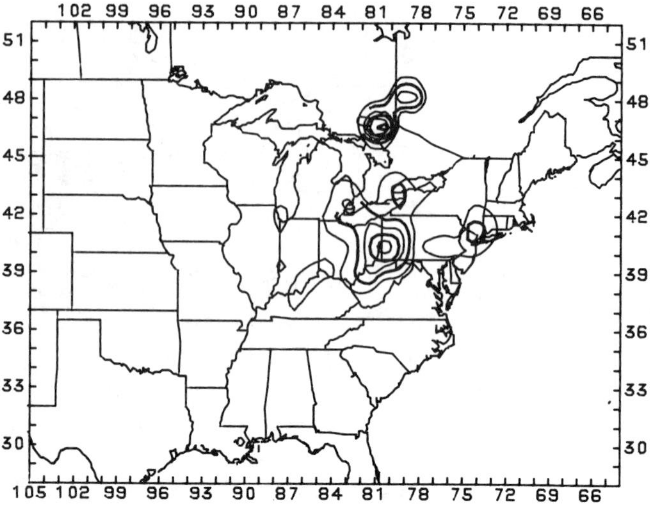

Figure 5. Spatial distribution of probable contribution to the 1978 annual ambient $SO_4^=$ concentration at Whiteface Mountain, New York. Isopleths are spaced every 10 μgm^{-3} per 10^6 km^2.

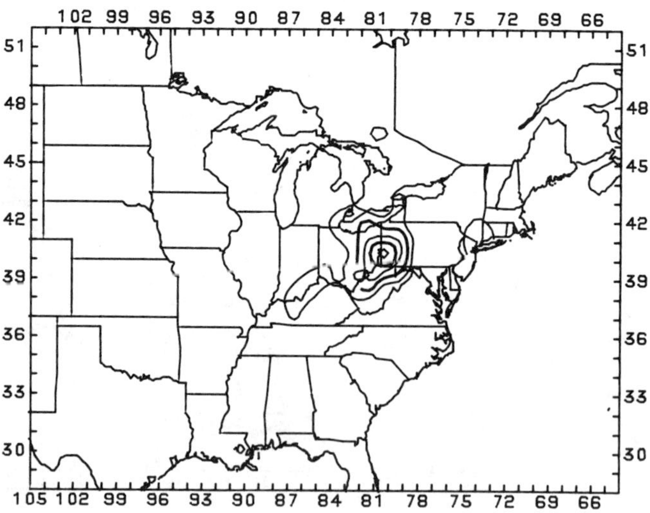

Figure 6. Spatial distribution of probable contribution to the 1978 sulfur wet deposition at Whiteface Mountain, New York. Isopleths are spaced every 5 kg ha^{-1} per 10^6 km^2.

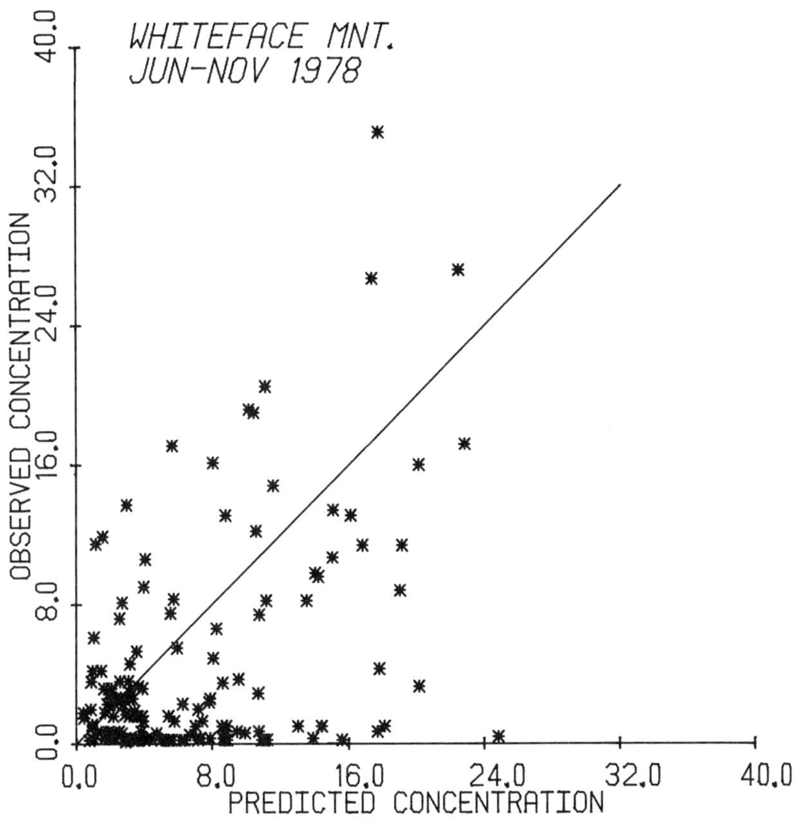

Figure 7. Scatter diagram of predicted daily $SO_4^=$ concentrations at Whiteface Mountain, New York versus those observed by the New York State Department of Health[11]. The samples were collected between June 15, 1978 and December 20, 1978 on a daily basis (r=0.44, N=170).

AMBIENT PARTICULATE MATTER CONTRIBUTIONS TO ACIDIC PRECIPITATION

Michael A. Stafford →
Texas Air Control Board
Austin, Texas

Howard M. Liljestrand
University of Texas, Civil Engineering
Austin, Texas

The alkalinity contribution of particulate matter to precipitation chemistry changes significantly from west to east in Texas with the transition in soil chemistry. Although this crustal component provides important buffering in rainwater, it is poorly delineated spatially with respect to quantities of local and long range sources, such as the magnitude of alkaline calcium carbonate soil dust transport from the west. A target transformation factor analysis is used to identify the source factors to total suspended particulate (TSP) air samples at eight sites in Texas. These sources are tested as factors for the rainwater chemistry. The target transformation factor analysis of the rainwater data from the same sites determines the contributions of aerosol sources to the chemical composition and net acidity or alkalinity of rainwater. The results provide a better definition of crustal alkalinity contributions than the commonly used chemical mass balance methods.

Introduction

The previous attempts at determining source contributions to alkalinity in Texas rainwater have used chemical mass balance methods.[1,2] In order to use the chemical mass balance (CMB) approach, accurate source profiles must be known. Unfortunately, the soil dust composition has not been chemically determined at many rainwater sampling sites, alkalinity sources may show chemical fractionation during precipitation, and long range transport of crustal components may contribute to the rainwater chemistry. To avoid these uncertainties, the CMB analysis of the major ionic species has presumed all the non-sea salt contributions result from unity sources composed of only one species in its acid or alkaline form.[3,4] Typically, three unity alkalinity sources - calcium alkalinity ($CaCO_3/Ca(OH)_2/CaO$), magnesium alkalinity and sodium alkalinity - are postulated to describe the soil contributions, and no distinction is made among the possible local and long range sources.[3]

This investigation uses target transformation factor analysis (TTFA) to determine empirically the elemental profiles and alkalinity contributions of local crustal components to rainwater chemistry in Texas. Soil alkalinity is an important buffering agent in Texas rainwater, and the background buffering capacity in rain changes from west to east as a result of soil composition[5] and ammonia contributions.[2] The weathered soils of East Texas are predominantly buffered by alumino-silicates, as opposed to the calcium carbonate soils of West Texas, as shown in Figure 1. The generalization of two soil types is only approximate, since surface soil chemistry is quite variable on a local scale. TTFA of total suspended particulate matter (TSP) in the air is used to define the local soil dust composition, including fractionation effects. The elemental profiles from the aerosol results are then applied to the target transformation factor analysis of rain chemistry data sets. This analysis provides a characterization of the local crustal contributions to rainwater.

Methods

Source of Aerosol Composition Data (TSP)

The TSP data used in this study was obtained from the Texas Air Control Board (TACB) for samples collected and chemically analyzed at the sites nearest the rain sampling sites shown in Figure 1 for 1980-81 and 1976 for Austin. Twenty-four hour high volume (HiVol) samples were collected at six day intervals as part of the TACB's ambient monitoring program. Samples were collected on cellulose acetate filters and analyzed by X-ray fluorescence (XRF) using four different X-ray sources to quantify the concentrations of thirty-two elements (Al, Si, P, S, Cl, Ar, K, Ca, Ti, V, Cr, Mn, Fe, Co, Ni, Cu, Zn, Ge, As, Se, Br, Rb, Sr, Zr, Mo, Cd, Sn, Sb, Ba, La, Tl and Pb).[6]

Sources of Precipitation Chemistry Data

Nine sets of rainwater from four sources were investigated. Firstly, Lopez-Gonzalez[6] determined the pH and trace metal composition of Austin rain in 1975. Cation-exchange filters and coprecipitation were used to pre-concentrate trace metals for XRF at the TACB. Samples were collected by storm event. Secondly, the WMO-NOAA-EPA global precipitation monitoring network[7-9] collected monthly aggregates of precipitation at Victoria and San Angelo. The major ionic species (Na, K, Ca, Mg, NH_4^+, F^-, Cl^-, NO_3^-, SO_4^{2-} and pH) as well as some trace metals, which were near the limit of detection, were measured. Thirdly, Feeley and Liljestrand[2] reported the major ionic species as well as trace species (Pb, Br, Mn, o-PO_4, and NO_2^-) for sequential 0.25-inch increments of rain at Austin,

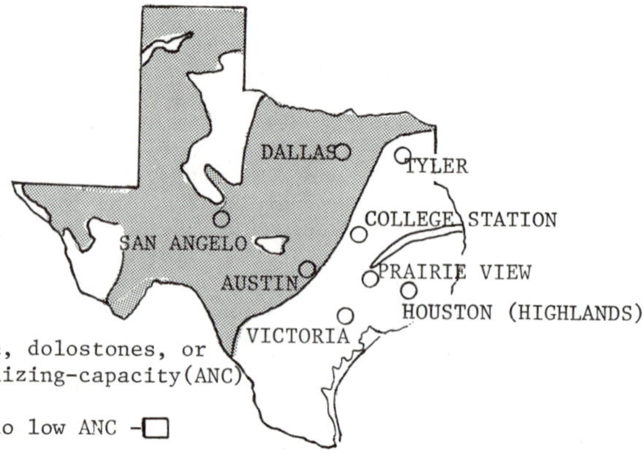

LEGEND

Bedrock of limestone, dolostones, or highest acid-neutralizing-capacity(ANC) materials - ▩
Beckrock of medium to low ANC - □

Figure 1. Locations of Rainwater Monitors and Bedrock Alkalinity[5]

Prairie View and Houston (Highlands) during 1980-81. Additional trace elements were determined for annual precipitation-weighted composites. Fourthly, Lopez-Gonzalez[10] collected sequential 0.014-inch increments of precipitation for storm events at Tyler, Dallas and College Station. The major ionic species, trace metals (Cr, Ni, Zn, Cu and Pb) and pH during collection were measured.

Screening of TSP Data

Before factor analysis of the raw TSP data sets, a screening of the elements was performed to reduce error in the data, hopefully without losing valuable information. Each element whose mean of observed values was less than a minimum detectable mean (MDM) for that element was removed from the data set. The MDM is estimated by a t-test to determine if the mean of the measurements is statistically different than zero, and it is defined by,[11]

$$MDM = t(0.001, n-1)\ s\ /\ (n-1)^{\frac{1}{2}} \qquad (1)$$

where t(0.001, n-1) = student's t for n-1 degrees of freedom and 0.001 confidence level in a two-tailed test,
n = number of samples, and
s = standard error of measurement, or one standard deviation of the random detector error at zero chemical concentration.[11]

In previous studies, measured values in the retained data which fell below the detection limit (usually assumed to be three times the detector error, s) were set equal to one-half the detection limit.[12] This approach was not used in this study, since slightly lower average errors were obtained in factor analysis using the unaltered data. Using the unaltered screened element data avoided the introduction of any bias when the mean of the replaced values is less than 3s/2.

Screening of Rainwater Data

In addition to the screening of elements below the limit of detection, the rainwater data for pH was converted to the corresponding net strong acidity/alkalinity for water open to the atmospheric partial pressure of carbon dioxide (P_{CO_2}). While pH represents a logrithmic scale (pH = $-\log_{10}\{H^+\}$), alkalinity and net strong acidity are conservative linear quantities. Alkalinity and strong acidity for aqueous systems open to $P_{CO_2} = 10^{-3.5}$ atm at 20°C are defined by,

$$[\text{Alkalinity}] = [HCO_3^-] + 2[CO_3^{2-}] + [OH^-] - [H^+] =$$

$$- [\text{Acidity}] = \frac{10^{-11.3}}{\{H^+\}} + \frac{10^{-21.25}}{\{H^+\}^2} + \frac{10^{-14}}{\{H^+\}} - \{H^+\} \qquad (2)$$

with a reference acidity of zero at pH 5.6. By engineering convention, these values are multiplied by 50,000 to report a weight concentration as calcium carbonate, although this is not a true weight in mass fractions.

Acidity and alkalinity in equation 2 apply only for strong acid-base, dilute solutions open to atmospheric P_{CO_2}. This approximation is good for Texas rainwater.[2] Weak acids with $-\log K_a$ values in the pH range of rainwater (3-7) would require corrections for their extent of dissociation to be included in equation 2. Ammonium can be treated as a cation rather than a weak acid, since it is less than 1% dissociated in rainwater. Solid phases may provide alkalinity from solubility controls or weak acidity from surface charge.[13] For dilute solutions, the strong alkalinity contribution is presumed to be dissolved and measured by the aqueous ion concentrations as in equation 3, which results from the charge balance constraint.

$$[\text{Alk}] = -[\text{Acid}] = [Na^+] + [K^+] + 2[Ca^{2+}] + 2[Mg^{2+}] + [NH_4^+] - [Cl^-] - [F^-]$$

$$- [NO_2^-] - [H_2PO_4^-] - [NO_3^-] - 2[SO_4^{2-}] \qquad (3)$$

Net strong acidity and alkalinity in rainwater as calculated by equation 2 are treated as being independent of the alkalinity calculated by the major ionic concentrations (equation 3) in this investigation, since charge balance and quality assurance checks have not been met in the older rainwater data sets in Texas.[2,14] Equation 3 also provides an estimate independent of TTFA for the alkalinity contributions to rainwater from aerosol TSP, assuming the particulate matter basic and acidic components are soluble in dilute rainwater. Even if the solid phases do not completely dissolve, their alkalinity is measurable as a buffering capacity.[13] Alternatively, this alkalinity can be expressed as the equivalent difference between cationic and anionic species (analogous to equation 3) and expressed as mg $CaCO_3$ solid. This is a conservative quantity in solid or aqueous phase, in the absence of oxidation-reduction reactions.[13]

Using net strong acidity in factor analysis presents some difficulty in data sets with rain samples both above and below pH 5.6 (net alkalinity as well as net acidity). Samples with pH values greater than 5.6 have a negative value on the strong acidity scale. However, these negative values are meaningful and should not be set equal to zero or one-half the detection limit during data manipulations. An alternative to handling positive and negative values for net strong acidity/alkalinity is to offset the values with respect to the minimum value and thus redefine the acidity and alkalinity reference point. This is analogous to the titration method for acidity with the endpoint outside the range of pH values in the data set.

Methods of Numerical Analysis

The method of numerical analysis used in this investigation was target transformation factor analysis (TTFA), because of its ability to determine source profiles from the data as well as to rotate the data toward pre-selected source profiles. TTFA has been described by Malinowski[15], Alpert and Hopke[16], and others. The computer algorithm used for TTFA was FANTASIA (Factor Analysis To Apportion Sources In Aerosols) which was developed by Alpert et al.[18] based upon a program written by Malinowski and co-workers.[17] Modifications of the FANTASIA algorithm at the University of Texas at Austin have been minor.

The results of TTFA depend on a series of subjective judgements made during data analysis. Each data set was analyzed in five steps. Firstly, the screening program removed species below the MDM from the raw data. Secondly, the screened data set was input to FANTASIA to determine the number of factors. The magnitudes of the eigenvalues for the correlation matrix, factors with Exner Psi functions greater than 0.1, as well as variations in the root-mean-square error, chi-squared and Malinowski's factor indicator function were examined to indicate the range of probable number of factors. Thirdly, TSP test vectors were input to FANTASIA with the best estimate of the number of factors (from step 2) to obtain iterated test vectors (vectors from the iterated rotation of the eigenvectors toward the test vectors) as well as iterated unity vectors (vectors from the eigenvector rotation toward each unique chemical species). This output was plotted for visual comparison of the initial and iterated vectors as well as to identify common source types among the iterated vectors. The determination of likely test vectors was based upon 1) the number of iterations required for convergence, 2) the maintenance of source vector characteristics before and after iteration, 3) the ease of eigenvector rotation with only one iteration, and 4) the common source types among the iterated vectors. Fourthly, the likely iterated test vectors (from step 3) were grouped according to similarity and twenty combinations of vectors were chosen by varying the most probable groups and substituting various vectors within each group. This process is repeated until reasonable fits are obtained with few source contributions calculated to be negative in the individual samples. Fifthly, the process is repeated with a different number of factors, to see if a better fit can be obtained within the range of probable number of factors.

Results

Seven TSP data sets were analyzed by TTFA to determine the crustal source composition and the major contributing sources at each site. Nine rainwater data sets from eight sites were then analyzed by TTFA, using the results from the aerosol study. The major TSP sources identified for each site are outlined in Table I. The mass fraction of each element in the crustal and road source profiles is provided in Table II. Blanks in Table II indicate that element was below the limit of detection and removed in the screening before TTFA. In most of the TSP data sets, two alkalinity sources are evident - road dust and crustal soil components. The crustal sources in both TSP and rainwater data are tentatively identified as West Texas or East Texas to reflect Al:Si:Ca:Fe ratios. The crustal component identified as West is enriched in calcium and shows higher contributions to West Texas TSP sites. In some rainwater data sets, two crustal soil components were evident, as shown in Table I.

For the rainwater data sets, the objectives were to determine the number of sources, determine if sources to the particulate matter data sets were also sources of elements in rainwater, and identify sources which contributed primarily only one element (the unity source vectors). For each site the factors identified are listed in Table I, and the compositions of the iterated crustal source vectors are given in Table III. While distinctions of mechanisms (primary vs secondary) and sources for sulfate, nitrate and other non-crustal component contributions to rainwater are beyond the objectives of this investigation, source vectors iterated from TSP sulfate vectors have been distinguished from sulfate, nitrate and sulfate+nitrate source vectors found in the rainwater which are distinct from the aerosol TSP contributions. This difference between TSP and non-TSP sources of nitrate and sulfate should be re-examined. Artifacts and losses in sulfate and especially nitrate TSP as well as sampling under non-precipitating (non-representative) conditions may bias the aerosol contribution of these acidic components to rainwater.

Few unity sources or sources primarily of one element showed distinct character in any of the rainwater data sets. All of the iterated unity vectors tended toward a common vector for each site, which may indicate a common precipitation scavenging mechanism rather than a common source. The iterated unity vectors other than sulfate and nitrate were not included in the final source apportionment and hence are not included in Table I. Those species which showed some distinct character in source profile included calcium at Austin[6], College Station[10], Victoria and San Angelo[7-9]; manganese at Houston, Austin and Prairie View[2]; arsenic at Austin[6]; nickel at Dallas[10]; and lead at Tyler[10].

Discussion of Results

The conclusions of this investigation are somewhat limited by the types of available data. The use of HiVol TSP samples limits the resolution of local crustal sources from anthropogenic sources. Data from dichotomous samplers could provide a more exact identification of the TSP sources as well as better input vectors for the crustal component to the rainwater factor analysis. However, the use of HiVol TSP is appropriate, since it is the only historical data available for the geographic distribution of Texas. Even so, gaps in the HiVol data (i.e., no XRF data for College Station) necessitate the use of the nearest monitor's crustal component to be iterated in the TTFA of rainwater.

Moreover, HiVol TSP and rainwater source contributions are both biased to the larger diameter aerosol particles. While HiVol includes both coarse and fine particles, the mass collected is dominated by the mass peak of the particle size distribution in the coarse size. Similarly, below cloud scavenging by raindrops via diffusion, interception and impaction mechanisms is most efficient for aerosol particles <0.005 μm and >10 μm in diameter[19]. Thus the greatest mass contributed to rainwater by local soil types resuspended below cloud level should be from the large aerosols which contribute most of the metal mass collected by HiVol samplers. If dichotomous data were available, the crustal component found for the coarse aerosol would be used for the analysis of rain data.

A second limitation in the interpretation of the results lies in the labels East and West for the crustal components. The use of calcium enrichment to aid in distinguishing soil sources is tenuous, since calcium is generally hard to fit in receptor modeling. However, two crustal components are expected in precipitation samples, because frontal systems entrain soil dust for long-range transport. The crustal components determined from factor analysis may represent the local soil type involved in below-cloud scavenging as well as a long-range transport factor scavenged both below- and in-cloud. Lopez-Gonzalez[10] documented dramatic shifts in rainwater composition (pH as well as metals) as surface winds and temperature shifted in the passage of a front through Dallas. Likewise, this investigation determined two crustal components for the two air masses.

A third potential limitation in the interpretation results from the rain sample collection and handling. While wet-only precipitation was attempted at all sites, precipitation samples are always a combination of wet deposition plus dry deposition during the storm. As sampler cleaning and mechanical and electronic maintenance become lax, the rainwater samples approach the composition of bulk deposition samples dominated by the local soil type. A large contribution from one factor makes the resolution of trace factors more difficult. Alternatively, TTFA provides an indication of quality assurance by distinguishing wet from bulk precipitation sampling.

The largest limitation is the degree of overlap between the TSP metals determined by XRF (atomic number >13) and the species routinely monitored in precipitation studies. More trace metal analysis is needed to define source vectors in the rainwater data sets. TSP data for Na, Mg, and alkalinity by titration would be useful. The Lopez-Gonzalez[6] trace metal analysis of Austin rainwater provides the best overlap of elements as well as a minimum difference in instrumental biases since the same XRF instrument was used for TSP and rain samples. There is fair agreement between TSP crustal sources and rainwater source components in Tables II and III. Unfortunately, this rainwater data set has insufficient major ionic concentration determinations to provide quality assurance for the pH measurements used to calculate the alkalinity.

Despite these limitations, the TTFA of the rainwater data sets provides several noteworthy results. First, the resolution of source components or scavenging mechanisms is improved by sampling increments of precipitation. The least number of factors are found for the monthly aggregate samples at Victoria and San Angelo. Alternatively, the low number of significant factors may result from a high dry deposition of the local soil component, which obscures other factors. These sites had the lowest precipitation. The probable number of factors is greatest for the Feeley and Liljestrand[2] data sets which sampled multiple storms by increment within each event. The added factors are associated with gaseous precursors (ammonium, nitrate and sulfate). In data sets with the fewest factors, the resolution of common sources from common mechanisms of scavenging becomes more difficult; the first-order removal of all chemical constituents from the atmosphere by precipitation scavenging provides a large covariance among all species. In data sets with the largest number of factors, the additional factors may represent distinct mechanisms rather than distinct sources for the gaseous species. The differentiation of common sources vs common mechanisms of scavenging is less of a problem for the non-volatile crustal alkalinity components.

Secondly, the sea salt component appears in all the rainwater data sets, even though it may not be evident in the TSP results for the same site. The resolution of a sea salt factor substantiates the postulate of the simple CMB model for rainwater.[3,4] Most of the sodium, magnesium and chloride in rainwater as far inland as Austin is attributed to the marine air mass.[2]

Thirdly, the resolution of strong acidity components requires a data set which is consistent with the charge balance constraint. Sulfate, nitrate and ammonium are evident as factors in both TSP and rainwater data sets, and the sulfate vector identified in the TSP results is preserved in the TTFA of rainwater data for the same site. Moreover, the rainwater data sets show nitrate factors in addition to the vectors identified for TSP. Unfortunately, the strong acidity (calculated from pH measurements by equation 2) associated with these sulfate and nitrate acid components as well as those for the crustal components are often inconsistent with the charge balance constraint (equation 3). Only those precipitation data sets which meet the charge balance quality assurance test show factors with consistent strong acidity fractions. For these data sets (Austin, Prairie View and Houston), strong acidity is not linearly independent from the major ionic species.

Some of the inconsistency for strong acid fractions within the source vectors can be attributed to the pH measurements used to calculate the alkalinity/acidity. Lopez-Gonzalez[10] recorded increases of up to 1 pH unit between immediate and successive measurements over a week following collection. This would correspond to decreases in acidity by an order of magnitude. Moreover, the component highest in strong acidity is often enriched in potassium chloride, the main component of the outer reference electrode solution.

Conclusions

1. The local crustal components identified by the TTFA of particulate matter appear useful in the interpretation of precipitation chemistry data. The target transformed vectors in rainwater data are consistent with those input from the TSP analysis. The interpretation of TSP data is a better approximation than the limited CMB models commonly used in rainwater interpretation, since fewer sources are presumed.

2. For an accurate assessment of local soil contributions to rainwater, more overlap is needed between the chemical species determined in TSP and precipitation. The additional determinations which would be useful include trace metals in individual rainwater samples and the major ions as well as alkalinity by titration for aerosol samples.

3. The sulfate vector from the total suspended particulate matter analysis appears to be an important source to the rainwater as well. A nitrate factor for rainwater appears in addition to the TSP factors.

4. The target transformation factor analysis of rainwater data confirms the sea salt (Na, Cl, and Mg) component at all sites, which supports the assumption used in chemical mass balance and tracer models.

5. The resolution of acidity/alkalinity factors in historical data for rainwater may be limited by the pH/acidity measurements. An alternative is to estimate the net acidity/alkalinity contribution with a charge relationship and the concentrations of the major ionic species.

Acknowledgments

This study was supported by the Texas Air Control Board and grants from the Texas Energy and Natural Resources Advisory Council as well as the U. S. Environmental Protection Agency.

References

1. H. M. Liljestrand, J. J. Morgan, "Error analysis applied to indirect methods for precipitation acidity," Tellus, 31: 421 (1979).

2. J. A. Feeley, H. M. Liljestrand, "Source contributions to acid precipitation in Texas," Atmos. Environ., in press.

3. L. Granat, "On the relation between pH and the chemical composition in atmospheric precipitation," Tellus, 24: 550 (1972).

4. C. V. Cogbill, G. E. Likens, "Acid precipitation in the northeastern United States," Water Resour. Res., 10: 1133 (1974).

5. S. A. Norton, J. J. Akielaszek, T. A. Haines, K. L. Stromborg, J. R. Longcore, "Bedrock geologic control of sensitivity of aquatic ecosystems in the United States to acidic deposition," and "Distribution of surface waters sensitive to acidic precipitation: a state-level atlas," National

Atmospheric Deposition Program, Colorado State University, Fort Collins, Colorado, 1982.

6. J. A. Lopez-Gonzalez, "X-ray fluorescence sample preparation for evaluating trace element scavenging by precipitation," Unpublished Masters Thesis, University of Texas at Austin (1975).

7. "Atmospheric turbidity and precipitation chemistry for the world 1973," Environmental Data Service, National Climatic Center, Asheville, North Carolina, 1975.

8. "Atmospheric turbidity and precipitation chemistry for the world 1974," Environmental Data Service, National Climatic Center, Asheville, North Carolina, 1976.

9. "Global monitoring of the environment for selected atmospheric constituents 1975," Environmental Data Service, National Climatic Center, Asheville, North Carolina, 1977.

10. J. A. Lopez-Gonzalez, "Determination of variations in the chemical composition of rainfall in selected regions of Texas through the use of an automated sequential sampler and analyzer," PhD Dissertation, University of Texas at Austin (1982).

11. S. L. Dattner, "Statistical estimation of the minimum detectable mean," Proceedings, 73rd National Meeting, APCA, June 22-27, 1980.

12. E. Kushner, "On determining the statistical parameters for pollution concentrations from a truncated data set," Atmos. Environ., 10: 975 (1976).

13. H. M. Liljestrand, "Equalization-neutralization basin design, discussion," J. Environ. Eng. Div. (ASCE), 108: 224 (1982).

14. S. Y. Tyree, Jr., "Rainwater acidity measurement problems," Atmos. Environ., 15: 57 (1981).

15. E. R. Malinowski, D. G. Howery, Factor Analysis in Chemistry, John Wiley and Sons, New York, 1980.

16. D. Alpert, P. Hopke, "A quantitative determination of sources in the Boston aerosol," Atmos. Environ., 14: 1137 (1980).

17. E. R. Malinowski, D. G. Howery, P. H. Weiner, J. M. Sorka, P. T. Funke, R. B. Swizer, A. Levinstone, "FACTANL - target transformation factor analysis," Program 320, Quantum Chemistry Program Exchange, Indiana University, Bloomington, Indiana (1976).

18. D. Alpert, "Quantitative apportionment of urban aerosol mass by factor analysis, " PhD Dissertation, University of Illinois at Urbana-Champaign (1980).

19. W. G. N. Slinn, "Some approximations for the wet and dry removal of particles and gases from the atmosphere," Water Air Soil Pollut., 7: 513 (1977).

TABLE I. COMPARISON OF MAIN SOURCES DETERMINED BY TARGET TRANSFORMATION FACTOR ANALYSIS OF ELEMENTS MEASURED IN TOTAL SUSPENDED PARTICULATE MATTER SAMPLES WITH THOSE FOR PRECIPITATION

Main Sources of Measured Elements in HiVol Samples

Data Set	Source Identities
Austin (1976)	Crustal (West Texas), Sea Salt, Auto
Austin	Road+Auto, Crustal (East Texas), Sea Salt, Sulfate TSP
Dallas	Road, Auto, Crustal (West Texas)
Houston	Crustal (East Texas), Road, Sea Salt, Sulfate TSP
San Angelo	Crustal (West Texas), Road, Auto
Tyler	Crustal (East Texas), Auto+Diesel, Sulfate TSP
Victoria	Crustal (East Texas), Auto, Sea Salt

Main Sources of Measured Elements in Rainwater Samples

Data Set	Identified Source Vectors	Range of Probable Number of Vectors
Austin (1975)	Sea Salt, Crustal (East Texas), Crustal (West Texas)	2-4
Austin	Sea Salt, Crustal (East Texas), Crustal (West Texas)	3-6
College Station	Crustal (East Texas), Ammonia, Sea Salt	2-5
Dallas	Crustal (East Texas), Crustal (West Texas), Sea Salt	2-5
Houston (Highlands)	Sea Salt, Sulfate TSP, Nitrate, Road	3-5
Prairie View	Crustal (East Texas), Sea Salt, Sulfate TSP, Nitrate	3-6
San Angelo	Sea Salt, Crustal (West Texas), Auto	2-3
Tyler	Crustal (East Texas), Sea Salt, Sulfate+Nitrate	2-4
Victoria	Crustal (East Texas), Sea Salt, Sulfate+Nitrate	2-3

Note: West Texas crustal component refers to soils high in calcium carbonate indicated by the shaded area of Figure 1. While East Texas refers to soils of lower alkalinity and higher clay content, the clear area of Figure 1. All soils differ locally.

TABLE II. ALKALINITY SOURCE PROFILES FOR HIVOL SAMPLES NORMALIZED TO % OF CALCULATED MASS

Element / Data Set	Pb	Br	Zn	Cu	Fe	Mn	Ti	Ca	K	Cl	S	P	Si	Al
Austin (1976) (West)	0.35	0.18	-	0.35	5.18	-	-	16.3	4.68	-	5.5	-	45.1	22.3
Austin (Road+Auto)	3.05	0.92	0.43	3.73	1.65	0.06	0.06	41.4	1.21	4.0	29.2	0.66	9.4	4.0
Austin (East)	1.23	-	0.054	-	10.7	0.16	1.1	13.0	5.42	-	10.2	0.34	45.4	12.0
Houston (East)	0.17	0.02	0.14	0.05	15.6	0.18	0.62	12.3	4.13	12.9	-	0.61	38.1	14.0
Houston (Road+Auto)	2.34	0.07	0.79	0.66	2.83	0.22	0.28	43.6	0.60	-	18.0	0.84	19.2	9.78
San Angelo (Road)	0.25	0.11	0.43	1.40	4.48	0.08	0.40	64.2	3.22	2.03	-	0.51	20.4	5.92
San Angelo (West)	0.02	-	0.15	2.55	5.56	0.13	0.60	23.4	4.16	0.50	14.3	0.34	35.4	12.9
Tyler (East)	0.15	0.12	0.03	0.60	10.1	0.12	0.81	4.65	4.25	2.94	-	-	47.4	23.7
Tyler (Road)	7.72	0.64	0.74	1.52	12.2	0.54	0.41	39.6	0.61	5.47	2.69	1.37	0.06	0.03
Dallas (Road+Auto)	2.50	-	0.57	2.15	1.44	0.10	0.01	29.6	0.71	2.61	49.7	1.0	6.78	1.66
Dallas (West)	0.89	0.28	0.15	0.38	8.21	0.16	0.91	23.6	5.0	1.72	0.95	0.55	43.0	14.2
Victoria (East)	0.22	0.02	0.12	-	9.5	-	1.24	17.7	5.1	-	1.9	-	51.0	13.3

TABLE III. ALKALINITY SOURCE PROFILES FOR PRECIPITATION SAMPLES NORMALIZED TO PERCENT OF CALCULATED DRY MASS

Element	Pb	Zn	Ni	Fe	Mn	Ti	Ca	K	Cl	SO_4^{2-}	P	Si	Al	Mg	Na	NO_3^-	NH_4^+	Alk_1[a]	Alk_2[b]
Data Set																			
Austin (East)	(1.74)[c]	(0.08)	–[d]	–	0.14	(1.56)	18.4	1.02	9.50	18.5	0.13	(29.7)	(7.86)	0.96	4.10	5.55	0.35	25.5	28.5
Austin (West)	(0.02)	(1.08)	–	(3.03)	0.19	(0.32)	11.6	7.68	16.5	11.7	0.07	(8.4)	(3.6)	1.07	7.33	6.27	2.97	32.6	36.1
College Station (East)	0.65	(0.06)	0.16	–	(0.08)	(0.48)	5.51	3.23	10.1	22.8	(0.27)	(17.1)	(6.28)	1.37	10.97	19.5	2.44	38.0	2.0
Dallas (East)	0.09	0.29	0.16	(0.33)	(0.04)	(0.25)	6.69	0.64	22.7	24.1	(0.27)	(12.1)	(4.02)	0.48	1.63	17.4	8.5	-98.3	-22.5
Dallas (West)	1.85	0.03	0.16	(2.08)	(0.14)	(0.01)	64.2	0.01	1.87	–	(1.45)	(9.81)	(2.40)	–	–	17.4	–	-8.75	150.
Houston (Road)	(0.53)	(0.18)	–	(0.64)	0.09	(0.07)	9.88	0.20	12.6	36.2	0.19	(4.35)	(2.22)	1.20	6.40	22.9	2.23	-19.6	-21.4
Prairie View (East)	(0.11)	(0.09)	–	–	0.16	(0.41)	5.36	3.29	20.7	7.8	0.81	(24.9)	(9.15)	1.82	10.6	5.20	7.13	20.6	26.5
San Angelo (Road)	(0.19)	(0.32)	–	(3.38)	(0.06)	(0.30)	55.1	1.13	12.4	–	(0.38)	(15.4)	(4.47)	0.78	–	5.61	0.52	<0	132.
San Angelo (West)	(0.02)	(0.14)	–	(5.12)	(0.12)	(0.55)	14.3	0.65	0.20	11.4	(0.32)	(32.6)	(11.9)	–	0.02	22.6	0.04	<0	21.4
Tyler (East)	0.50	0.14	–	–	(0.07)	(0.49)	7.57	12.5	25.5	5.7	–	(28.7)	(14.4)	1.04	–	–	3.93	-22.8	9.74
Victoria (East)	(0.08)	(0.04)	–	(3.53)	–	(0.46)	6.57	4.38	25.1	13.5	–	(18.9)	(4.93)	1.43	14.1	2.14	4.78	60.8	31.1
Austin (East-75)	0.12	0.006	–	5.71	–	0.81	7.2	1.9	–	–	–	52.1	32.1	–	–	–	–	–	–
Austin (West-75)	0.13	0.04	–	3.61	–	0.42	43.0	1.89	–	–	–	39.4	11.2	–	–	–	–	–	–

[a] Alk_1 is the alkalinity expressed as $CaCO_3$ and determined from pH measurements (equation 3). [b] Alk_2 is the alkalinity expressed as $CaCO_3$ and determined by the charge balance (equation 5). Both Alk_1 and Alk_2 are not true weights, are not included in the mass fraction calculation, and represent relative concentrations. [c] ( ) implies an elemental composition from the target transformation factor analysis of the HiVol total suspended particulate matter which has not been rotated in the TTFA of rainwater data sets. [d] – implies not determined or less than zero.

ACIDIC PRECIPITATION SOURCE IDENTIFICATION
BY CHEMICAL MASS BALANCE METHODS
EMPLOYING FRACTIONATION FACTORS

Howard M. Liljestrand
The University of Texas at Austin
Department of Civil Engineering, ECJ 8.6
Austin, Texas

Source-sink relationships in rainwater quality are subject to greater fractionation effects than in aerosol studies. The gaseous acid precursors (NO_x and SO_2) are inefficiently scavenged relative to the metal tracers used to identify sources. Moreover, the tracer species are preferentially scavenged according to their mass median diameter within the aerosol size distribution. The fractionation factor, α_{ij}, is used to correct the mass fraction of each chemical species (i) at the receptor relative to the composition at the source (j). During long range transport and precipitation scavenging, α_{ij} for acid precursors increases relative to α_{ij} for the tracer species, making source contributions difficult to estimate.

A physical-chemical model is applied to the unsteady mass-continuity equation to identify the important atmospheric processes as well as fundamentally estimate the effect of precipitation scavenging on the fractionation factors. The fractionation factor is found to depend primarily on the cumulative precipitation that has scavenged a parcel of air in a Lagrangian framework, and to a lesser degree, on the average precipitation rate. The fractionation effects are most pronounced over the first half-inch of rainfall, and an α_{ij} approaches a constant value with further precipitation. With the predicted fraction factors (α_{ij}), the chemical mass balance method is used to estimate the contributions of five main source types (sea salt, crustal soil dust, mobile, stationary, and ammonia sources) to nine sites in southern California.

Introduction

The chemical mass balance (CMB) method for source apportionment for aerosol samples as proposed by Miller, Hidy and Friedlander[1] includes a fractionation factor, α_{ij}, for the preferential loss or enrichment of the ith element in transit between the jth source and the receptor. In general, the CMB simultaneous equations are given by

$$P_i = \sum_{j=1}^{n} \alpha_{ij} P_{ij} C_j \tag{1}$$

where P_i is the percentage of the ith species in the receptor, P_{ij} is the percentage of the ith species in the emissions from the jth source and C_j is the fractional source contribution from the jth source. The fractionation factor has been used to account for elemental variations in aged aerosols, differences between resuspended dusts and top soils, and gas to particle conversion. In the absence of empirical measurements, α_{ij} is usually assumed to be unity.

For precipitation samples, the secondary pollutants involved in gas-to-particle processes account for approximately 75% of the total solids. The reference state required for conservation of mass between the source and receptor (rainwater) is the sum of particulate and soluble gaseous emissions. For this reference, the fractionation factors (α_{ij}) can no longer be approximated as unity, since α_{ij} represents the relative scavenging efficiency of the ith species to that for the total emissions (gaseous and particulate). The α_{ij} can be empirically determined, estimated with a statistical analysis or theoretically predicted from the physical-chemical mechanisms, as in this investigation. Once the fractionation factors have been estimated, the source apportionment can be determined within the uncertainties of the source emissions as well as the fractionation effects.

Methods

The sampling methods, sampler locations and analytical techniques have been described in detail elsewhere.[2] Briefly, wet-precipitation-only samplers were employed at nine sites in the Los Angeles area separated by >16 kilometers. Sequential increments of precipitation were collected for inorganic analysis. The chemical analyses included pH, anions by ion chromatography, metals by atomic absorption spectroscopy (flame and carbon rod), organic carbon by total organic carbon analysis as well as standard colorimetric techniques for ammonium, phosphate, fluoride and chloride.

Results

Theoretical Model for Estimating α_{ij}

Since the CMB approach assumes a mass balance, a consideration of the conservation of mass in the air is required for the estimation of fractionation factors. The unsteady-state mass continuity equation after application of the Boussinesq approximations for density, viscosity and conductivity effects can be written as,[3]

$$\frac{\partial X}{\partial t} = \nabla \cdot (K \nabla X) - \nabla \cdot \vec{U} X + P + Q - R \tag{2}$$

where

- X is a pollutant concentration in a given aerosol size range;
- $\nabla\cdot(K\nabla X)$ is the diffusivity term with K an eddy diffusivity tensor;
- $\nabla\cdot\vec{U}X$ is the advection term with $\vec{U}$ the wind vector;
- P is the rate of production of X by chemical reactions;
- Q is the primary emission rate into volume of air; and
- R is the removal rate by physical chemical mechanisms not included in the P-production term.

For the source tracer species, P is assumed to be the net rate of coagulation of species X into the aerosol size range being considered. R is approximated by the rate of precipitation scavenging, where $R = -\gamma(D_p)\,X$ and $\gamma(D_p)$ is the first order scavenging rate coefficient as a function of particle size.

This fundamental equation can be solved in restricted cases to give an analytical solution. Peterson, Gelbard and Seinfeld[4] determined the similarity solutions for aerosols including coagulation, condensation, nucleation and first order removal of particles, with the rate constant independent of particle size. Since the precipitation scavenging rate is highly dependent on particle diameter, their solutions are not applicable for the changes in aerosols during rainfall.

To make the mass continuity equation more tractable, approximations for the main features can be made to simplify the terms as well as deletions of the negligible processes. Using the approach of Brock,[5] the characteristic times of each term are compared with the hydraulic residence time (τ_{HYD}), the characteristic time for the advection process, which is given by

$$\tau_{HYD} = \frac{L}{|\vec{U}|} \tag{3}$$

or the ratio of the air parcel length to the wind speed. Under rainfall conditions, τ_{HYD} is less than one hour. For the area averages measured by only nine sampling sites in southern California, the diffusion processes have a half-life much greater than one hour. Likewise, the coagulation of aerosols is relatively slow, unless in-cloud droplet scavenging is treated as coagulation rather than precipitation scavenging. The characteristic time for condensation depends on the aerosol size. Empirical results by Stein and Liljestrand[6] indicate condensation effects are most important for aerosols of diameter less than 3 micrometers with the characteristic time much less than one hour or the time to reach 100% relative humidity in the air mass. Likewise, precipitation scavenging is nonnegligible.

The first order removal rate constant (γ) for precipitation scavenging can be estimated by a physical-chemical model. As outlined by Englemann,[7] the scavenging rate coefficient, $\gamma(D_p)$, for a given particle diameter (D_p) can be expressed as

$$\gamma(D_p) = \int A(D)E(D_p,D)V(D)N(D)dD \tag{4}$$

where $A(D)$ = the aerodynamic cross-sectional area of a raindrop of diameter D;
$E(D_p,D)$ = the collection efficiency of an aerosol particle of diameter D_p by a raindrop of diameter D;
$V(D)$ = terminal fall velocity of a raindrop of diameter D; and
$N(D)dD$ = the number concentration of raindrops with diameters between D and D + dD.

The collection efficiency $E(D_p,D)$ has been semi-empirically described by Slinn[8] for diffusion, interception and inertial impaction processes by

$$E(D_p,D) = \frac{4}{Pe}[1 + 0.4Re^{1/2}Sc^{1/3}] + 4K[K+\frac{(1 + 2vK)}{1 + vRe^{-1/2}}] + (\frac{S-S*}{S+C})^{3/2} \quad (5)$$

where

Re = Reynolds Number;
Sc = Schmidt Number;
Pe = Peclet Number;
K = interception parameter = D_p/D;
v = absolute viscosity of water/absolute viscosity of air;
S = Stokes Number;
$S*$ = Critical Stokes Number = $\frac{12/10 + 1/12 \ln(1 + Re)}{1 + \ln(1 + Re)}$; and
$C = 2/3 - S*$

Pilat and Prem[9] indicate thermophoretic and diffusiophoretic forces are larger than the Brownian diffusion effects. Thus the scavenging efficiency for a 0.1 micrometer aerosol may be underestimated by an order of magnitude by restricting the physical mechanisms of scavenging to only diffusion, interception and impaction. The result of Slinn's approximate collection efficiency model is a dramatic decrease in scavenging efficiency for aerosols below ~ 2 micrometer in diameter. The relative collision efficiencies of 10 micrometer and 0.1 micrometer aerosol particles are about 1:0.001. Theoretically, raindrops are biased collectors that preferentially scavenge species in aerosols particles greater than 2 micrometer in diameter. This conclusion is substantiated by the variation in the rainfall scavenging ratio with the mass median diameter for the chemical species, in empirical field studies[10].

Even with these approximations, the unsteady mass continuity equation is difficult to solve numerically, since the precipitation intensity must be known accurately as a function of time. From the precipitation intensity, the raindrop size distribution (and thus $N(D)$ required for the calculation of the scavenging rate $\gamma(D_p)$) can be estimated from the Marshall-Palmer distribution, the Best distribution, or other empirical distributions of raindrop size. However, the rate of precipitation is poorly known from the standard rain monitoring efforts, especially at low precipitation intensity. A more convenient approach is to parameterize the fractionation processes as a function of the cummulative precipitation, which is known fairly accurately.

An example of the influence of precipitation intensity on the rate of aerosol particle scavenging is presented in Figure 1. These results assume a Best distribution of raindrop size as a function of rainfall intensity as well as Slinn's semi-empirical scavenging efficiency. A family of similar curves can be calculated for all particle sizes; the shape of the curve is characteristic, although the magnitude of $\gamma(D_p)$ shifts for a different diameter particle. Since γ is a non-linear function of precipitation rate, the amount of material scavenged is highly dependent upon the precipitation intensity. The fractionation factors (α_{ij}) are less dependent upon the rainfall rate, since they depend upon the ratio of the mass of the ith species scavenged to the total mass of pollutants removed. Moreover, precipitation samples are volume-weighted or precipitation-intensity-weighted aggregates. Thus, the average precipitation intensity controls the fractionation factors. Finally, the fractionation factor at a given receptor depends upon the cumulative precipitation that has scavenged or "weathered" the emissions between the source and sink. While the history of each air parcel need not be known, the amount of precipitation which has fallen through a parcel in transit between

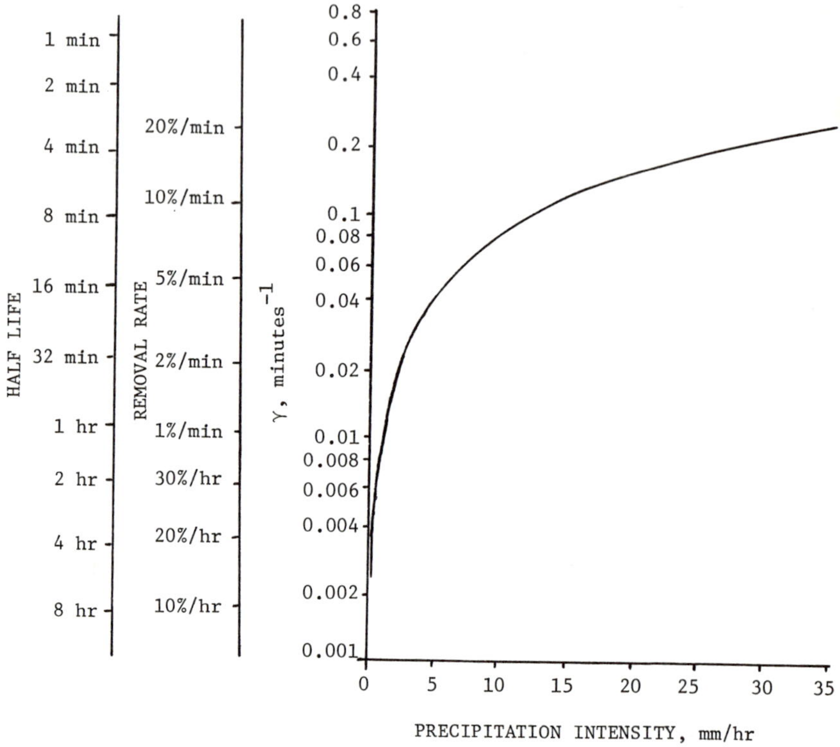

Figure 1. The First Order Scavenging Rate Coefficient for 7.5 Micrometer Diameter Particles (γ(7.5 μm)) vs Precipitation Intensity. The assumptions are 1) the raindrop size distribution follows Best's distribution as a function of precipitation intensity and 2) the efficiency of collection by each raindrop can be approximated by Slinn's semi-empirical algorithm. For comparison, the first order scavenging rate is expressed as the removal rate and half life or time to reach 50% of the initial concentration.[6]

source and receptor does need to be known.

An Application of Aerosol Scavenging Models to Estimate Fractionation of Tracer Elements

The fractionation of tracer elements depends upon the distribution of the species within the particle size distribution. To account for condensation effects, this chemical distribution with respect to particle size should be known over the range in relative humidity from emission conditions to rainfall conditions. Hygroscopic aerosols which grow to sizes greater than ~2 μm will be efficiently scavenged.

A case analysis of the magnitude of fractionation effects is presented for the steel mill emissions shown in Figure 2. The chemical distribution within the particle size distribution as measured by Davidson and Chu[13] shows most of the iron mass is distributed in particles >2 μm, while most of the lead mass is associated with particles <2 μm. As precipitation scavenges these emissions, the lead to iron ratio in the air will rise as the iron is preferentially scavenged. Figure 3 compares the ratio of $\alpha_{Pb,steel}$ to $\alpha_{Fe,steel}$ for rain as a function of the extent of precipitation for two average rainfall rates, again assuming a Best raindrop size distribution, Slinn's scavenging mechanism approximations and Englemann's scavenging rate algorithm. This example is an extreme case since the mass distributions of the tracers with respect to particle size differ greatly from each other and from that which is typical for total mass, a bimodal distribution with peaks at ~0.3 μm and ~10 μm.[14] A comparison of α_{ij} for two source tracers having similar mass distributions shows smaller fractionation effects than implied by Figure 3.

The results of the steel mill example (Figure 3) lead to five generalizations. First, the relative fractionation of tracers can be approximated from the cumulative rain and chemical distribution with respect to particle size. The dependence on rainfall rate is of lessor importance. Second, scavenging by >0.5 inch rain changes relative tracer concentrations by factors of 2-4X. For larger relative changes, extensive deposition is required. Third, fractionation occurs with any preferential removal process, including dry deposition which is likewise dominated by diffusion for D_p <0.5 μm and impaction/sedimentation for D_p >0.5 μm. Fourth, the fractionation is strongly dependent on total rainfall <0.3 inch, in the manner indicated in Figure 3. Fifth, if the total mass had the same distribution as iron mass in Figure 2, Figure 3 would directly show α_{Pb}.

Gas Scavenging

While the previous discussion has focused on the relative scavenging of source tracer elements in the particulate phase, the fractionation of the acid gases with these nonvolatile elements must also be considered. Gas scavenging can be controlled by kinetic processes or equilibrium constraints. An equilibrium solubility appears to be attained for the following reactions:[15,16]

$$SO_{2(g)} + H_2O_{(\ell)} \overset{K_H}{=} SO_2 \cdot H_2O_{(aq)} \overset{K_1}{=} H^+ + HSO_3^- \overset{K_2}{=} 2H^+ + SO_3^{2-} \qquad (6)$$

$$HNO_{2(g)} \overset{K_H}{=} HNO_{2(aq)} \overset{K_1}{=} H^+ + NO_2^- \qquad (7)$$

$$HNO_{3(g)} \overset{K_H}{=} HNO_{3(aq)} \overset{K_1}{=} H^+ + NO_3^- \qquad (8)$$

$$NH_{3(g)} + H_2O_{(\ell)} \overset{K_H}{=} NH_4OH_{(aq)} \overset{K_b}{=} NH_4^+ + OH^- \qquad (9)$$

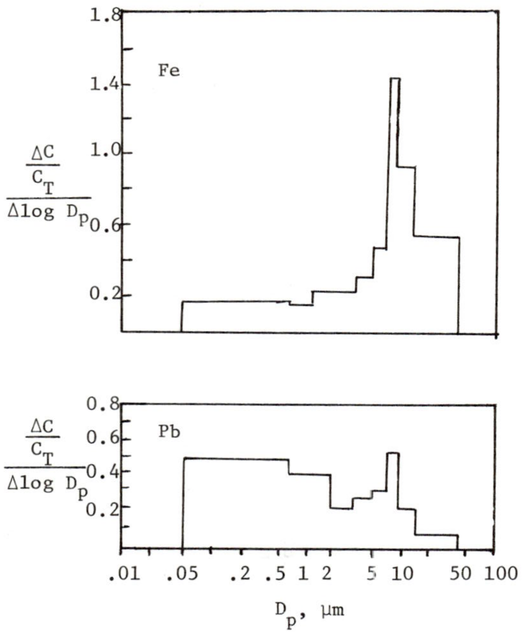

Figure 2. Mass Distributions of Iron and Lead Within the Particle Size Distribution For Steelmaking Facilities.[13] The fractional mass within each particle range corresponds to mass mean aerodynamic diameters of 8 and 0.6 μm for iron and lead, respectively. ΔC is the mass of a species in the size range $\Delta \log D_p$. By dividing by the total mass of the species C_T, the fractional mass in each size range is given above.

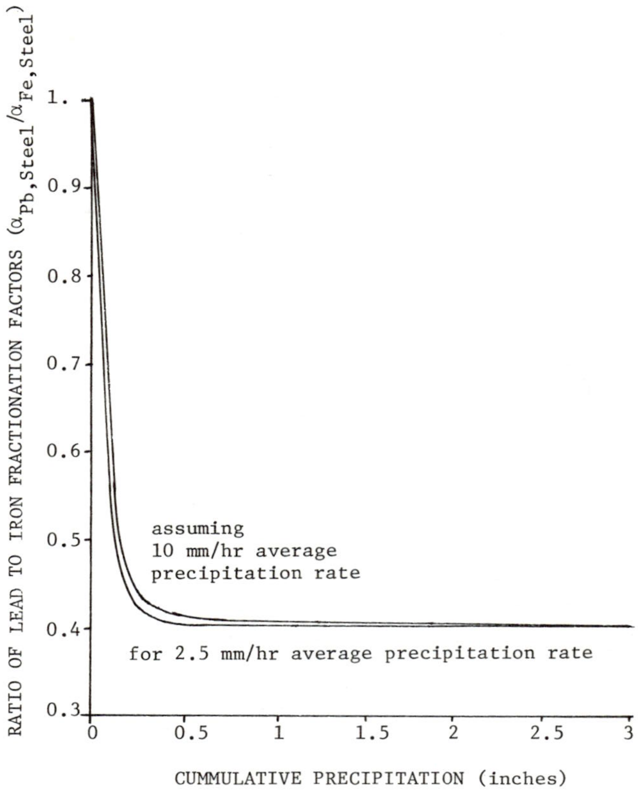

Figure 3. The Relative Fractionation of Lead to Iron From Steel Mill Emissions by Rain Scavenging. The chemical distribution with respect to particle size shown in Figure 2 and the Best raindrop size distribution are assumed to calculate these relative scavenging efficiencies.

An equilibrium model for acid gas scavenging has two important features. Firstly, the scavenging rate is pH dependent or buffer capacity dependent for the open air-water system. Charlson and Rodhe[17] discuss the effect of ammonia and background acidity/alkalinity on the partitioning of sulfur dioxide into raindrops. Once the buffering capacity of the rain is known, the maximum solubility of the gas can be determined from the equilibrium pH. For SO_2 solubility, the total dissolved sulfite is given by

$$[SO_2 \cdot H_2O] + [HSO_3^-] + [SO_3^{2-}] = \frac{P_{SO_2} K_H}{\left(\frac{K_1 K_2}{[H^+]^2} + \frac{K_2}{[H^+]} + 1\right)^{-1}} = \frac{P_{SO_2} K_H}{\alpha_o} \qquad (10)$$

where α_o is the fraction of total dissolved sulfite in the undissociated form. Secondly, a thermodynamic model implies a function of state as opposed to a path or time dependent mechanism and resulting concentration. Thus, an apportionment of the sulfur dioxide gas contributions to the partial pressure of sulfur dioxide (P_{SO_2}) at the receptor also apportions the source contributions to the total dissolved sulfite. The fraction contributed to the gas phase, after corrections for the extent of chemical conversion/oxidation, is also the fraction of dissolved acid of the same oxidation state.

The removal of gaseous pollutants which reach a reversible equilibrium solubility can also be described by a first order rate expression. The rate of removal, R, is given by

$$R = -\gamma \, P_{gas} \qquad (11)$$

where P_{gas} is the partial pressure of the gaseous species and γ is the first order scavenging rate. A mass balance approach to that for the particulate matter determines γ to be defined by

$$\gamma = \frac{K_H p'}{H \alpha_o} \Bigg/ \left(\frac{1}{RT} + \frac{V_\ell K_H}{V_g \alpha_o}\right) \qquad (12)$$

where

V_ℓ / V_g is the average ratio of liquid water volume to air volume,
K_H is the Henry's Law solubility constant,
p' is the precipitation rate,
H is the mixing depth or height of the cloud top,
R is the ideal gas law constant, and
T is the absolute temperature at ground level.

In southern California, the scavenging rate for sulfur dioxide/sulfite (γ_{SO_2}) is pH dependent as a function of α_o, and it is typically 1-10 x 10^{-4}/minute for a rainfall rate of 10 mm/hr, which is comparable to the first order removal rate constant for aerosol particles of 0.5 μm diameter.

Empirically, the scavenging of NO_x and SO_2 primary emissions is not a simple first order process, making scavenging rates between acids and tracers difficult to compare.[18] Oxidant concentrations and oxidation reactions appear to control the rates of SO_2 and NO_x scavenging as sulfate and nitrate, respectively. Mechanistically, nitrogen dioxide dissolution by the reaction

$$2\, NO_{2(g)} + H_2O_{(\ell)} = HNO_{2(aq)} + H^+ + NO_3^- \qquad (13)$$

is a second order process and kinetically slow, even at diffusion limited rates in raindrops.[19] If a pseudo-first order reaction rate can be empirically determined for precipitation conditions, Friedlander's approach[20] of comparing the hydraulic residence time (τ_{HYD}) with the chemical half-life (τ_{CHEM}) to yield a fractionation factor α_{ij} given by

$$\alpha_{ij} = 1/\left(1 + \frac{\tau_{HYD}}{\tau_{CHEM}} \right) \tag{14}$$

is a convenient method. Empirical estimates of τ_{CHEM} are available for many gaseous species, and fundamental estimates of τ_{CHEM} for NO_x and SO_2 primary emissions are currently being investigated.

CMB With Fractionation on a Regional Scale

The application of the chemical mass balance method with fractionation is possible in southern California because it involves regional scale transport with negligible long range sources. The sources are well characterized, including some information regarding chemical distributions with respect to particle size. The fractionation factors for gaseous and particulate emissions are estimated in three steps - aerosol tracer fractionation with extent of of precipitation, gaseous acid precursor oxidation/scavenging with empirical pseudo-first order reaction rates, and gas scavenging as a function of pH. The later two processes must also satisfy the boundary conditions provided by the observed spatial distribution of P_{SO_2}, P_{NO}, P_{NO_2}, aerosol sulfate and aerosol nitrate under precipitation conditions. The largest uncertainty in these estimates lies in the determination of source contributions to in-cloud scavenging processes, where the effective extent of precipitation "weathering" of the air pollutant concentrations is much larger than for below cloud processes.

Within these uncertainties and the limitations of available tracer measurements, five main source types are assumed to determine their contributions to acid precipitation chemistry. Figure 4 presents the charge contributions in microequivalent/liter (μN) from sea salt (primarily sodium chloride and magnesium sulfate), crustal components (primarily calcium alkalinity), mobile sources (non-cross hatched area of sulfate and nitrate) stationary sources (cross hatched area of sulfate and nitrate) and ammonia. For lack of source emission characteristics, no distinction is made between anthropogenic and natural sources of ammonia. The results for the annual precipitation-weighted-mean concentrations for 1978-1979 in Figure 4 are not representative for every storm. The approximately log-normal distributions of concentrations for all species in rainwater[21] results in mean concentrations which are weighted towards the samples with highest concentrations. These samples coincide with the precipitation events having the least rainfall and least fractionation. Thus, the contributions to mean concentrations of rainwater reflect the primary emission proportions from the major sources.

Conclusions

1. A fundamental model has been proposed for the estimation of fractionation effects between source emissions and rainwater compositions. In an Eulerian framework, local emissions, advection, condensation and precipitation scavenging control the spatial average chemical concentrations for the tracer elements.

2. The scavenging model requires data for the chemical distribution with respect to particle size. While scavenging depends both on rainfall

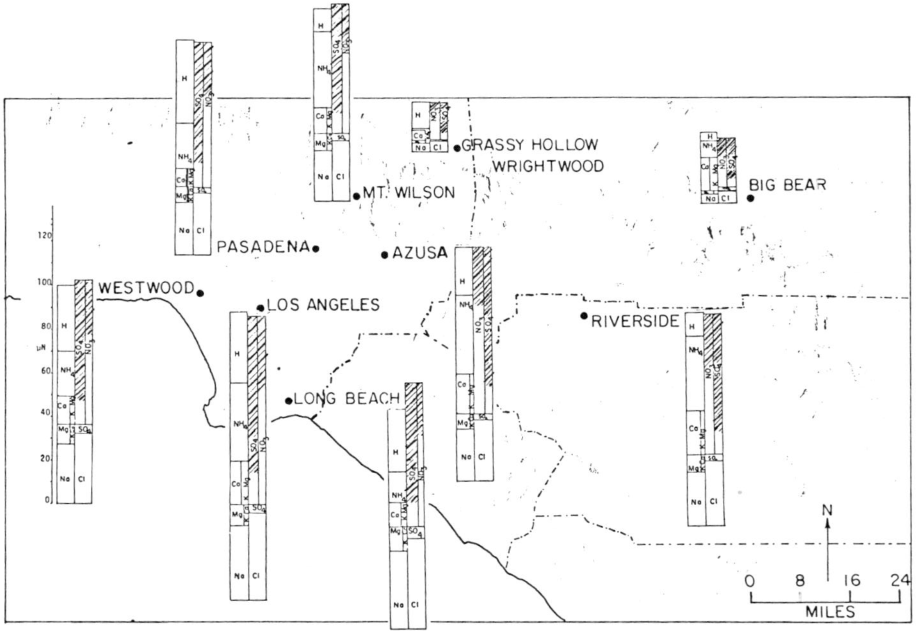

Figure 4. Source Contributions to the Precipitation-Weighted-Mean Chemical Composition of Acid Rain. The five major source types are, in ascending order, sea salt ($NaCl + MgSO_4$), crustal component (Ca alkalinity) neutralized acids, ammonia neutralized acids, nitric and sulfuric acids from mobile sources (open) and sulfuric and nitric acids from stationary sources (cross hatched area).[2,19] Concentrations are presented on a charge basis as μN, microequivalents/liter.

rate and total rainfall, the fractionation effects are most pronounced over the first one-half inch of precipitation as material in particles greater than 2 micrometers is removed from the atmosphere. The influence of average precipitation intensity is of secondary importance. Thus, source profiles can be parameterized into two classes - those influenced by >0.5 inch of precipitation in transit to the receptor and those scavenged by <0.5 inch of rainfall in a Lagrangian framework. The trajectories need not be explicitly calculated but rather the cumulative precipitation which has interacted with the emissions.

3. Gas scavenging can be considered with two mechanisms - linear equilibrium solubility and oxidation-reduction kinetics. While equilibrium solubility (Henry's Law solubility as well as acid-base dissociation corrections) yields a first order removal similar to aerosol tracer species scavenging, the kinetic controls are more important for sulfur dioxide and especially nitrogen oxides scavenging. If the chemical conversion rates to sulfate and nitrate are empirically known, Friedlander's method of comparing the reaction half life or chemical residence time with the hydraulic residence time is a convenient approach for the estimation of fractionation effects.

4. The reference state required for the conservation of mass in CMB models applied to rainwater as the receptor is the sum of particulate and soluble gaseous emissions.

5. Application of these models to mean concentrations in precipitation in southern California, a regional scale test with little long range transport, yields the estimates of source apportionments given in Figure 4. The approximately log-normal character of all rainwater species concentrations yields precipitation-weighted-mean concentrations which are biased towards the most concentrated samples. These samples, in turn, are dominated by the major source of each species.

Acknowledgments

This study was partially supported by a grant from the U. S. Environmental Protection Agency. The cooperation of the California Air Resources Board and the South Coast Air Quality Management District is greatly appreciated.

References

1. M. S. Miller, S. K. Friedlander, G. M. Hidy, "A chemical element balance for the Pasadena aerosol," J. Colloid Interface Sci., 39: 165 (1972).

2. H. M. Liljestrand, J. J. Morgan, "Spatial variations of acid precipitation in southern California," Environ. Sci. Tech., 15: 333 (1981).

3. R. B. Bird, W. E. Stewart, E. N. Lightfoot, Transport Phenomena, Wiley, New York, 1964.

4. T. W. Peterson, F. Gelbard, J. H. Seinfeld, "Dynamics of source-reinforced, coagulating, and condensing aerosols," J. Colloid Interface Sci., 63: 426 (1978).

5. J. R. Brock, "Processes, sources and particle size distributions," in Fogs and Smokes, The Faraday Division, Chemical Society, London, pp. 198-208, 1973.

6. D. A. Stein, H. M. Liljestrand, "Precipitation scavenging of aerosols >0.5 micrometer in diameter," under review.

7. R. Englemann, "The calculation of precipitation scavenging," in Meteorology and Atomic Energy, U.S.A.E.C Report TID-24190, pp. 208-221, 1968.

8. W. G. N. Slinn, "Some approximations for the wet and dry removal of particles and gases from the atmosphere," Water Air Soil Pollut., 7: 513 (1977).

9. M. J. Pilat, A. Prem, "Calculated particle collection efficiencies of single droplets including inertial impaction, Brownian diffusion, diffusiophoresis and thermophoresis," Atmos. Environ., 10: 13 (1976).

10. D. F. Gatz, "Wet deposition estimation using scavenging ratios," J. Great Lakes Res., 2(Supplement 1): 21 (1976).

11. J. S. Marshall, W. McK. Palmer, "The distribution of raindrops with size," J. Meteor., 5: 165 (1948).

12. A. Best, "The size distribution of raindrops," Quart. J. Roy. Meteor. Soc., 76: 16 (1950).

13. C. I. Davidson, L. Chu, "Scanning electron microscope study of iron-containing particles on foxtail," Environ. Sci. Technol., 15: 198 (1981)

14. K. T. Whitby, R. B. Husar, B. Y. H. Liu, "The aerosol size distribution of Los Angeles smog," J. Colloid Interface Sci., 39: 177 (1972).

15. J. M. Hales, "Fundamentals of the theory of gas scavenging by rain," Atmos. Environ., 6: 635 (1972).

16. W. G. N. Slinn, L. Hasse, B. B. Hicks, A. W. Hogan, D. Lal, P. S. Liss, K. O. Munnich, G. A. Sehmel, O. Vittori, "Some aspects of the transfer of atmospheric trace constituents past the air-sea interface," Atmos. Environ., 12: 2055 (1978).

17. R. J. Charlson, H. Rodhe, "Factors controlling the acidity of natural rainwater," Nature, 295: 683 (1982).

18. H. M. Liljestrand, J. J. Morgan, "Chemical source, equilibrium and kinetic models of acid precipitation in southern California," in Energy and Environmental Chemistry, Vol. II, Ann Arbor Science, pp. 103-122, 1982.

19. Y.-N. Lee, S. E. Schwartz, "Evaluation of the rate of uptake of nitrogen dioxide by atmospheric and surface liquid water," J. Geophys. Res., 86: 11971 (1981).

20. S. K. Friedlander, "New developements in receptor modeling theory," in Atmospheric Aerosol: Source/Air Quality Relationships, ACS, pp. 1-19, 1981.

21. H. M. Liljestrand, J. J. Morgan, "Measurement and interpretation of acid rainfall in the Los Angeles basin," Proceedings, California Symposium on Acid Precipitation, California Air Resources Board, Sacramento, CA, 1982.

STATISTICAL CONSIDERATIONS IN THE EMPLOYMENT OF SAX RESULTS FOR RECEPTOR MODELS

David L. Johnson →

John P. Twist

Department of Chemistry
State University of New York
College of Environmental Science and Forestry
Syracuse, New York

Hundreds of thousands of individual particle measurements may be accumulated in a receptor model study employing Scanning electron microscopy with Automated image analysis and X-ray energy spectroscopy (SAX). At present, the summaries of these data are utilized in apportionment calculations. Two basic approaches have been taken to the sorting of these observations: variable algorithms which are sample specific, and fixed algorithms which have temporal and inter-airshed continuity. We report here the use of a "fixed" sorting scheme developed for one area as it is applied to an airshed where no source specific information was available. Discriminant analysis, as applied to the individual particle characterizations, was shown to be very useful in improving the quality of the mathematical solution using a Q mode Factor/Vector analysis. It is concluded that statistical manipulation of the individual object vectors is more efficient than dealing with data summaries. Use of a "fixed" sorting algorithm is shown to be feasible when unknown ambient aerosol samples are to be apportioned to sources on the basis of their comparison with library "source signatures."

Introduction

Scanning Electron Microscopy (SEM) and Analytical Light Microscopy (ALM) techniques, as applied to aerosol samples, can extract much valuable information from the heterogeneous nature of individual particle assemblages. Utilized along with traditional receptor models such as the CMB, they improve source resolution. In fact, given appropriate sample collection and/or preparation, microscopy-based analytical methods of individual particle characterization present viable alternatives to bulk chemical analysis models. Scanning electron microscopy interfaced with Automated image analysis and X-ray energy spectroscopy (SAX) methods create and store hundreds or thousands of object vectors from the aerosol particles of each sample obtained[1]. General descriptions of the methodology can be found in the works of Johnson[2], Hanna _et_ al[3], Johnson et _al_[4], Casuccio and Janocko[5], McIntyre _et al_[6], and Johnson and McIntyre[7]; therefore we will not present any operational details of the technique here. Instead, we will concentrate on the problems of, and approaches to, sorting and classifying the individual particle analytical data for use in SAX receptor models.

If SAX characterizations of aerosols could be absolute, with unique identification of each particle coming from a particular source, then the model would be absolute and no further manipulation of the results would be required. But sources are variable and SAX suffers certain inherent limitations[8] which preclude it from being quantitative (for a given particle) at the present time. Therefore, practitioners of this technique must employ summaries of the data which result from some sorting or classification scheme. No doubt some information is lost in this compression of the observations, so it is important to examine the applicability of the two major approaches to particle classification and subsequent data use.

Currently SAX observations are sorted by either a "_fixed_" or a "_variable_" algorithm (we mean these in a generic sense). In the "_fixed_" case, two major approaches are taken. In a _linear_ file structure, the user defines the number of categories desired with chemical and physical specifications for each. An object vector (particle with certain morphological and relative X-ray intensity chemical observations) is sifted through the file until it comes to the category for which it meets all the requirements. The other alternative for the "_fixed_" file structure is to sift the particle using best-fit criteria for the defined classification structure[9].

A "variable" sorting algorithm builds classes of particles from the object vector data of the sample. It may employ a relative X-ray intensity "ranked clustering" approach to identify unique groups of particles. Thus the classifications and the resulting data summaries may be different in structure from sample to sample. Results from a "variable" sorting algorithm are employed in the mathematical aspects of receptor models in a different way than are those of "fixed" schemes. "Variable" sorting results are used in a "Fingerprint/Radio" receptor model[10]. "Tracer" particle types from specific sources are identified in ambient aerosol samples and their dilution ratios fitted by a least squares procedure to estimate source contributions. Such an approach has the specific advantage of higher information entropy[11], since unused classes are not reported, and it shows excellent resolution of sources. However, it requires that airshed-specific source information be available. That is, ambient samples may contain "tracer" classes of particles which are not recognized as such since no source analyses pointed them out and were identified with them. Thus, the "variable" sorting algorithm and the Fingerprint/Ratio SAX receptor model are difficult to apply to airshed aerosol collections which are depauperate in specific source samples.

The "fixed" sorting algorithm takes a more generalized approach in SAX receptor models. A fixed classification scheme for all ambient and source sampling allows a long term (or historic) internal consistency in the comparing of results. In this sense it is more like a bulk chemical analysis - library source signatures can be applied to receptor models for airsheds in which little or no source information is available. This approach is a cost effective preliminary receptor model for source apportionment, and used along with bulk analysis, can increase the resolution of CMB models.

Of course, there are disadvantages in the blanket of a "fixed" particle classification scheme. What is optimum for one airshed may not be for another. In this report we show how statistical treatment of the SAX observation vectors can improve the source apportionment of an unknown airshed when employing "library source signatures." The specific example is that of discriminant analysis of Houston aerosols to optimize their classification. This report represents an advance in interpretation over that presented at the Quail Roost II Conference.

Methods

SAX analyses of aerosol samples were carried out on 25 coarse fraction samples of Houston aerosol collected by dichotomous sampler on Nuclepore® membrane filters during September of 1980. A detailed description of sample collection and background history can be found in Dzubay *et al*[12] and in the Quail Roost II report by Stevens *et al*[13]. We have chosen 4 samples from this set of 25 upon which to base our examples of object vector manipulation. Table I lists the filter identifications and the sampling times for the 4 samples considered here, and presents the results of the original SAX

TABLE I. Houston Nuclepore Filters used in Experiment. Times and dates of collection and original SAX analyses for population and weight are shown.

	124-502C		124-510C		124-516C		124-522C	
	9/10/80 7am		9/12/80 7pm		9/14/80 7pm		9/16/80 7pm	
Particle Class	#%	Wt%	#%	Wt%	#%	Wt%	#%	Wt%
1. Low counts	8.9	1.4	14.3	4.9	6.5	0.5	0.9	0
2. Marine	0	0	0	0	8.6	6.7	17.3	13.7
3. S only	0.3	0	0	0	0	0	0.4	0
4. Ca & S	2.0	2.5	2.6	2.7	2.5	1.2	1.5	0
5. Wood ash	0.5	1.5	0.6	1.2	0.2	0	0.4	0.2
6. Refuse ash	0.9	0.3	0.1	0	0	0	0.2	0
7. Auto	3.6	3.1	4.8	1.8	3.6	4.8	4.3	1.1
8. Oil FA	0	0	0.4	0.7	0.6	0.2	0	0
9. Fly Ash, lg.	1.5	3.3	0.4	2.0	0.4	3.4	0	0
10. Fly Ash, sm.	4.4	1.2	7.1	2.5	6.3	2.2	2.8	1.1
11. Metal Rich	11.8	6.7	2.6	0.6	3.0	2.3	2.8	0.9
12. Mg-Ca	0.3	0.4	0.3	0.1	0.8	0.5	0.9	0.3
13. Ca only	19.6	17.2	19.3	42.8	30.3	32.0	36.6	60.3
14. Construction	7.0	12.7	5.0	4.5	4.4	10.1	5.4	3.4
15. Talc like	1.0	0.1	0.8	0.6	0.2	0	0	0
16. Illite like	0.1	0.2	0.3	0	0.4	0.3	0	0
17. Kaolinite like	2.5	0.5	1.7	0.8	1.1	0.1	0.2	0
18. Feldspar like	0	0	0	0	0	0	0.2	0
19. Biotite like	0	0	0	0	0	0	0	0
20. Muscovite like	1.1	1.5	2.1	2.2	1.5	1.9	0.6	1.1
21. Al rich	0.6	0.4	0.4	0	0	0	0	0
22. Si only	5.6	10.7	5.5	3.0	3.0	2.9	1.3	1.0
23. Misc. Litho.	20.6	32.6	22.5	25.0	16.0	25.3	13.4	7.8
24. Misc. Salts	2.6	0.4	2.7	2.7	4.2	0.7	3.2	2.1
25. Unknown	5.2	3.3	6.2	2.6	6.3	5.0	7.8	7.1

classification (population percent and weight percent) as determined by the "*fixed*" sorting algorithm employed for analysis of all source and ambient samples.

All of the apportionment experiments carried out for this work have used the same mathematical treatment of the data - namely the Q-Model Factor/Vector analysis as described by Johnson and McIntyre[7] from the work of Full *et al*[14]. Having selected a set of 33 possible source signatures from our signature library, which might reasonably apply to Houston aerosol, we have carried out a series of signature modification experiments to assess the efficiency of the original particle classifications. We wished to know how well source signatures from a "world list" could be applied to an unknown ambient aerosol using a "*fixed*" sorting algorithm. The signature library contained non-size fractionated SAX weight percent results for 52 sources (mostly from Syracuse, N.Y.) and 3 Houston Soil samples or road dusts. By using a Chi Square Minimum Marginal Homogeneity test[15], 30 sources (types) were selected as potential end members for the Q-Model Factor/Vector analysis. These 30, along with the 3 Houston soils and the 25 ambient samples, were used consistently in the Q-Model apportionments. The experiments consisted of adjusting some of the signatures as described below. Figure 1 shows a block diagram for how these experiments were conducted.

The Q-Model apportionment is based upon a vector description of the position of the ambient samples within the extreme boundaries of sample space. The particular extremes (potential source types) might not be appropriate for the Houston airshed; yet they might allow a mathematically correct solution for the apportionment. We are not using the Factor Analysis portion of Q-Model as the sole mechanism for choosing the source types which go into the apportionment. Chi Square Minimum Marginal Homogeneity was employed to select those signatures from our library which might reasonably represent, in some mixture, the Houston aerosol. The test employs an $\underline{X}$ $\underline{\log}$ $\underline{X}$ transformation of the number percent or weight percent by particle type categories (according to Kullback[16]). The 25 ambient samples were compared one at a time to each of the 52 library signatures and a homogeneity value calculated. The marginal or by-particle-class, contributions to the total homogeneity were computed. Any library source signature for which the cumulative marginal X^2 homogeniety for 2/3 of the 25 particle class types was less than the critical value (95% confidence level) was retained for inclusion in the Q-Model apportionment calculations. Certain generic signatures (5 in all) were also retained; these included: secondary particles, sea salt, calcium sulfate, limestone, and organic debris.

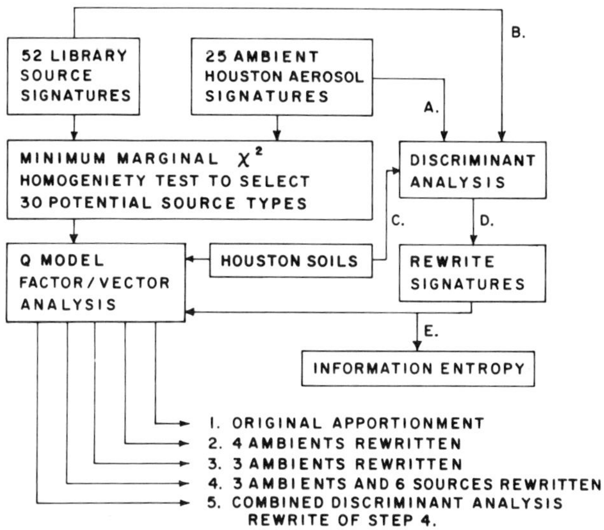

A. AMBIENTS: -502,-510,-516,-522 DISCRIMINATED INDIVIDUALLY
B. SOURCES: 6 LISTED IN TABLE III DISCRIMINATED INDIVIDUALLY
C. HOUSTON SOIL I: DISCRIMINATED INDIVIDUALLY
D. DISCRIMINATED SIGNATURES REWRITTEN FROM NUMBER PERCENT TO WEIGHT PERCENT - SEE TEXT
E. INFORMATION ENTROPY CALCULATED FOR NEW SIGNATURES

Fig. 1. Flow diagram for the steps in the statistical manipulation of the SAX results.

The 4 example aerosol samples as well as 6 prominent "source" samples were subjected to discriminant analysis (avg. 500 object vectors, or particles, per sample) using SAS[16] programs. Each particle object vector contained the nominal or original linear chemistry definition file classification, the image area, the total net x-ray count for the particle, and the relative x-ray intensity for up to 22 elements. These were punched on cards (approximately 5250 total) for creation of a SAS data set which could also be used for Factor and Cluster analyses.

After discriminant analysis for each sample of 500 particles, the discriminant functions indicated some reclassification of the original object vectors. From the discriminated results, a new weight percent by-particle-class signature was created and substituted in the Q-Model run to see how it affected the apportionments. The new weight percent signatures were created from the discriminated set of object vectors using the image area, assuming that each particle was spherical, and utilizing the specific gravity assigned to the particle type[7]. The "efficiency" of the restructured signatures was evaluated by the information entropy approach of Full et al[11]. Shannon[18] defined entropy in the following fashion:

$$E = -\sum_{i=1}^{n} P_i \log P_i$$

where Pi is the probability of finding an observation in class i and n is the number of classes in the sorting algorithm (25 in this work). Entropy, as applied to either the number percent or weight percent by particle class histograms, is a measure of the contrast between classes. The entropy of the "fixed" classification scheme is a measure of the potential of the signature histograms to provide reasonable solutions when further analysis is performed on that set of summary data.

Figure II contains the original results of the SAX procedure as well as the discriminant analysis results for ambient sample 124-502C as a histogram of the original weight percent by particle class signature along with the new signatures that were created by the discriminant procedure. The classes, by number, are those given in Table I. The classification procedure provided by the discriminanted particle data is advantageous as it places each particle in a class by its best fit. In the original procedure, the particle must meet strict requirements before being placed in a certain class. The miscellaneous class percentage is reduced by the discriminant procedure and class thirteen (limestone) is increased. The two histograms for 124-502C have only slight differences and yet the percent of maximum possible information entropy is increased. The discriminant procedure is also advantageous as it uses all the information provided by the analysis to make a classification.

Results and Discussion

Q-Model Factor/Vector analysis was used as an apportionment tool for this series of investigations. We have operated it in a consistent

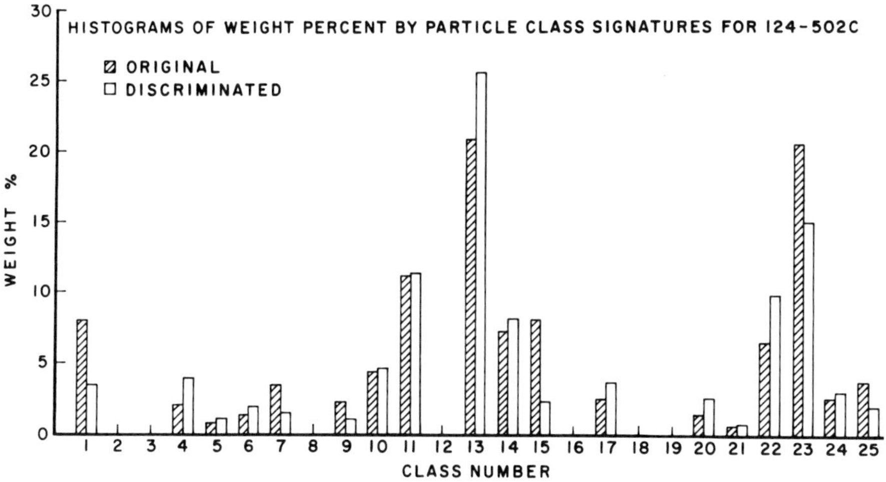

Fig. 2. Original and adjusted Signature.

manner, with the same set of source signatures and ambient sample signatures, so that our results are internally consistent. Our purpose was to adjust the "signatures" of certain ambient and source samples, after statistical manipulation of the object vectors, in order to judge whether the refinements improve or worsen the apportionment results. Reclassification of original observations, by discriminant analysis, leads to a restructuring of the "signatures" used in an apportionment model. This process implies an increase in information "entropy" which means more efficient use of the data contained in the summaries object vector characterizations. However, the drive towards maximum information entropy leads toward the minimum in source resolution. Maximum information entropy for a "fixed" number of categories would be obtained by an equal distribution of observations in each of the possible classes. Maximum source resolution would result from unique classification in one of the possible particle type categories. Both resolution and information entropy can be increased by increasing the number of classes in the categorization scheme. However, "counting statistics" for maximum precision would then dictate more data acquisition for the sample. All these tendencies must be balanced at some optimum intermediate value of categories. At present, that number would seem to be between 15 and 20.

The use of a "fixed" sorting algorithm allows one to compare source signatures from a library against the analysis results for samples from an unknown airshed. We have assumed that a classification scheme developed for one reason will not be optimum when applied to another region. Our "fixed" chemistry definition file for particle sorting was developed on the basis of "standard" materials from the Syracuse AQMA. When individual particles (object vectors) from several samples were subjected to discriminant analysis, we found that 15% of them were reclassified. That is, our step functions in the linear sorting algorithm were only 85% optimum. When Houston aerosol object vectors were sifted with the same scheme and subjected to discriminant analysis, we found a 30% reclassification rate on a sample by sample basis. Since the "fixed" file structure was only 70% optimum for Houston aerosol, the question arose, "How reasonable are the apportionment results for Houston based on library source signatures?"

It would seem that if classification schemes are not optimum, there is a high probability that the apportionment mathematics employed are also not optimally efficient. There is much value in examining the negative apportionment vectors - especially when library signatures are to be applied in a case where airshed-specific source information is lacking. Study of the negative apportionment vectors can reveal some structure of the potential sources. In operating Q-Model, after signature modification, we have employed two criteria for assessing optimum data treatment: we have looked at the change in information entropy after signature adjustment, and we have examined the sum of the negative apportionment vectors in the common case where we specified a 12 end member (potential source) solution.

Q-Model begins with a factor analysis of the samples and observation variables submitted to it. For a particular run or application there may be different dimensionalities to the mathematical solution of the vector analysis depending upon variance of the input data. When utilizing the 58 entries for this work, we decided to cut off the number of factors (end members) at the point where the eigen values fell below 1.00. This resulted in the specification of a 12 end member (factor) solution, and explained 95% of the variability in the complete data set. The Q-Model computations were carried out with summaries of the SAX data (weight percent by particle class histograms). When the individual object vectors were subjected to factor analysis (500 at a time), higher degrees of dimensionality were indicated. The number of factors seemed to be related to the number of elements monitored by the X-ray spectrometer. When unusual, or trace, elements like I, Sn, Cd, Mo, and Se were eliminated from the variables for factor analysis, 10-12 factors routinely had eigen values greater than 1.00. Thus, all Q-Model experimental runs were made using 12 possible end members (factors identified with specific sources). Table II shows the "initial" and "final" source apportionment results for the Houston aerosol samples which we studied in detail.

TABLE II. Comparison of Initial and Final Q-Model Source type fractional apportionments for three Houston aerosol samples.*

	124-502C		124-510C		124-516C	
Source Type	Init.	Final	Init.	Final	Init.	Final
Coal FA 1	.50		.56		.34	
Coal FA 2	-.10		-.10		-.05	
Steel Mill	-.13	.05	-.20	-.01	-.12	0
Limestone	.06		.26		.21	
Sea Salt	.01	0	.02	.01	.06	.05
Cement Plant	.01	.09	.01	.62	0	.37
Organic Debris	.03	0	.05	.01	.02	0
Auto (Pb)	.05	.01	.05	-.01	.06	.04
$CaSO_4$	.09	.05	.10	.03	.06	.03
Construction	.54	.49	.33	.15	.43	.35
2° Particles	.01	0	.01	-.01	.01	0
Dirt 1 (Syr)		-.03		.03		-.01
HST Soil 2		.37		.29		.21
Misc.	-.07	-.02	-.08	-.09	-.02	-.03

* Initial apportionments as submitted to Q-Model. Final apportionments after discriminant analysis reclassification for three ambient and six source signatures.

The signature readjustment experiments were conducted by within-sample discriminant analysis in a series of steps. Step 1 was the Q-Model apportionment carried out on the original signatures. In

Step 2, four of the Houston ambient aerosol samples were discriminated individually and the resulting adjusted (or best fit) particle class population signatures converted to weight percent by particle class histograms for substitution in the Q-Model apportionment. In Step 2, Houston ambient sample 124-522 was selected as a potential source type after discriminant analysis and signature rewrite. Therefore, in Step 3, sample 124-522 was eliminated from the Q-Model run. In Step 4, six of the prominent potential source type signatures were discriminated, rewritten, and subjected to the Q-Model apportionment as replacements for the existing signature entries. In the final step of the experiment, Number 5, all of the individual particle object vectors were simultaneously subjected to discriminant analysis for signature rewrite and inclusion in the Q-Model apportionment.

Table III shows the general success (or non-success) of the discriminant analysis prior to Q-Model apportionments. What is displayed is the summation of the negative apportionment vectors for each step of the experiment as outlined above. Also shown (in parentheses) are the calculated information entropy values as a percentage of the maximum possible for the original and rewritten signatures. The values presented are for the rewritten weight percent-by-particle class signatures.

It would appear from these results that a generic or source type apportionment of unknown ambient aerosol samples can be made using a source type signature library and employing SAX methods. From an examination of the emission inventory for the Houston AQMA, the types of apportionments shown in Table II would appear reasonable. Of course, the approach utilized here is dependent upon the textural quality of the source type signature library - potential major source types must be on file. In the case of the Houston aerosol samples used in this example, the apportionment results indicate that much of the aerosol has calcium rich sources (cement plant like, construction/destruction, limestone, and $CaSO_4$). Based upon this type of SAX apportionment, it would seem that subsequent source samplings (as may be required in Houston) should concentrate on appropriate Ca rich sources.

It can be seen from the example results for Step 5 (Table III) that as information entropy increases, resolution of source types decreases. Clearly the application of discriminant analysis to the individual object vectors within a sample of aerosol material can sharpen the signature histograms which result from SAX characterizations. This helps to circumvent the difficulty of not having an "optimum" sorting scheme for each airshed. As long as the source signature library contains the minimum of appropriate signature types, best-fit manipulation of the raw SAX data would seem to show great promise in generic source apportionments of unknown samples.

TABLE III. Summary of total negative apportionment vectors.

[STEPS]	1	2	3	4	5
Sample	Original	New Ambients	New Ambients▲ (Not 552)	All New Signs.* (Not 522)	Total Discrim.‡
124-502C	-30.4 (63)**	-19.9	-19.3	-5.9	-7.7 (80)
124-510C	-37.2 (56)	-31.7	-25.4	-12.6	-25.5 (62)
124-516C	-19.5 (62)	-14.4	-14.6	- 4.9	114.0 (70)
124-522C	- 3.5	0	- 6.8	-23.8	

▲ Sample 124-522 chosen as "extreme" in step 2. Did not use Discrim. Signature in later steps.

* New Discrim. signatures for 502, 510, 516, Limestone, Syracuse Dirt 7, Foundary Sand, Houston Soil 1, Lime Kiln, Coal fly ash 2.

‡ New Discrim. signatures, as in step 4, but all 9 signatures submitted to the same discriminant analysis calibration simultaneously.

** Percentage of maximum possible information entropy.

There are some limitations in the use of SAX data for source apportionment. One particular problem is when the individual particle being characterized is an aggregate, possibly representing particle emission for several sources. Such particles obviously exist in nature but no quantified description in ambient aerosol is apparent. Through the use of multiple source samplings and through "features" in the SEM image, the presence of aggregates can be determined. These particles can be divided among two or more categories, or eliminated for consideration in the statistical summary of the total sample. In the case of the results we present here, software limitations precluded these options as the x-rays collected were from the geometric center of the particle feature. Manual quantification of the sample we used indicated a low incidence of compositionally distinct aggregates. We feel, therefore, that the characterization of large numbers of particles (greater than 600) in each specimen will tend to average out the uncertainties. Further work is needed to quantify this SAX model uncertainty.

Conclusions

SAX analyses of unknown aerosol samples can be used in a generic receptor model apportionment scheme which utilizes a library of source type signatures. "Fixed" particle type sorting algorithms developed for one airshed may not be optimum when applied to other airshed ambient samples. However, the results of the experiments performed here with the raw analytical data for individual particle observations show that the quality of the mathematical solution can be improved to acceptable levels. Unknown ambient aerosol samples, characterized by SAX methods, can be apportioned by using a "world list" of source type signatures. The results can serve as a guide to improving the resolution of CMB receptor models and can indicate where source sampling efforts can be most efficiently pursued. It is clear that statistical treatment of the raw data (object vectors for individual particles) is more efficient than dealig with the summaries of such observations.

Acknowledgements

This work was supported under a cooperative agreement with the U.S. Environmental Protection Agency: Number CR-809760011. Special thanks are due to Don McLean and Gwen May of the CESF Computer Center staff for expediting operation and maintenance of our variours computer programs.

REFERENCES

1. R.J. Lee, J.F. Kelly, "Overview of SEM-based automated image analysis," Scanning Electron Microscopy/1980/I: p. 303 (1980).

2. D.L. Johnson, "Monitoring Particulate Pollutants; Air and Water: A Breakthrough in Individual Particle Analysis, "Proc. Conf. on Toxic and Hazardous Chemicals, K. Pilon (ed), Central New York Water Quality Management Program, Syracuse, N.Y. (1980).

3. R.B. Hanna, K.J. Karcich, and D.L. Johnson, "Determination of Particle Identities via Computer Assisted SEM/EDXA System," Scanning Electron/Microscopy/1980/I: p. 323, (1980).

4. D.L. Johnson, B. McIntyre, R. Fortmann, R.K. Stevens, and R.B. Hanna, "A Chemical Element Comparison of Individual Particle Analysis and Bulk Analysis Methods," Scanning Electron Microscopy/1981/I: p. 469 (1981).

5. G.S. Casuccio and P.B. Janocko, "Quantifying Blast Furnace Cost House Contribution to TSP using a Unique Receptor Model", Proc. 74th APCA Meeting, Philadelphia (1981).

6. B.L. McIntyre, M.M. Minnis, D.L. Johnson, "An individual particle analysis method and its application to atmospheric transport." Proc. Symp. on: Intermediate range atmospheric transport processes and technology assessment, DOE Gatlinburg, Tn., pp. 435-443, (1981).

7. D.L. Johnson and B.L. McIntyre, "A Particle Class Balance Receptor Model for Aerosol Apportionment in Syracuse, N.Y." (these proceedings) (1983).

8. D.L. Johnson, "SEM/XES Image Analysis Characterization of Particulate Inclusions in Biological Tissues," Scanning Electron Microscopy/1982/I: in press (1982).

9. K.J. Karcich, M. Moore, R. Morton, R. Klein, J.C. Russ, "Utilization of Alternating Chemistry Definition File Structures for Materials Characterization," Scanning Electron Microscopy/1982/I: p. 463 (1981).

10. P. Janocko, G. Casuccio, "Final report to TACB", in press.

11. W.E. Full, R. Ehrlich, S.K. Kennedy, "Optimal Definition of Class Intervals of Histograms or Frequency Plots," Int. J. Math Geol., (in press), (1983).

12. T. Dzubay, R.K. Stevens, C.W. Lewis, D.H. Hern, W.J. Courtney, J.W. Tesch, and M.A. Mason, "Visibility and Aerosol Composition in Houston, Texas," Environmental Science and Technology 16: p. 514 (1982). *

13. R.K. Stevens, et al, USEPA Quail Roost II Final Report (in press).

14. W.E. Full, R. Ehrlich, J.E. Klovan, "Extended Q-Model - Objective definition of external end members in the analysis of mixtures," Int. J. Math. Geol. 13: p. 331 (1981).

15. E. Gallaher (personal communication), Author Computer Program. Director Computer Services, Phil. Acad. Nat. Sciences, Phil., Pa.

16. R. Kullback, "Information Theory and Statistics", Wiley Interscience, New York, N.Y. (1959).

17. Statistical Analysis System, N.C. (1979).

18. E. Shannon, Bell System Tech. Journal, 379, p. 623. Reprinted in C.E., Shannon and Weaver, The Mathematical Theory of Communication, Univ. Ill., Press (1963).

A PARTICLE CLASS BALANCE RECEPTOR MODEL FOR AEROSOL APPORTIONMENT IN SYRACUSE, N.Y.

David L. Johnson
Department of Chemistry
State University of New York
College of Environmental Science and Forestry
Syracuse, New York

Brian L. McIntyre

Scanning electron microscopy interfaced with automated image analysis and x-ray energy spectroscopy (SAX) can be employed for individual particle characterization of aerosol samples in a time efficient manner. Such an individual particle analysis partitioning of the bulk chemical data provides a new perspective on the heterogeneity of ambient aerosols. As particles are individuals, it is possible to increase the source resolution of conventional receptor models which are based on bulk chemical analyses. This paper presents a particle class balance (PCB) which is directly analogous, in mathematical approach, to the CMB types of receptor models.

Samples of ambient aerosol and potential source materials from Syracuse, N.Y. were characterized by SAX. Particles from both types of specimens were sorted by a fixed classification algorithm wherein the number percent and weight percent contribution to each of 25 categories were recorded. The resulting "signatures" of the source standards served as the basis for "unmixing" the composite ambient signatures into their fractional components. This was accomplished by both a linear least squares method and a Q-mode factor analysis/vector analysis technique, for comparative purposes. In both cases, sources were apportioned to 12 generic source types - an improvement in resolution over most bulk chemical analysis models. Results were consistent with what is known of area and point source types and strengths in Syracuse.

Introduction

Receptor models for aerosol source apportionment may take a variety of forms. Each is based upon a chemical analytical characterization of atmospheric particles, and each may employ one or more mathematical tools in the treatment of the data (or lack of data) for ambient aerosols and their potential source materials. At present three basic types of receptor models are in use: those based on bulk chemical element analyses, models based upon crystalline component identification and quantification, and models which rely on individual aerosol particle class or type characterizations. Organic molecular compound analysis and quantification will no doubt provide a new type of model.

Bulk chemical element composition models such as the chemical element balance (CEB) or chemical mass balance (CMB) are the most commonly used and are usually based on x-ray fluorescence, neutron activation, or atomic absorption spectroscopy analyses. Several general descriptions of bulk chemical analysis receptor models are available.[1-4] The mathematical data manipulation techniques which are employed with bulk elemental analysis results include linear least squares,[5] multiple regression,[6] ridge regression[7] analysis and factor analysis.[8]

Receptor models based on aerosol characterization other than elemental composition are relatively new and have not been extensively described in the scientific literature. A provenence analysis utilizing the crystalline component composition (by XRD) of urban aerosol in Rapid City, SD, was carried out by Davis.[9] Casuccio and Janocko[10] utilized computer controlled scanning electron microscopy (CCSEM) for an aerosol particle type based receptor model, but did not describe in detail its operation, applicability, or limitations.

We report here an individual aerosol particle characterization model for source apportionment based on particle analysis by scanning electron microscopy with automated image analysis and x-ray energy spectroscopy (SAX). The model structure is equivalent to the CMB except that ambient and source aerosol bulk elemental composition is replaced by the number or mass contribution of particle types in the aerosol samples. Hence, we have termed this approach a particle class balance (PCB) model. A variety of potential source and ambient atmospheric particle samplings for 1980 and 1981 in Syracuse, New York, serve as the basis for the introduction of the PCB receptor model.

Methods and Results

SAX Particle Characterization

All particle analyses were conducted by SEM-XES using an ETEC Autoscan scanning electron microscope coupled to a Kevex 5100C x-ray spectrometer, and interfaced to a LeMont Scientific B10-E image analysis system. Automation of the processes associated with these types of analyses was done using an Interdata 8/16E microprocessor, which also was responsible for subsequent data manipulation. The use of the computerized system made it possible to analyze significant numbers of particles in a reasonable length of time -- 400-600 particles/hour.[11-14] SAX is performed in three discrete operations: 1) morphological determinations, 2) elemental composition determination, and 3) classification of particles. To arrive at morphological measurements of individual particles, image analysis procedures were imposed upon the sample. A back scattered electron (BSE) image of particles can be resolved into areas of high contrast and low contrast. Such contrast difference is directly proportional to the atomic number of the atoms producing the BSE image. Thus, particles composed of atomic number elements greater than 11 can be distinguished from standard collection filter media. Threshold gray level selection allows for differentiation of particles from filter substrate. The result of this process is a binary image of the particles in question; the particles appear as a white signal, the filter as a black background. Since binary representations of particles are recognizable to a digital computer, control of image analysis can be automated. Figure 1 shows a secondary electron image of 2 aerosol particles, its BSE binary signal, and the diameter analysis of those images.

Through electron beam and sample electron interaction, the elemental composition of the particle being measured can be acquired. With the software package available on the system we used, the electron beam is positioned at the geometric center of the particle image. x-rays are then collected for 3 seconds. The net X-ray counts above the background channels on either side of each x-ray energy spectrum elemental region of interest are recorded for later use along with the lengths of the four (or optionally, 8) diameters determined for the particle image. These measurements constitute the raw data for each particle, and are utilized for particle classification as well as chemical and physical summaries for the populations of particles in each user defined category.

Ambient Aerosol Sample Collection

The accuracy and time efficiency of individual particle analysis is quite sensitive to aspects of the aerosol mass loading on the collection substrate. Since samples analyzed in the SEM are generally examined at magnification of 300-1000X, a relatively small field of view (or frame area) results. Typically an area of the filter of 1 to $5 \times 10^{-4} cm^2$ is examined for sample characterization at the optimum loading range. This small area dictates that uniform particle deposition on the filter surface is required to minimize bias in the analysis/sample characterization process.

Aerosol loading on the filters to be subjected to SEM/XES image analysis techniques is important for several reasons. Too high a loading results in overlapping particle images, thus confusing the individual particle characterization. Too light a loading requires excessive operator time in the selection of fields of view for analysis. Particles should be separated from each other about 2-4 μm in order to minimize fluorescent x-ray emission from particles adjacent to the one under analysis. We find that these conditions are satisfied by a loading range of 30-80 $\mu g/cm^2$ of filter surface.

Samples of ambient particulate matter were collected in the Syracuse metropolitan area over a one-year period from May 1980 to May 1981. Small spot samplers were employed for this. Teflon cylinders with a 1 m diameter axial bore were clamped over Nuclepore membranes for sample collection. They were operated either by a small diaphram pump or by a source of bottled vacuum. In either case, samples were generally collected for a 3-6 hr. period depending upon the conditions of atmospheric particle loading. Since the amount of material deposited on the samplers was too little to weigh, the source apportionment results reported represent percentage contributions from the various source types to the average particulate matter loading in Syracuse.

During the complete study period, 60 samples were collected from a variety of locations within the Syracuse air quality maintenance area (AQMA). However, the present results were from one urban residential site, which was sampled over a two-week period at 12-hour intervals. This produced a set of 23 consecutive samples during March and April of 1981.

Characterization of Emission Source Materials

In order to use the source apportionment procedures developed by the PCB model, a source inventory for the study area was conducted. Table I summarizes the number and types of source materials which were characterized for this report. Some of the source standards were not obtained from the Syracuse AQMA, but were felt to be representative of similar materials found there.

Bulk-point and-area source standard materials were deposited on membrane filters for analysis by shaking the material in a plastic bag, letting the large (>15 μm) particles settle for about 10 seconds, and then inserting an air sampler into the bag for 5-15 seconds. This sampling procedure attempts to collect particles of the source material representative of that in the IP range which could be transported from source to receptor site. Replicate samplings of this nature show greater than 85% similarity. In preparation for the SEM analysis, portions of the membrane filters were mounted on carbon stubs using colloidal graphite in isopropanol. The samples were then coated with 20-40nm carbon in a high vacuum evaporator.

Particle Classification

A critical step in the individual particle analysis procedure as applied to the PCB model for source apportionment is the user defined classification of particles on the basis of their measured characteristics. Each category in the chemistry definition file (CDF) is specified by the operator and saved for future use. The CDF used here contained 25 separate classes of particles based on the observed relative net x-ray contributions for up to 27 elements, as well as a variety of physical parameters (i.e. width/length ratio, avg. diameter, etc.). Details of the classification scheme are given in Johnson, et al.

The individual classes in the CDF were based upon analyzes of standard materials that describe each class. For example, the marine aerosol category was defined by analyzing air samples obtained from a sampling expedition in the Carribean Sea; the wood combustion category was derived by analyzing ash from domestic as well as industrial wood burning operations; the combustion cars category was based on samples from auto exhaust, and so on. Source materials are not pure; in this sense; for instance, some real fly ash might be classified as lithophilic. (a general alumino-silicate type material).

The data collected from a population of discrete particles was summarized in a variety of ways. The weight % in each class was based on an estimate of individual particle weight. Particle volume is computed from the image assuming it to be a prolate spheroid rotated around the major axis. These were broken down into user defined size categories (based upon average particle image diameter). For our work, both numbers % and weight % of the total sample in each chemical category were used as input data for the mathematical apportionment manipulations.

Analytical Reproducibility

Several experiments were conducted to determine the variability of the analytical results obtained from SAX. Similarity coefficients[15] were used as an index since they are sensitive to small differences in the classification results of individual particle analyses. (The similarity coefficient is the sum, over all user-defined categories, of the minimum value that two samples in a pairwise comparison have in common). With skilled operators, and the analysis of 1000 particles each from the same sample, similarity coefficients of 90% were achieved for between-operator comparisons for weight percent by particle class signatures. For single operator replicate analyses of fields from a sample of ambient aerosols, we also have found 90% similarity coefficients. As the number of particles analyzed increases, the similarity of replicates increases in direct accordance with counting statistics.

Real sample variability, as observed from replicate samplings with samplings in the same location, exceeds the analytical uncertainties described above. In two experiments with triplicate samplers whose inlets were placed within 10 cm of each other, an average similarity coefficient of 81% for weight percent histograms was observed for pair-wise comparison. In a third experiment in which 6 spot samplers were deployed synoptically at 30 meter spacings, a mean similarity coefficient of ≅ 74% for replicates was observed. In all these tests, the variability resulting from individual particle class differences was observed to be randomly distributed. Thus, we take the over all uncertainty of our analyses to be approximately 25% where we mean to relate particle classification to reproducibility.

Mathematical Apportionment Techniques

For the first SAX data treatment a linear least squares procedure was used to fit the source data to the observed ambient particle class distributions. Due to matrix inversion problems in the programs used, the most efficient number of sources to be fit were 12-17. Both weight and number percent (of the total sample) apportionment vectors were calculated. After obtaining information as described for particle class distributions within the samples, a routine very much similar to the previously mentioned CMB technique was used to arrive at a best-fit of individual source magnitudes to the observed data.

A PCB replaces the elemental distribution of CMB with a particle class distribution generated by SAX. The replacement occurs in the source signature matrix. From this point the same mathematics apply to both PCB and CMB techniques. The steps involve: Calculation of a predicted value for the proportion of Class k for all sources j

(Equation 1)

$$P_k = (M_j)(C_{jk})$$

where

P_k = predicted proportion of class K

C_{jk} = proportion of class K in source j

M_j = fractional abundance of source j

and minimization of the value of X^2 with a linear least squares regression that minimizes the value of $(O_k - P_k)$ in Equation 2 below:

(Equation 2)

$$X^2 = (O_k - P_k)2/\delta K$$

where,

O_k = observed proportion of class k

δK = total error term associated with data

The result of PCB is a vector of mathematically correct individual source magnitudes for a given receptor (sampling) location. Mathematically correct does not imply absolutely correct, since a best-fit procedure may project obviously erroneous results (i.e. >100%, or < 0%).

Each ambient aerosol sample and specimen of source materials was characterized by SAX under the same SEM operating conditions, and utilizing the same CDF algorithm for particle type sorting. From the base of 47 source signatures an average 12 source type by 17 particle class source signature matrix was created for an unweighted linear least squares fit source apportionment of the 23 ambient aerosol samples collected at the site in Syracuse. This amounted to 11,500 ambient particles analyzed. Since, in the SAX data summary routines, both population percent by particle class and total aerosol weight percent by particle class histograms are produced, least squares apportionment solutions for both types of signatures were calculated. The results, expressed as a fractional mass contribution to the total ambient aerosol sample, are shown in Table II.

Table II. Source apportion for 23 Syracuse aerosol samples collected at one site, March and April, 1981. Results are average wt.% contributions for Q-Model.

PARTICLE SOURCE TYPE	LEAST SQUARES NO.%	LEAST SQUARES WT.%	Q-MODEL WT.%
Organic material	2	0	3
Salt resembling	0	0	2
Secondary sulfate	1	0	0
Ca and S rich	8	5	21
Wood flyash	1	-7	0
Auto emissions	4	2	2
Oil flyash	1	1	3
Coal flyash	12	-17	5
Steel mill emissions	3	7	5
Limestone/calcite	10	12	13
Other Lithophilic	56	89	44
Miscellaneous	3	1	2

Also shown in Table II are the source apportionment results from the PCB model as obtained by a different mathematical treatment of the same SAX data - a factor analysis/vector analysis approach called CAB-

FAC/EXTENDED Q-MODEL as described by Full, et al.[16] For Q-Model, data on total weight percent contribution for each of the 25 particle types were used in the solution as opposed to 17 particle types in the linear least squares solution described above. The weight percent-by-particle type signatures for 47 possible sources and each of the 23 ambient aerosol samples were combined and submitted to a Q-Mode factor analysis which employed an orthogonal (VARIMAX) factor axis rotation. The computer program employed identified the most extreme samples (now in Varimax factor space) as the boundary points defining a hyperellipsoid. The number of such boundary was user specified at 15. Solutions were sought which did not include any of the ambient aerosol samples as end members. The apportionment of the ambient samples was then taken as the vector composition description of the ambient specimens in terms of the selected extreme or end member source samples. The 15 specific end members identified were then collapsed to the 12 generic categories shown in Table II.

Discussion

The chemistry definition file, or sorting algorithm, used for the SAX results in this work may not be optimum for general application as a descriptive tool. Every set of particles collected at a sampling site represents a subset of the particle types which historically might occur there. That is, during a discrete sampling period, not all potential sources would be contributing to the receptor site aerosol. As this phenomenon has geographic variations as well, optimum particle sorting schemes might be different for samples collected in different locations. A sample specific (or variable, tailor made) particle sorting scheme would no doubt be the most cleanly descriptive as a microscopic technique. If accurate specific particle type and/or source identities could be established for each feature observed in the SEM, then an absolute source apportionment would be possible. There are two difficulties in attaining this goal: 1) particle sources are not pure - they generally contain a variety of particle types similar to other sources, and 2) the current state of the art for SAX technology is insufficient for unique feature discernment - uncertainties in both x-ray data acquisition and the feature image analysis lead to uncertainties in classification. To the extent that unique particle types from specific sources can be identified by SAX, then a variable sorting algorithm may be employed. Since sources are variable and complex, such an approach in a PCB receptor model appears to have limited applicability; especially when airshed - specific source information is not available.

We have elected to employ a fixed particle sorting scheme to be uniformly applied to source materials as well as ambient samples. The particle classifications used are diverse and intended to represent a variety of generic source types. We recognize that in application of this method we lose some purely descriptive capabilities of the data summary as well as increase the possibility of particle class overlap when comparing signatures of similar sources, but these disadvantages are outweighed by the advantages of internal or relative sample and source analytical consistency. The signature overlap we experience can be sorted out by a variety of statistical methods.

The source apportionment results for the Syracuse aerosol, as shown in Table II, seem reasonable based on our knowledge of the emission source inventory for the local AQMA. Of course, verification of these results will only be possible by comparison to results of other receptor and dispersion models.

Considering the linear least squares results it seemed, at first, surprising that such differences in average apportionment results for coal flyash and lithophilic materials could result from the use of SAX particle class population percent signatures as compared to the particle class weight percent signatures. However, the least squares generic source signature matrix is an average for a number of specific sources. In the analysis of the ambient samples, a few large particles could distort the signature (or significantly weight it) towards non-average values. Thus the coal fly ash/general lithophilic colinearity shown by a -17% and a + 89% in the least squares weight percent apportionment of Table II might be resolved if the SAX results were size fractionated in the apportionment model. Preliminary results show promise and will be presented as a theoretical approach to size-specific apportionment and source identification in a subsequent publication. SAX procedures make it very easy to peruse size specific results since the image analysis of each individual aerosol particle contains detailed morphological information.

The Q-Model apportionment results shown in column three of Table II are very similar to the number percent results obtained from the least squares solution. The total particle class weight percent contribution signature was employed in the Q-Model calculation. While a few large particles in a given source or ambient sample may also have distorted these results, the Q-Model apportionment was based on a population of 47 specific source signatures, and not directly on generic signatures resulting from averaging several sources.

Conclusions

Individual aerosol particle characterizations by SAX are a sufficient data base for the development of a new type of receptor model for source apportionment. Computer control of SEM/XES functions speeds the acquisition of such data so as to make individual particle analysis an economically feasible technique. Since hundreds of particles can be recovered per hour, large data bases for source and ambient aerosol samples can be assembled. This increases the potential source type resolution of SAX based receptor models by increasing the number of degrees of freedom available for multivariate statistical treatments of the data. SAX receptor models would appear to be most useful in situations where conventional CMB applications have trouble separating similar sources because of multicolinearities. To be sure, SAX results may suffer from colinearities in particle definition, but used in conjunction with other techniques, different perspectives may be obtained. Where unique particle class source tracers are present in an aerosol, receptor models based on particle type classification offer an extremely efficient method for the validation of dispersion models. The automatic size fractionation available through image analysis of SEM signals could provide a new and valuable perspective for receptor based source apportionment models.

Acknolwedgements

A portion of this work served as the basis for an MS Degree for B.L. McIntyre. portions of this research were funded under USEPA Cooperative Agreement No. CR 809F60011.

Special thanks are due to the staff of the SUNY-ESF A&TS unit for sampler construction and to members of the NC Brown Laboratory for Ultrastructure Study for support and advice relating to the use of the Scanning Electron Microscope.

REFERENCES

1. G.E. Gordon, "Receptor models - determining signatures for sources of emissions," Env. Sci. and Technol., 14: p. 792 (1980).

2. J.A. Cooper, J.G. Watson, Jr., "Receptor oriented methods for air particulate source apportionment," J. Air Poll. Contr. Assoc., 30: p. 1116 (1980).

3. J.E. Core, T.G. Pace, "Receptor models - How great thou art!(?)." 74th Annual Mtg. of the A.P.C.A., Philadelphia, Pa., (1981).

4. S.K. Friedlander, "Chemical element balances and identification of air pollution sources," Env. Sci. and Technol., 7: p.235 (1973).

5. T. Dzubay, "Chemical element balance method applied to dichotomous sampler data," Ann. N.Y. Acad. Sci. (Aerosols: Anthropogenic and Natural), 338: p. 126 (1980).

6. T.J. Kneip, R.P. Mallon, M.T. Kleinman, "Multiple regression models for source apportionment," Proceedings 74th. APCA Meeting; Philadelphia, Pa., (1981).

7. D.W. Wichern, G.A. Churchill, "A comparison of ridge estimators," Technometrics, 20: p. 301 (1978).

8. P.K. Hopke, "The application of factor analyses to urban aerosol source resolution." In: Atmospheric Aerosol: Source/Air Quality Relationships, ACS Symposium No. 167. Amer. Chem. Soc., Wash. D.C. pp. 21-49, (1981).

9. B.L. Davis, L.R. Johnson, M.J. Flannagan, "Provenance factor analysis of fugitive dust produced in Rapid City, S.D.," J. Air. Poll. Contr. Assoc., 31: p.241, (1981).

10. G.S. Casuccio, P.B. Janocko, "Quantifying blast furnace cast house contribution to TSP using a unique receptor model." Proc. 74th APCA Mtg., Philadelphia, Pa., (1981).

11. D.L. Johnson, B.L. McIntyre, R. Fortmann, R.K. Stevens, R.B. Hanna, "A chemical element comparison of individual particle analysis and bulk analysis methods," Scanning Electron Microscopy/1981/I: p. 469, (1981).

12. B.L. McIntyre, M.M. Minnis, D.L. Johnson, "An individual particle analysis method and its application to atmospheric transport." Proc. Symp. on: Intermediate range atmospheric transport processes and technology assessment, DOE Gatlinburg, Tn., pp. 435-443, (1981).

13. R.B. Hanna, K.J. Karcich, D.L. Johnson, "Determination of particle identities via computer assisted SEM/EDXA system," Scanning Electron Microscopy/1980/I: p. 323, (1980).

14. R.J. Lee, J.F. Kelly, "Overview of SEM-based automated image analysis," Scanning Electron Microscopy/1980/I: p. 303 (1980).

15. J.R. Bray, J.T. Curtis, "An ordinationm of the upland forest communities of southern Wisconsin," Ecology Monthly, 27: p.325 (1957).

16. W.E. Full, R. Ehrlich, J.E. Klovan, "Extended Q-Model - objective definition of external end members in the analysis of mixtures," Int. J. Math. Geol. 13(4): p. 331 (1981).

Table 1. Syracuse Study Particle Sources

SOURCE TYPE	CHEMICAL PARTICLES	NUMBER OF SAMPLES
Cement Kilns	Ca containing	4
Foundries	Fe and Al-Si materials	9
Chemical Plants*	Ca, Na, Cl, Al-Si cont.	5
Steel Mills*	Fe, Cr, Mn containing	6
Lumber Yard	Ca, K, S containing	2
Plastics Plant	Al-Si containing	2
Refuse Incinerator ǂ	Pb, Zn etc. containing	2
Oil Burner ǂ	S, V, Al-Si rich	4
Lithophilic Mat. (area source)	Al-Si, Ca, Ti, Fe cont.	9
Automobiles	Pb, Br, Al, Fe containing	2
Wood Burners	Ca, K, S rich	2
Vegetative Debris (area source)	Ca, K, S, organic cont.	1
Secondary Parts. (area source)	S containing	1
		47

* Includes Coal Fly Ash Samples

ǂ Obtained from outside the Syracuse AQMA

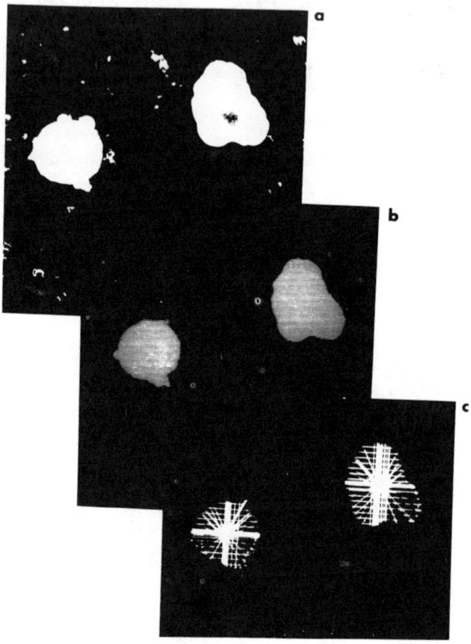

Figure 1. Example of normal secondary electron image of particles at a magnification of 700X (a), their image from the back-scattered electron signal (b), and the LeMont Diameter (Image) analysis of the resultant signal.

THE EL PASO AIRSHED; SOURCE APPORTIONMENT USING COMPLEMENTARY ANALYSES AND RECEPTOR MODELS

Philip B. Janocko,

Gary S. Casuccio,
Energy Technology Consultants,
Murrysville, Pa.

Stuart L. Dattner,
Texas Air Control Board,
Austin, Tx.

David L. Johnson,
State University of New York,
Syracuse, N.Y.

E.R. Crutcher,
Microlab Northwest,
Redmond, Wa.

The Texas Air Control Board (TACB) is currently evaluating the ambient air quality of the El Paso airshed. A major part of this evaluation is directed toward identifying the sources that contribute to total suspended particulate (TSP), with special emphasis on sources of particulate lead. The TACB elected to use receptor oriented modeling for the particulate source identification. A combination of microscopic and bulk chemistry methods have been employed to acquire the data used in the receptor models. The ambient samples analyzed were both cellulose and glass fiber filters collected by high volume (Hi-Vol) samplers in 1981.

All source and ambient samples were analyzed by computer controlled scanning electron microscopy, proton induced x-ray emission and ion chromatography. To augment these techniques, selected samples were anlayzed by optical microscopy, x-ray fluorescence, low temperature ashing and atomic absorption.

This paper presents examples of data from the various analytical procedures as well as preliminary results from the receptor models employed.

Introduction

The Texas Air Control Board (TACB) is currently evaluating the overall air quality of the El Paso airshed. A major part of this evaluation is directed toward identifying and quantifying the sources that contribute to total suspended particulate (TSP), with special emphasis on sources of particulate lead. The TACB elected to use receptor modeling of TSP as one of the techniques to obtain the information that would allow future control strategies to be developed with confidence. Various sample collection, sample analysis, and receptor modeling methods were chosen to provide a means for comparing and evaluating the results. Although the El Paso study is nearly concluded, the relatively short duration of the project permits presentation only of early results. While they are preliminary, they may be considered as representative of the techniques employed.

Background

The city of El Paso is located at the tip of the Texas western panhandle, with New Mexico to the west and north. The Rio Grande River flows along the southern edge of the city, forming the boundary between the United States and Mexico. The Mexican city of Ciudad Juarez lies immediately to the south.

The El Paso airshed is a river basin situated southwest of the Rocky Mountains, with the Franklin Mountains intruding into the basin from the north. The climate is semi-arid, with most of the annual precipitation occurring during a single rainy season. In the dry seasons, dust storms may occur. Winter nightime temperatures are cold enough to require space heating. Temperature inversion trapping conditions occur frequently.

Local industrial particulate sources include a lead smelter, a copper refinery, hard rock and limestone mining and processing, slag processing, and portland cement and brick manufacturing. Non-industrial sources include vehicular traffic and wind-blown soil. Open burning of refuse is common in Mexico.

Particulate monitoring in El Paso is conducted at 24 monitoring sites by three separate agencies. The principle sampler employed is the high volume sampler (Hi-Vol). TACB operates seven sites, El Paso City-County Health Department (EPCCHD) operates fifteen sites, and U.S. EPA operates two sites. Many of these sites have additional Hi-Vols or size selective samplers co-located with the primary Hi-Vol. The TACB collects TSP samples on cellulose filters, whereas the EPCCHD and EPA used the standard glass fiber filter. All Hi-Vol samples collected by TACB are routinely analyzed for 32 elements by XRF. Filters from nine of the EPCCHD sites are routinely analyzed for arsenic, cadmium, lead and zinc by atomic absorption spectroscopy. The EPA El Paso sites are operated as part of its inhalable particulate network. Size selective inlet (SSI) Hi-Vols and/or dichotomous samplers are co-located with Standard Hi-Vols at these sites. The Hi-Vol and SSI Hi-Vol filters are analyzed for sulfate, nitrate and ten elements. The dichotomous sampler filters are analyzed for eleven elements, as well as sulfate and nitrate.

The El Paso airshed has been studied using conventional dispersion models. These studies have been hampered by the complex nature of the terrain, as well as the arid conditions that result in large quantities of wind eroded and vehicularly entrained soil. TACB staff have been developing receptor models to take advantage of the large data base from XRF analysis, which began in 1972. While results have been promising, the staff desired additional resolution of the soil or crustal component, as well as comparison of their model with an alternate receptor model or models.

Design of the El Paso Receptor Modeling Study

The goal of this study was to identify and quantify sources of TSP and particulate lead in the El Paso airshed. To accomplish this goal, the following objectives were established. Source contributions would be determined by receptor modeling. For increased confidence in the results, several receptor models would be employed. The use of multiple models would provide a comparison of source resolution, precision and cost.

These goals and objectives resulted in the following project guidelines. The existing and extensive XRF data base was used, first as a guide for filter selection, and then for comparison with other analyses. Other analytical methods were selected because they could be compared with XRF as well as provide additional information. Additional receptor models used a different method, different analytical data input, or both. Microscopic analyses were employed to enhance source resolution. Analysis methods capable of analyzing both cellulose and glass fiber filters were desired, to take advantage of additional archived glass fiber filters, thereby avoiding the cost and time delay of special sample collection. Sampling sites and filter samples were selected to provide a characterization of the problem areas in the airshed as well as the extent of background contribution. A part of the project was performed as a blind study. An interlaboratory comparison of microscopic methods was a part of this blind study.

Several filter analysis methods were selected. These methods were to be employed either on all filters (where practical), or only on the filters in the microscopic comparison study. The methods selected were as follows:

Bulk Methods:

- o Radioisotope excitation x-ray fluorescence (XRF)
- o Proton induced x-ray excitation XRF (PIXE)
- o Atomic absorption spectroscopy (AA)
- o Ion chromatography (IC)
- o Low temperature ashing (LTA)

Microscopic Methods:

- o Computer controlled scanning electron microscopy (CCSEM)
- o SEM with automated image analysis and x-ray energy spectroscopy (SAX)
- o Quantitative optical light microscopy (OLM)
- o Qualitative optical light microscopy (OLM screening)

The receptor models were selected to use the XRF and CCSEM data. In addition, receptor modeling was performed on the microscopic method comparison filters using an OLM-based model. The models used were designated, by their input data, as follows:

XRF - TACB Linear Regression
CCSEM - Fingerprint Ratio (Least Squares)
OLM - Morphological Tracer Quantification

The monitoring sites and the filters to be analyzed were chosen by TACB. The sources were selected and sampled by TACB. Monitoring sites and source locations are identified in Figure 1.

Comparison of Sample Analysis Methods

The XRF analyses were performed by TACB, using radioisotope excitation and energy dispersive x-ray detection.[1,2] Concentrations of 32 elements from aluminum through lead were determined. For glass fiber filters, only heavier elements (iron or heavier) were reported, because of the large interferences from the glass matrix. PIXE analyses provided 53 elemental concentrations from magnesium through uranium. The PIXE method also used energy dispersive x-ray detection, but sample excitation was through a high energy proton beam.[3,4] Results for all elements were reported from both cellulose and glass fiber filters, although the glass fiber blank correction for interfering elements was only moderately successful. Concentrations of zinc, cadmium, arsenic and lead were determined by AA for all glass fiber Hi Vol filters.[5] All filters were analyzed by ion chromatography (IC) to determine concentrations of seven ion species. Low temperature ashing (LTA) was performed on all glass fiber filters to monitor the levels of carbon and hydrocarbon on the filters. Because the energy dispersive x-ray analyses were not sensitive to lighter elements, the mass fraction consisting of oxides (oxygen), carbonates, carbon and hydrocarbons was undetectable by PIXE and XRF. LTA was used to estimate the carbonaceous contribution to this light element fraction.

Electron microscopy and more recently, automated scanning electron microscopy (SEM), have been used extensively for particulate characterization.[6-13] Two automated SEM techniques, CCSEM and SAX, were used in this study to acquire size, shape and elemental chemistry information from individual particles within each sample. CCSEM was used on all samples in this study, while SAX was employed on the 10 interlaboratory comparison filters. Although not originally designed to do so, both CCSEM and SAX have been modified to report the cummulative particle data as bulk elemental composition.[14,15] This capability was crucial to the intermethod comparisons of this study. While similar in concept, CCSEM and SAX employed completely different hardware and software for automation of the SEM. CCSEM monitored 19 elements, reported as up to 65 particle types. SAX monitored 27 elemnts, reported as 36 particle types. The CCSEM analyses had an upper size limit of 80 um while the upper limit of the SAX analyses was 10 um.

Quantitative optical light microscopy (OLM) was performed on the 10 interlaboratory comparison samples. Appearance and optical properties of individual particles were used to identify species, while the size and number of each species were used to determine concentrations.[16,17] A less rigorous optical analysis (OLM screening) was employed on every sample, to provide a qualitative basis of comparison with the other analyses.

A total of 110 samples were analyzed during this study, including co-located and replicate filters. Many of the 77 ambient filters belonged to several groups of samples collected at different sites on the same day. Some of the 27 ambient filters in the blind study consisted of unidentified replicates as well as filters from areas outside the El Paso airshed. Several of the 33 source samples were replicates of the same filter. This combination of samples and analytical methods provided opportunities for comparison within a method (replicability) as well as between methods.

Tables I and II present data obtained from two co-located Hi-Vols sampling simultaneously over a twenty-four hour period, one employing a glass fiber filter, the other a cellulose filter. The cellulose fiber filter was submitted for analysis twice, producing a replicate analysis. In Table I, relative concentrations of the particle types determined by CCSEM are presented. For all particle types the agreement is generally good. The Pb-rich particle type does show poorer agreement between the glass fiber and cellulose filters. The results are expressed as percent of TSP because of the significant difference in TSP between the two co-located filter samples. For five sampling events in this study where co-located glass fiber and cellulose filters were collected, the cellulose filter always reported a lower TSP value, averaging 75% of the glass fiber filter TSP. Table II presents concentrations of selected elements determined by three methods for the same filters reported in Table I. For the XRF method, only heavy element results from the glass fiber filter were reported because of the previously mentioned interferences from the glass. Replicate XRF analyses were not performed. Consequently, the number of opportunities for comparison with XRF data was reduced and the determination of XRF replicability was affected.

However, Table II shows reasonable agreement of the results for Ca and Fe between samples and methods. The PIXE results for Si illustrate the difficulties encountered in correcting for the interferences from the elements present in the glass filters. The microscopic methods, however, excluded the glass fibers from the analyses. The levels of sulfur reported by CCSEM were generally slightly lower than those reported by either PIXE or XRF. The XRF sulfur values were often the highest. The results for lead again show good agreement except for the CCSEM value from the glass fiber filter. As in Table I, the lead level from the glass fiber filter is lower than that of the cellulose filter. A review of the CCSEM results provided the reason for the difference. While the cellulose results showed that well over half, by weight, of the Pb-rich particles were larger than 10 um (geometric), all of the Pb-rich particles on the glass fiber filter sample were less than 10 um. The cause of this problem has yet to be determined, but may be associated with differences in sampling, selection of filter area for preparation or sample preparation itself.

Results obtained from analysis of a glass fiber filter by three bulk and two microscopic methods are given in Table III. The agreement is quite good with the exception of the SAX value for Fe and the values for the light elements, Na and Mg. The exact cause of the high Fe value from SAX, relative to the other Fe results, is unknown, but may be caused by secondary excitation of the SEM chamber. It may also be significant that SAX did not analyze particles larger than about 10 um (geometric), while CCSEM analyzed up to 80 um.

In Table IV, results are presented for a cellulose filter analyzed by the same methods as in Table III, with the exception of AA, which was not performed. These results show considerable variation between methods for the lighter elements (below sulfur). For the rest of the elements, all methods show reasonably good agreement, with the exception of the zinc and lead values reported by SAX.

Table V presents a comparison of three microscopic analyses from the same filter as in Table III. The particle types of CCSEM and SAX were assigned to the appropriate classes described by OLM. These results indicate good agreement when particle types are classified by generic sources.

While the comparisons presented here are limited, they illustrate many of the general observations from the overall study. Specifically, the individual tables presented may imply performance by an analytical method that was outside of that typically experienced in this study. Taken as a whole, however, these tables should provide an indication of both the variability and the agreement between analytical methods. A detailed statistical treatment of the data base is in progress at this writing and will be reported in the future.

Receptor Models Utilized in the El Paso Study

Source apportionment of selected ambient filters in the El Paso study was accomplished using three independent receptor modeling techniques: TACB linear regression model based on bulk XRF results; ETC Fingerprint/Ratio least squares model using microscopic CCSEM results; and Microlab Northwest morphological tracer quantification model based on optical microscopic results. A brief explaination of each technique is presented below.

The Fingerprint/Ratio receptor model was developed to obtain additional resolution of specific sources by using CCSEM data from individual particle size and chemistry analysis. An important assumption of this model is that the conservation of mass from source to receptor is valid only within narrow size ranges. By independently determining source contributions for each size range, the effects of particle deposition between source and receptor are minimized. However, this model still assumes that each source emits a characteristic set of particles in approximately fixed proportions.[18]

Source apportionment was performed on the ambient TSP filters using key particle types and a series of "tracers" for each source. A key particle type was selected for each source sample based on its uniqueness and rank. If the key particle type appeared on the ambient filter, then the associated source was incorporated in the apportionment process. If the key particle type did not appear on the ambient filter, the associated source was rejected. Sources with the same key particle type were grouped together and apportioned against the ambient filter using tracers specific to each source. Once a tracer for a specific source was identified on an ambient (receptor) filter, the balance of particle types constituting the source fingerprint were tested against the ambient particle types for best fit in each size range and apportioned accordingly. This calculation was performed for each source tracer. A chi-square test was used to determine the best fit from the results obtained from each tracer.

In order to maximize the total amount of filter mass attributable to specific sources, particle types that individually accounted for less than 1.5 weight percent of a size range were combined in a miscellaneous category. Any particle containing lead and/or zinc was not included in the creation of the miscellaneous particle type. With the exception of the lead-rich (Pb) and lead-bromine (Pb-Br) particle types, all particle types containing lead and/or zinc were assigned to either a lead-bearing or zinc-bearing particle category, if the original particle type accounted for less than five weight percent of the size range.

The final determination of specific source contribution was obtained through standard matrix manipulation techniques. The apportionment used a least square solution to a set of simultaneous equations, minimizing the sum of the squares in the deviation between the calculated and expected particle concentrations. All sources selected during the initial source appotionment were analyzed in this fashion until a calculated concentration distribution, approximating the observed, was obtained. The sources that contributed to this "best fit" were printed out with contribution from each particle type reported for each source. Table VI presents the output from the Fingerprint/Ratio model, treating the entire sample as a single size range (0-80 um).

One of the advantages of the use of microscopic analyses data in receptor modeling is the ability to discern chemical (particle type) information as a function of size, within a single sample. In the Fingerprint/Ratio model, the calculations exemplified in Table VI were made for individual aerodynamic size ranges. Table VII shows an example of the contribuiton from each source to each size range. The sum of the contributions to all size ranges from each source is shown in the TOTAL row. For comparison purposes, the source contribution calculated by treating the sample as a single size range is listed below the TOTAL row. It can be seen that while the overall estimated source contributions are similar (Table VIII), the contributions from individual sources show some variation.

In addition to performing a specific source apportionment, a generic source apportionment was produced by combining residual particle types and apportioned particle types within generic classifications (soil, highway, smelter and unknown) for TSP, zinc and lead. An example is presented in Table X.

The TACB receptor model combined factor analysis with chemical mass balance (CMB) in a two step process. First, the XRF data were treated to determine the variability of elements over a large number of ambient filters. This factor analysis established loading factors which may represent source characteristics. After associating each factor with the most likely source, the individual ambient filters were source apportioned with a multiple least squares fit CMB model. The tracer for each source was the element with the highest factor score. In the El Paso airshed, silicon was the tracer for soil, zinc for the ASARCO smelter, and bromine was the tracer for automobile emissions. The assumptions upon which this model relies include the following:

- The relationship among elements emitted from each source is constant.[19]
- Each major emission category has a unique emission source profile.
- Measurements of source and ambient air concentrations have good precision.

An example of results obtained by the source apportionment program used by the TACB is presented in Table XI.

The identification of source contribution by optical microscopy was performed on ten ambient filters as part of the blind study, interlaboratory comparison. The optical microscopic properties used to identify and source apportion particles included morphology, refractive index, and crystallography.[20]

The properties observed in particles from the ambient filter were compared to those of known particle types. This comparison, carried to completion, gave a positive identification, often including a specific origin or source. Generic source information was augmented by analysis of specific source (emission) samples. The inherent resolution of this method required reduction of the number of observed particle types to a manageable level, generally less than fifty. The grouping of these particle types, combined with the quantitative analysis, produced results stated in terms of source contribution. An example of this optical analysis source apportionment is presented in Table XII.

Advantages of a Multimethod Approach

Several benefits were realized by employing multiple analyses in this study. The comparison of methods provided reassurance when results were in agreement and warning when they were not. The different types of information produced by each method increased our understanding of the airshed source contribution, beyond the level possible through any single method. Because of the additional insight provided by the multimethod approach, each method will continue to be refined and improved. But, most importantly, the reliability of the results has been better defined. This will allow decisions based upon these results to be made with increased confidence.

Acknowledgments

The authors are grateful to those contributing to the El Paso receptor model study. Although we cannot list everyone, some merit special mention. John Zalar and Joseph Costantino of Bituminous Coal Research, Inc. provided valuable assistance with the statistical treatment of the data. Gary Burns of the NUS Corporation coordinated the ion chromatography and low temperature ashing analyses, and Henry Kaufmann of Element Analysis Corporation provided the PIXE analyses. We are especially indebted to Lin Cole, who devoted many long nights to the preparation of this manuscript.

References

1. Rhodes, J.R., Pradzynski, A.H., Hunter, C.B., Payne, J.S., and Lindgren, J.L., "Energy dispersive x-ray fluorescence analysis of air particulates in Texas," Environmental Sci & Tech. 6 (10),922 (1972).

2. Rhodes, J.R., "Energy-dispersive x-ray spectrometry," American Laboratory , July (1973)

3. Nelson, J.W., "Proton-Induced Aerosol Analyses: Methods and Samplers," X-ray Fluorescence Analysis of Environmental Samples , T.G. Dzubay, (ed.), Ann Arbor Science, Ann Arbor, MI, 1977.

4. Kaufmann, H.C., Akselsson, K.R. and Courtney, W.J., "REX: A Computer Program for PIXE Spectrum Resolution of Aerosols," Advances in X-Ray Analysis , Vol 19, R.W. Gould, C.S. Barrett, J.B. Newkirk and C.O. Ruud, (eds.) (Dubuque, Iowa: Kendall/Hunt Publishing Co., 1976), pp. 355-366.

5. Hubert, J., Candelaria, R.M., Applegate, H.G., At. Spectrosc ., 1:90 (1980).

6. Lee, R.J and Fisher, R.M., "Quantitative Characterization of Particulates by Scanning and High Voltage Electron Microscopy," National Bureau of Standards Special Publication 553 (1980).

7. Armstrong, J.T. and Buseck, P.R., "Applications in Air Pollution Research of Quantitative Analysis of Individual Microparticles With Electron Beam Instruments," Electron Microscopy and X-Ray Applications to Environmental and Occupational Health Analysis , P.A. Russel and A.E. Hutchings, eds., Ann Arbor Science, Ann Arbor, MI, 1978, p. 211.

8. Hoover, M.R., White, E.W., Lebiedzik J., and Johnson, Jr., G.G., "Automated Characterization of Particulates and Inclusions by Computer Controlled SEM/Probe," Proc. 10th Ann. MAS Conf. , 1975, p. 54A.

9. Lee, R.J., Fasiska, E.G., Janocko, P.B., McFarland, D.,and Penkala, S.J., "Electron-beam Particulate Analysis," Industrial Research and Development , June 1979, 25-28.

10. McCarthy, J.J., "Applications of a Digital Scan Generator to EDS Analysis," Microbeam Analysis , D.E. Newbury, ed., San Francisco Press, San Francisco, CA (1979).

11. Stott, W.R. and Chatfield, E.J. "A Precision SEM Image Analysis System With Full Feature EDXA Characterization," Scanning Electron Microscopy 2: 53 (1979).

12. Lee, R.J. and Kelly, J.F., "Overview of SEM-Based Automated Image Analysis," Scanning Electron Microscopy 1: 303 (1980).

13. Casuccio, G.S. and Janocko, P.B., "Quantifying Blast Furnace Cast House Contribution to TSP Using a Unique Receptor Model," 74th Annual Meeting of the Air Pollution Control Association , Philadelphia, PA (1981).

14. Casuccio, G.S., Janocko, P.B., Lee, R.J., and Kelly, J.F., "The Role of Computer Controlled Scanning Electron Microscopy in Receptor Modeling", 75th Annual Meeting of the Air Pollution Control Association , New Orleans, LA (1982).

15. Johnson, D.L., McIntyre, B.L., Fortmann, R., Stevens, R.K., and Hanna, R.B., "A Chemical Element Comparison of Individual Particle Analysis and Bulk Analysis Methods," Scanning Electron Microscopy , 1981 , I: p. 469, (1981).

16. Crutcher, E.R., "Analysis of Airborne Particulate Collected on Glass Fiber Filters," Report for the Environmental Protection Agency: Region X, #WR-8-0321-B, Reference AHM-78, May 18, 1978.

17. Crutcher, E.R., "A Standardized Approach to Airborne Particulate Analysis Using Quantitative Light Microscopy," Proceedings, Institute of Environmental Sciences , Seattle, WA., (1979).

18. S.K. Friedlander, "Chemical Element Balances and Identification of Air Pollution Sources," Environ. Sci. Technol. 7: 235 (1973).

19. Stevens, R.K., et al, "Mathematical and Empirical Receptor Models:Quail Roost II", U.S. Environmental Protection Agency (in print).

20. Crutcher, E.R., "The Selection of Optical Criteria for the Identification of Airborne Particulate", BOEING Company Report.

TABLE I. Comparison of CCSEM Results From Co-located and Replicate Filters (%TSP)

Particle Type	TACB #14 Glass Fiber (156 ug/m^3)	TACB #13 Cellulose (111 ug/m^3)	TACB #13R Cellulose (Rep.) (111 ug/m^3)
Si-Rich	6.7	6.7	7.1
Ca-Rich	23.2	19.0	20.6
Fe-Rich	2.8	2.6	2.1
Ca-Si	23.3	23.2	21.4
Pb-Rich	.8	6.1	6.3
Pb-Br	2.4	1.3	2.3
Pb-Bearing	4.3	3.0	3.4

TABLE II. Comparison of Elemental Concentrations Between Analyses and Across Co-located and Replicate Filters (% TSP)

	TACB #14 Glass Fiber (156 ug/m^3)			TACB #13 Cellulose (111 ug/m^3)			TACB #13R Cellulose(Rep.) (111 ug/m^3)		
Element	XRF	PIXE	CCSEM	XRF	PIXE	CCSEM	XRF	PIXE	CCSEM
Si	a	0.0	10.5	6.9	11.8	10.1	b	10.3	9.6
S	a	3.1	1.7	4.2	3.2	2.8	b	2.9	2.1
Ca	a	10.3	12.5	11.9	10.7	10.8	b	9.6	10.8
Fe	a	2.1	3.3	2.3	2.6	3.0	b	2.4	2.3
Pb	2.2	2.3	1.3	2.6	2.6	2.5	b	2.6	2.7

a, not reported b, not analyzed

TABLE III. Intermethod Comparison of Elemental Analyses of a Glass Fiber Filter (ug/m^3)

Q.A. Sample: TACB #96 (159 ug/m^3)

Element	XRF	PIXE	SAX	CCSEM	AA
Na	b	b	0.3	0.7	b
Mg	b	3.7	0.2	0.9	b
Al	a	10.4	13.0	12.0	b
Si	a	34.4	34.7	34.2	b
S	a	2.8	2.9	2.0	b
K	a	2.0	2.3	3.0	b
Ca	a	8.5	7.8	5.7	b
Fe	a	5.0	18.0	9.0	b
Zn	a	0.4	0.5	1.0	1.3
Pb	1.1	1.1	1.5	1.3	1.4
Br	0.1	0.1	trace	0.1	b
TOTAL	2.2	68.4	81.2	69.9	2.7

a, not reported b, not analyzed

TABLE IV. Intermethod Comparison of Elemental Analyses of a Cellulose Fitler (ug/m^3)

Q.A. Sample: TACB #97 (76 ug/m^3)

Element	XRF	PIXE	SAX	CCSEM	AA
Na	b	b	0.1	0.6	b
Mg	b	5.0	0.5	1.6	b
Al	1.2	3.9	2.8	2.6	b
Si	5.1	11.2	8.8	9.3	b
S	3.1	2.6	2.0	2.6	b
K	0.8	0.9	0.9	1.1	b
Ca	11.6	12.2	13.3	11.6	b
Fe	1.7	2.3	2.6	1.6	b
Zn	0.6	0.8	3.4	0.7	b
Pb	1.6	1.6	2.9	1.5	b
Br	0.1	0.1	trace	0.2	b
TOTAL	25.8	40.6	37.3	33.4	b

b, not analyzed

TABLE V. Interlaboratory Comparison of Microscopic Analyses (Wt.%)

Q.A. Sample: TACB #96 Glass Fiber Filter (159 ug/m^3)

Particle Class	OLM	SAX	CCSEM
Minerals	73.0	77.3	72.3
Biologicals	0.1	-	-
Combustion	0.2	4.6	1.6
Industrial/ Transportation	26.8	18.2	24.6

TABLE VI. Example of Fingerprint/Ratio 0-80 um Particle Type Source Apportionment (ug/m^3)

Sample: TACB #19 Cellulose Filter (36 ug/m^3)

Type	S12	S24	S30	S31	S32	S 6	S 8	S28	SRC	AMB	EST	RES
Ca	0.0	0.2	0.1	0.2	3.4	0.0	0.0	1.5	7.0	5.4	5.4	0.0
Mix-Clay	0.0	0.6	0.3	0.5	1.3	0.0	0.0	2.7	5.8	5.4	5.4	0.0
Ca-Si	0.0	0.3	0.1	0.1	3.5	0.0	0.0	0.9	5.5	5.0	5.0	0.0
Carbon	0.3	0.0	0.0	0.0	1.1	0.0	0.0	0.3	1.7	3.2	1.7	1.5
Si	0.0	0.2	0.3	0.2	1.3	0.0	0.0	1.0	3.5	3.0	3.0	0.0
Ca-S	0.1	0.0	0.0	0.1	0.1	0.0	0.0	0.3	0.7	1.8	0.7	1.1
Pb-Br	0.0	0.0	0.0	0.0	0.8	0.0	0.0	0.0	0.8	1.8	0.8	0.9
Pb-	0.4	0.0	0.0	0.0	0.1	0.0	0.0	0.0	0.6	0.9	0.6	0.3
Pb-Bearing	0.2	0.0	0.0	0.0	0.0	0.0	0.0	0.2	0.7	0.5	0.5	0.0
Zn-Bearing	0.2	0.0	0.0	0.0	0.0	0.0	0.0	0.0	0.4	0.2	0.2	0.0
Misc.	0.1	0.3	0.1	0.3	1.5	0.0	0.0	1.4	3.8	7.9	3.8	4.0
Other	0.1	0.0	0.1	0.1	0.6	0.0	0.0	0.4	1.3	1.8	1.3	0.5
TOTAL	1.5	1.7	1.1	1.6	13.6	0.1	0.1	8.7	31.7	36.9	28.4	8.5

S 6 - S32 = Source number
SRC = Combined Source Fingerprint
AMB = Ambient Concentration
EST = Estimated Source Contribution
RES = Residual

TABLE VII. Fingerprint/Ratio Source Contribution by Size Range

Sample: TACB #19 Cellulose Filter (36 ug/m^3)

Aerodynamic Size Range (um) Apportioned	Source Number 5	6	8	9	11	12	13	24	28	30	31	32	Estimated Source Contribution
0-10	0.1	-	-	-	-	0.3	0.1	0.8	-	0.8	1.1	3.6	6.8
10-30	-	0.1	-	0.8	0.1	-	-	-	12.2	-	-	5.1	18.2
30-80	0.9	-	-	-	-	-	-	-	-	-	-	2.7	3.7
TOTAL	1.0	0.1	-	0.8	0.1	0.3	0.1	0.8	12.2	0.8	1.1	11.4	28.7
0-80	-	0.1	0.1	-	-	1.5	-	1.7	8.7	1.1	1.6	13.6	28.4

TABLE VIII. Fingerprint/Ratio Source Apportionment Totals by Size Range (ug/m^3)

Sample: TACB #19 Cellulose Filter (36 ug/m^3)

Aerodynamic Size Range (um) Apportioned	Combined Source Fingerprint	Ambient Concentration	Estimated Source Contribution	Residual
0-10	7.6	8.7	6.8	1.9
10-30	20.0	22.5	18.2	4.3
30-80	4.1	5.7	3.7	2.1
TOTAL	31.7	36.0	28.7	7.3
0-80	31.7	36.0	28.4	7.6

TABLE IX. Source Samples Contributing to TSP Concentrations in Tables VII and VIII

Sample: TACB #19 Cellulose Filter (36 ug/m^3)

Highway Emissions
Zack White School Soil
Soil at IB & WC Pumps
IB & WC Monitor Soil
Dirt Road at Ascarate Park
ASARCO Sinter
ASARCO Sinter Plant
ASARCO Lead Sinter
ASARCO Coke Storage
ASARCO Reverberatory
ASARCO Convertor
ASARCO Lead Blast Furnace

TABLE X. Example of Fingerprint/Ratio Generic Source Apportionment (ug/m^3)

Sample: TACB #19 Cellulose Filter (36 ug/m^3)

Source Type	TSP	Zn-Bearing*	Pb-Bearing*
Soil	12.9	-	0.2
Highway	13.5	-	0.9
Smelter, Process	-	-	-
Smelter, Fugitive	1.4	0.2	0.6
Soil (Residual)	4.0	-	-
Highway (Residual)	0.9	-	0.9
Smelter (Residual)	-	-	-
Unknown	3.4	-	0.3
TOTAL	36.0	0.2	3.1

*Mass in particle types, not elemental Zn or Pb

TABLE XI. Example of TACB Elemental Source Apportionment (ug/m^3)

Sample: TACB #19 Cellulose Filter (36 ug/m^3)

Source Type	TSP	Pb
Earth Crustal	17.0	-
Automotive	7.5	0.5
Lead Smelter	2.6	0.3
Unknown	8.9	-
Total	36.0	0.8

TABLE XII. Example of Information Obtained with Optical Microscopy Source Apportionment (Wt.% TSP)

Sample: TACB #96 Glass Fiber Filter (159 ug/m^3)

OLM CLASS/Category	% by Class	% by Category
MINERALS	73.0	
Quartz		8.2
Calcite		6.0
Melilite		9.5
Clay		7.8
Brick and Slag Minerals		41.6
BIOLOGICALS	0.1	
Trace Biologicals		.1
COMBUSTION	0.2	
Fused Flyash		.1
Soot		.1
INDUSTRIAL/TRANSPORTATION	26.8	
Tire Wear		.1
Black Slag		15.4
Hematite, Furnace		5.5
Hematite, Natural		2.8
Secondary Industrial		1.4
Calcium Oxide to Carbonate		.1
Heavy Metal Oxides		1.5

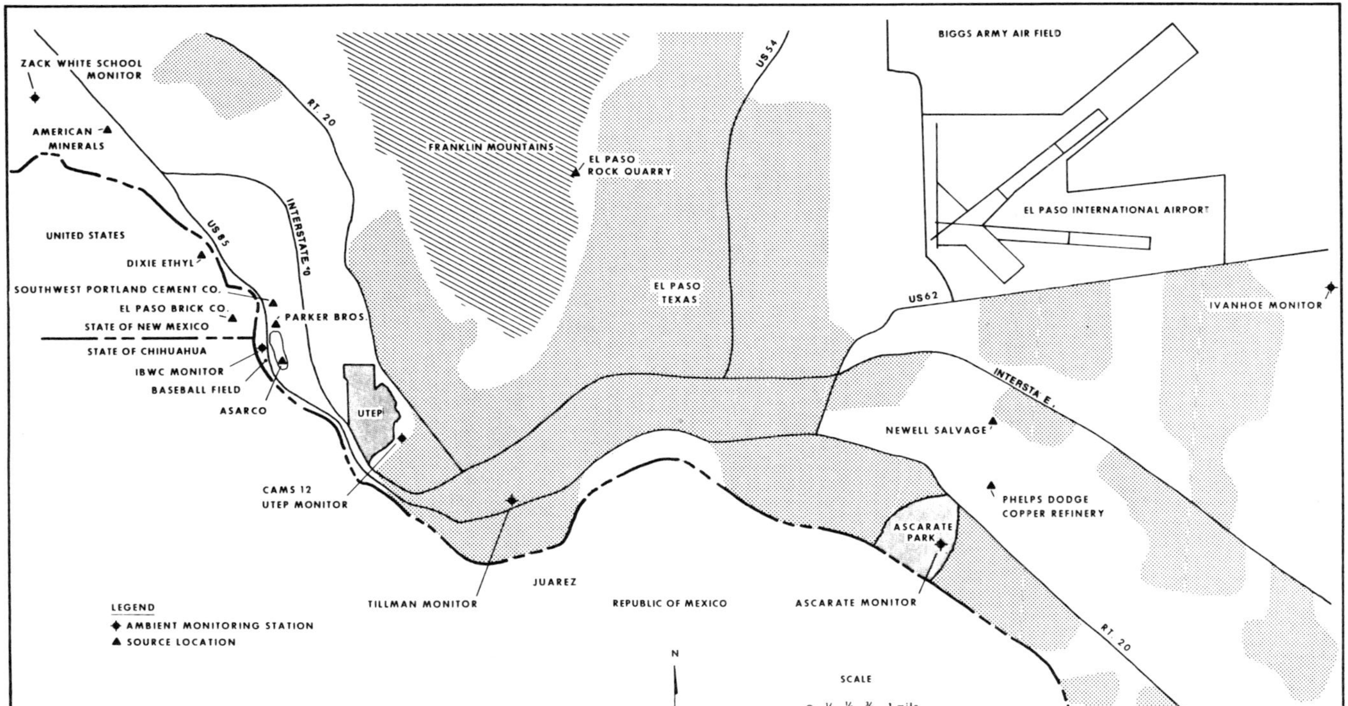

FIGURE 1

MAP OF EL PASO ILLUSTRATING THE LOCATIONS OF AMBIENT MONITORS AND SOURCES USED IN THE RECEPTOR MODELING STUDY

LIGHT MICROSCOPY AS AN ANALYTICAL APPROACH TO RECEPTOR MODELING

E.R. Crutcher
Microlab Northwest
7609 140th Pl. NE
Redmond, Washington

This paper has been written for those of you who would like to use or have used laboratories that provide light microscopical analysis. The purpose of this paper is to explain the type of properties used to identify particles and to serve as a reference against which such work can be evaluated. This is not a step by step guide to this analytical methodology, but it is hoped that it will help distinguish rhetoric from the hard science of light microscopy.

Why Light Microscopy?

The light microscope is the single most powerful analytical tool available for the qualitative analysis of particles with a single physical dimension greater than one micrometer. This includes all particles with an aerodynamic diameter greater than two micrometers and probably fifty percent of the mass below two micrometers. It characterizes compounds and sources on a particle by particle basis and is capable of detecting lot to lot variation in the same compound from the same process line. It is also the least understood and one of the most misused analytical tools.

The image created by the microscope contains two different types of information. The first type of information is form or morphology. Many objects can be uniquely identified by their form alone. Pollen grains, diatoms, and other biological materials are examples. Morphology is not sufficient for the identification of most airborne particles. Optical data is required. It is this second type of information that is most often missing in a poorly done analysis. The special techniques used to collect optical data, along with the mathematics and protocol of quantitative analysis, place a number of very specific constraints on the analysis. Among these constraints is the progression from qualitative analysis, to particle classification, to quantitative analysis, to final source analysis.

The type of optical data collected involves the interaction of light with the bonding electrons in matter. This interaction is discussed in the attached appendix. To collect this data, a polarized light microscope with at least an oblique top light is mandatory. The particles to be analyzed must be sufficiently free of surrounding particles and fibers to behave as separate optical entities. This generally requires removal from the filter substrate. Optically "clearing" a fiber filter is not adequate but some membrane filters can be satisfactorily "cleared." What follows is a discussion of the microscopical properties of particles and how these characteristics can be used to identify sources.

Characterization of Particle Types

The first step in any optical microscopic analysis of a particulate sample is the identification of the particle types in the sample. This phase of the study is qualitative. Individual particles are examined in great detail to establish their identity. The light microscopical properties used to identify these particles include both morphology and optical properties.

Morphology

The shape of any object is the result of the internal and external forces that have acted upon the object. Such an obvious statement, trivial in the macro-world, takes on a new significance in the micro-world of airborne particles. The structural unit of a material is modified in the bulk by distortions such as voids or misalignments whose frequency is a function of the number of unit volumes. As linear dimensions increase, the number of defects increases as the cube of the dimension. As a result, smaller particles are more likely to exhibit a form that indicates the structural units of their compositional material. These structural units may be a biological cell or possess a characteristic crystal symmetry.

Another effect of miniaturization is the increased importance of surface active forces over those dependent on volume or mass. Surface tension and surface chemistry become dominant over gravity and bulk chemistry. This has a profound effect on morphology as well as some analytical approaches.

Different types of objects often exhibit morphologies that have the greatest significance only for that type of object. A discussion of the cleavage planes for biological materials makes no more sense than a discussion of cell wall structure for a crystal. The terms used to describe these different materials must be different. For these reasons the following discussion is divided into three broad categories: crystals, biologicals, and morphological artifacts.

Crystals. The shape of a crystalline solid may result from its molecular crystalline structure in two general ways. The first, and most characteristic, is in its growth habit. This is a reflection of the molecular arrangement in the compound and can be sufficient in and of itself to establish the identity of the compound. The second is through the arrangement of the cleavage planes in the material. Cleavage planes reflect the relative bonding energies through the lattice and are generally less characteristic of specific compounds.

The two crystalline-based morphological types also imply the generation mechanism of their source. To exhibit a growth habit, the compound must grow free of physical constraints. Examples of such conditions include the growth of crystals from the vapor phase and from solution by precipitation or evaporation. Typically, the slower the growth rate, the larger and more perfect is the crystal′s structure. Cleavage is the result of force or stress applied to the material. Crushing or grinding operations typically produce cleavage fragments of those materials that tend to exhibit cleavage. In some materials such as the micas, the bonding energies are so different along different planes in the crystal, only cleavage fragments are ever found. In other materials, such as feldspars, cleavage may or may not be present.

Each crystal class has a set of symmetries associated with it. These symmetries are divided into axial, planar and point type. Many of these symmetries are immediately apparent in a crystal and are useful in helping to confirm an identity as well as a source type. A crystal habit is an outward form that expresses the internal order of the compound. It exhibits many of the symmetry elements of the specific crystal class to which the compound belongs and occasionally possesses unique interfacial angles which are sufficient to establish the compound's identity.

A subdivision of crystal habit is crystal pseudomorphs. These are forms whose outward habit reflects an earlier molecular order or an apparent order. An excellent example of this is alumina. Pseudohexagonal gibbsite ore is actually a monoclinic crystal. When calcined to aluminum oxide, it retains its pseudohexagonal shape. The result is a hexagonal-shaped particle of aluminum oxide (corundum) whose apparent hexagonal symmetry is not aligned to the actual hexagonal structure of the corundum produced by the calcining. This property and the presence of typical calcination-induced patterning uniquely identify calcined gibbsite.

Cleavage is the tendency of a crystal to part along specific lattice planes in response to stress. The extent to which a crystal will exhibit cleavage is dependent upon the relative energies or bonding forces between the ions in the lattice. If the forces are nearly the same in all directions, no cleavage fragments will form. Quartz is an example of a mineral that does not form cleavage fragments. Mica, on the other hand, has very weak bonds between hydroxyl ions along one plane in its lattice. The result is a single cleavage plane that is evident in all fragments of mica. The number of specific cleavage planes exhibited varies with the distribution of the energies through the lattice. Calcite exhibits three planes of cleavage and often one additional, less perfect cleavage plane diagonal to the others. Feldspars occasionally exhibit two cleavage planes. The angles of intersection for the cleavage planes of a material are often very characteristic. When combined with other optical crystallographic properties, a rapid and unique identification can often be made. The mica minerals and the divalent metal carbonates are examples.

Biologicals. Many biological entities or fragments can be defined in strict morphological terms. The morphology of biological materials is considerably more varied and complex than that of crystals, but they also exhibit texture, internal structure, wall patterning, functional subunits, characteristic terminus, characteristic associate structure, etc. Every division of biology has a unique language to describe the significant forms pertinent to that form of life. Typical biologicals encountered during an analysis of airborne particles include insect parts or fibers, herbaceous plant parts or hairs, charred herbaceous plant parts, cereal plants, wood parts, charred wood, pollens, spores, conidia, bacteria, algae (including diatoms), and animal hair. There are also numerous nondescript biological masses that defy simple classification. Rotted bacterial mats form a type of light absorbing mass with no distinct morphological features. This type of material can occasionally be at least identified as an organic by its interaction with or staining of the toluene soluble mounting media often used to mount these specimens.

The morphological identification of biological particles is an important part of any optical microscopic analysis of airborne particulate matter. The contribution of these biological particles to the total airborne burden in an area can be substantial.

Morphological Artifacts. Morphological artifacts is a term used here to describe all of those particles whose form reflects something specific about their source. Combustion-related particles and weathered soil particles are two typical examples but there are many more. Charred biological particles have already been mentioned but fossil fuels also produce many characteristic forms. Materials that burn as a liquid produce oil cenospheres. Oil cenospheres are carbonized small droplets of the original liquid fuel. They have been typically identified as oil cenospheres in the literature although plastics, resins in coal, tars, and similar materials with melting points well below the boiling point will form oil cenospheres when combusted.

Oil cenospheres cover a range of forms. They are all roughly spherical with a highly carbonized surface structure that may or may not be obvious. One extreme is a black solid sphere with some surface broad indentures. These particles are cenospheres which still contain a high proportion of volatile organics. These are characteristic of particles that pass quickly through the combustion chamber. If the particle remains exposed to high heat in an oxygen deprived atmosphere, the volatile organics bake out leaving a lacy, hollow sphere. In an actual sample, the two extremes and all intermediate forms would be found. Oil cenospheres are an indication of liquid fuels and do not represent the total carbonaceous impact of the combustion of the fuels. Much, if not most, of the carbonaceous particles produced by the combustion of liquid fuels consist of submicron "soot" particles.

The combustion of solid fuels such as coal produces another set of characteristic particles. This characteristic flyash consists of the fusion products of mineral impurities in the fuel. The flyash from a pulverized coal burner reflects the mineral impurities of the coal and the conditions of the firing chamber during use. The resultant flyash can be identified as having a calcite, clay, feldspar, pyrite or quartz origin. Calcite ($CaCO_3$) changes to $CaO + CO_2$. The carbon dioxide is trapped in the lime (CaO) matrix producing a frothy glass sphere. Clays are colored brown to yellow with a few small gas inclusions from released water. The feldspars and quartz tend to form spheres of glass without inclusions. Pyrite becomes a hemitite sphere.

Weathering produces another unique morphology which simplifies the characterization of roadside minerals as distinct from fresh sources. Weathering results in siliceous gels and clays adhering to or forming on the surface of clean particles. The exact mechanism of its formation is not well understood and the time required for its formation varies from area to area but its presence indicates that the particle has been essentially incorporated into the soil. When this happens to industrial particles it helps to distinguish between original emissions and secondary re-entrainment from road surfaces or parking lots.

Optical Properties

Optical properties include all the manifestations of the light's interaction with the material in the particle. A brief explanation of the source of these properties is presented here. The appendix contains a more detailed discussion. Light is actually a traveling force field that exerts force on the electrons that bond elements into molecules and molecules into larger structures, such as crystals. This force normally is directed to the direction of wave propagation. The way the electron reacts to the force of the light is dependent upon the type of bond, the orientation of the bond, and the direction the light is forcing the electron. The microscopist sees the effects of the number of electrons encountered, the bond absorption frequencies (if they are near the visible wavelengths), and the orientation of the electron bonds relative to the direction of the force field in the light beam. These effects are interpreted as the relative refractive index, color, and anisotropy, respectively. These three types of properties represent a somewhat arbitrary division of the observed effects. This division is intended to simplify an explanation of these effects. A brief definition of terms relative to each of these subheadings will be presented in each section.

Relative Refractive Index: The refractive index of a particle can only be determined relative to the media in which it is mounted. The observable effects include relief, Becke line motion, lens effect, dispersion, and relative phase shift.

Relief: The degree of contrast, seen as a sharp dark border at the edge of the particle, is called relief. High relief, a wide black border, indicates the particle's refractive index is much higher or much lower than the media's. Low relief, a very thin black border, indicates a smaller difference in refractive indices.

Becke Line Motion: Relief is caused by refraction and reflection at the interface between the particle and the mounting media. Refraction at the interface is always toward the material with the highest refractive index. This creates a bright band of light called the Becke line. As the plane of focus is moved from the edge of the particle to a position above the particle this bright band will either move inward, if the particle has the higher refractive index, or outward, away from the particle, if the mounting media has the higher refractive index.

Lens Effect: For particles too small to have visible Becke lines, the lens effect is very useful. If a particle has a higher refractive index than the mounting media, it behaves as a positive lens and focuses transmitted light to an intense bright spot above the particle. If it has a lower refractive index it diverges the beam creating a dark area above the particle.

Dispersion: Dispersion is the variation of refractive index with wavelength (see appendix). All materials exhibit dispersion, but to see its effects using the microscope the particle must be mounted in a media with a refractive index that matches that of the particle at some wavelength near or in the visible range. It is seen as two colored Becke lines. One is for the wavelengths for which the particle has the higher refractive index, and one is for the wavelengths for which the liquid has the higher refractive index.

Dispersion Staining: Dispersion effects can be enhanced to become a very powerful analytical technique. The methods used to enhance these effects are all referred to as dispersion staining techniques. They are all subject to the limitation of the particle being in a media with a matching refractive index near or in the visible range.

Phase Shift: When white light passes through two different materials a particle and the media it is mounted in, even if they share the same refractive index for some wavelenth, a relative phase difference is created in the different beams. Using phase contrast microscopy this phase shift can cause color effects that can provide relative refractive index information over a range of plus or minus 0.1 refractive index units.

During the qualitative part of the analysis, a particle′s refractive index can be determined very accurately because the mounting liquids can be changed until a match is found. During the quantitative part of the analysis, the mounting media cannot be changed and the degree to which the refractive index can be determined is a function of particle size and the amount of difference between the refractive index, of the particle and the mounting media. The larger the particle and the greater the difference in refractive index the less precise the determination.

The selection of a mounting media for the quantitative part of the analysis becomes very important when refractive index information is important. The technique with the greatest range is phase constrast. With a refractive index of 1.65 for the mounting media, minerals from quartz to corundum (1.55 to 1.76) fall within the useful range. This includes most slag minerals, many industrial minerals, and most natural minerals. When filter "clearing" techniques are used the mounting media must match the filter, about 1.51. This very significantly reduces the number of industrially relevant minerals that refractive index data can characterize.

Color. The spectral transmission of an object is the result of the angle of incidence for the illuminating beam of light, the angle of observation, the wavelength-dependent absorption coefficient for the object, and the object's crystallograhic orientation. The perceived color is dependent on the background intensity, the wavelength-dependent sensitivity of the sensor, the intensity of the light from the object toward the observer, and all of the other conditions mentioned for the spectral transmission of an object. Color can be a very sensitive criteria, but it can be equally misleading if misused. As an example, silicon carbide, a common industrial abrasive, has six forms with a color variation from colorless to blue to yellow to brown to black. These colors correspond to slight changes in electron vibration frequencies with different crystal phases in pure silicon carbide and not to the presence of impurities. Potassium bromide is another case in point. If it is radiated with x-rays it turns from colorless to pale blue. This change in color is the result of single electrons occupying sites normally occupied by bromine ions. The resultant low energy electron is a strong absorber in the red end of the spectrum. Different colors do not always indicate different compounds.

For the more common airborne particles, color can be a useful indicator of the source of a material. Fe_2O_3 is a good example. This material has two basic etiologies: oxidation through burning or oxidation through corrosion (hydration). The Fe_2O_3 from burning has made the transition from iron to FeO to Fe_3O_4 to Fe_2O_3. This transition is from a very high absorption coefficient to a final moderate absorption coefficient with very high absorption in the blue-green. The resultant particle is deep blood-red. When produced by corrosion, the transition is from iron to the whole Fe-O-OH-H_2O phase diagram ending in Fe_2O_3. Iron corrosion products vary in color from green-yellow to orange to red and black. They typically exhibit more than one phase in this series and have a morphology which is also different. The colors in the corrosion products are characteristic of their progression through this series of reactions.

Anisotropy. Anisotropy is a property that is immediately apparent using cross polarized light and indicates the presence of a number of additional properties that can be used to characterize the anisotropic particle. Any anisotropic substance will appear bright while the background and isotropic materials, such as glass, become black between crossed polars. Anisotropy indicates a directionally dependent difference in the material's ability to respond to the incoming light. A stretched piece of plastic, glass under stress, crystals, aligned polymers, etc., all have the property that light polarized in one direction will move with a phase velocity greater (lower refractive index) than light polarized at right angles to that direction (higher refractive index). The phase velocity difference introduces a relative retardation in one wave front with respect to the other. This retardation results in an interference effect when these beams are recombined as they pass through a linearly polarizing filter. The vocabulary associated with these properties is extensive, but a few of the more common terms are presented below.

Anisotropy: An anisotropic material is bright when viewed between cross polars. It indicates that the material has more than one refractive index.

Birefringence: The difference between the maximum and minimum refractive indices exhibited by this material in any possible orientation. This is a fundamental property of crystalline materials. The term "birefringent" is synonymous with "anisotropic".

Biabsorption: The difference between the maximum and minimum absorption exhibited by this material in any possible orientation. This is a fundamental property of crystalline materials. Pleochroism is a term used to identify absorption differences in the visible wavelengths that produce different color effects in the crystal when viewed while rotating a single polarizing filter.

Polarized Light: The term polarized light used in this paper will refer to linear polarized light. Light is polarized if all of its field vectors are aligned parallel to one another.

Polarizing Filter: A polarizing filter only allows light polarized parallel to one line to pass through. A polarizing light microscope contains one such filter below the stage, called the polarizer, and one above the stage called the analyzer.

Cross Polarized Light: The term cross polarized light used in this paper will refer to the use of a linear polarized filter below the subject and another above the subject with the angle between the polarization direction for the two filters at 90 degrees to one another. The result is that no light can pass through both filters except where modified by an anisotropic material.

Optic Axis: The optic axis of an anisotropic material is an imaginary line through the material normal to which all optical properties are the same. Viewed down this axis with a narrow, parallel light beam, an anisotropic material would not appear to be anisotropic.

Retardation Color: An interference color effect seen in anisotropic materials viewed between crossed polars, produced by a combination of the thickness of the particle and the difference between the two refractive indices measurable in that specific orientation of the material. The change of colors with increases in retardation follow a modified Newtonian interference color scale known as a Michel-Levy chart sequence.

Extinction Position: The position in which the particle disappears as it is rotated against the black field of view under cross polarized light. Extinction occurs every 90 degrees of rotation for anisotropic crystalline materials. Three types of extinction positions exist for particles exhibiting a characteristic axis or angle: parallel, symmetric and oblique.

Uniaxial Crystal: A crystal that has only one optic axis is uniaxial. The hexagonal, tetragonal, and trigonal crystal systems are uniaxial.

Biaxial Crystal: A crystal that has two optic axes is biaxial. The orthorhombic, monoclinic, and triclinic crystal systems are biaxial.

Sign of Elongation: Materials that have an obvious long axis can be easily tested to determine if their high refractive index is oriented more closely to that axis. If it is, the material has a positive sign of elongation. If not, it is negative (do not confuse the sign of elongation with the optical sign; the optical sign is outside the scope of this paper).

Interference Figure: This is the pattern at the back focal plane of the objective used to determine the orientation of the crystal on the stage and the optical crystal class, uniaxial or biaxial, to which it belongs.

Many different types of materials exhibit anisotropy: all crystals except the cubic system, stressed materials, aligned long chain polymers, and particles with certain special geometries. The appearance of every anisotropic material is unique if considered in sufficient detail. The partial list of attributes above gives some indication of the degree to which such a material can be characterized. A full listing and explanation of all of the attributes used to characterize industrially relevant particles would require a text many times the size of this article.

Particle Classification and Quantitative Analysis

The amount of information that can be gathered for any given particle is considerable, but often not relevant to the main purpose at hand, and much too time consuming to be part of a quantitative analysis of a complex sample. The quantitative application of the light microscope is not an extended qualitative analysis. There are special sample restrictions and statistical requirements. These impose restrictions on the optical properties that can be used to classify the particle types present in the sample. The manner of presenting the data and error analysis are also important considerations for the final report. The greatest single failing of most so called quantitative light microscopical analyses is in the transition from qualitative analysis to quantitative analysis. Two major differences are ignored, statistical considerations and the optical restrictions that result.

Statistical Considerations and Optical Restrictions

In a quantitative analysis of a particle sample there are a number of conditions that should be imposed. First, the sample should be one that reflects the sampled population in such a way that an additional sampling will produce a sample with the same properties. To analyze individual particles, these particles must be free of other particles. The particles must also be fixed in position so that they do not move while being analyzed. A sample population should be large enough to minimize the impact of statistically unreliable events. How do these conditions impact an analysis of particles collected on a high volume filter?

The subpopulation selected for analysis must be collected in a standardized fashion designed to minimize bias. The use of a cleared filter passes this requirement, tapping the back side of the filter to collect the dust does not. Ultrasonic collection techniques are good because the efficiency of the removal mechanism is not size dependent. A liquid carrier is required which presents possible solubility problems, similar to those involved in filter clearing. The ultrasonic technique also collects particles from both sides of the filter, another advantage over approaches involving the picking of particles from one surface. Whatever approach is used, the inherent variability of the population across the filter surface must also be considered.

The next restriction is that the particles must be free from one another. The probability of particles being isolated optically decreases rapidly as the area covered by particles increases above 10%. Twenty-four hour high volume filters generally fail this requirement with loading equivalent to ambient conditions of 50 micrograms per cubic meter. Cleared fiber filters in fact never meet the requirement of free particles because even if the particles are sufficiently separated, the underlying fibers will effect the observations of the particle.

During a quantitative analysis, individual particles are examined and recorded. If these particles are capable of moving in the liquid mounting media, the possibility of analyzing the same particle twice becomes a very real consideration. To prevent this, the mounting media must fix the particles in position, oils are not satisfactory. Aroclor 5442, Caromount, or other permanent mounting media will meet this requirement. Fixing the particles fixes the orientation of the particle and precludes changing the mounting media. The impact of these optical restrictions will be presented shortly.

The next requirement is that the sample size be large enough to minimize the impact of statistically rare events. A quantitative analysis involves the measurement of the particle, assigning the particle to a class of particles, multiplying the volume of the particle by the density assigned particles in that class, and finally calculating the percent of the total mass represented by that particle class. A single very large particle out of a sample population of more than a thousand may represent as much as 20% of the total mass. This is an example of a statistically rare or unreliable event. There are a number of ways to deal with this problem. If this problem is not mentioned and the manner of correction not presented, the data presented must be considered suspect.

Particle Classification

Typically 30 to 40 different particle classes are needed for a given analysis. Each of these classes may contain more than one type of particle. The category of melilite, a common slag mineral, will contain all of the different minerals in this family provided their elemental composition does not modify their optical properties beyond the classing criteria used. The pollen category may contain 15 different types of pollen as well as three or four different spores and a few conidia. There are two very practical reasons for combining these individual particles into classes. First, the addition of subcategories adds time and cost to the analysis. Second, the addition of irelevant categories clouds the relevant data. If all anisotropic particles are classed together, none of the sources in the area will be resolved by the data.

An essential part of any analysis is that the method be sufficiently described to be repeatable by another analyst skilled in this method of analysis. In the case of light microscopy this means that the classing criteria must be well defined. The term "slag" is not enough. There are hundreds of different slags and many are mineralized, meaning that they contain glasses and different minerals in a complex matrix. It would be nearly impossible to write a single optical criteria that would unambiguously identify such a mineralized slag and not include many materials from other sources. The solution is to establish individual classes of particles, each well defined by their own optical and morphological properties, that can be evaluated as to their source after the quantitative particle analysis has been completed. The definition of these classes should indicate how a particle is placed into this group, such as high relief, and not just a text value for the refractive index.

Presentation of Data

Not all sources emit at a constant rate so percentage data alone is not sufficient to describe the airshed. Both pecentage data and mass loading normalized data are needed to see distinctions between inversion effects when the mass contribution of all sources increase, and wind direction changes or local emission peaks. Seeing only percentage data can lead to erroneous conclusions. The total particle population used, the number and size distribution of those particles by class or category, and the criteria for inclusion in any category must be included or at least available on request. This data is critical to the proper evaluation of the conclusions. Many studies presented as based on quantitative microscopy cannot support their conclusions with any hard numbers. There is nothing wrong with a semiquantitative estimate for the major sources in an area provided it is presented as an estimate and an estimate was paid for. Estimates can be very useful under the conditions presented in the next section of this paper. An estimate of data reliability should also be provided along with the quality assurance procedures used and the results of the quality assurance tests.

Source Identification

The purpose of this type of analysis is two-fold: to identify emission sources and to quantify their impact on the environment at a specific site. It is often not necessary to perform a complete quantitative analysis to determine the major sources impacting a site. An estimate of the contribution from the dominant sources may suffice for days of abnormally high loading. Often these types of samples represent an out of control condition at a local industry. A source inventory and the location of the sample site under these conditions will be sufficient to produce a very reasonable estimate of the source contribution. The danger of an estimate is that it represents the opinion of the analyst. The value of that opinion rests in the skill of the analyst, something very difficult to evaluate. The advantage of an estimate is that is relatively inexpensive.

The quantification of the amount of material from a specific source impacting a specific site can only be based on quantitative analytical results. The transition from the analytical results to the final source apportionment requires the use of a model. Specific particles can often be attributed to a specific source but they are not the only emission from that source. Similarly many particle types have more than one source and must be distributed among these sources using some modeling criteria. There are a number of limiting factors acting at the site that will add scatter to the data. An incomplete list of sources can cloud the data and lead to erroneous conclusions. A poorly placed sampler has often indicated a problem where none actually existed. Most new state implementation plans include the relocation of sampling sites for this reason. There is also the unusual local event that can impact a site with unsuspected particle types. An example would be a landfill that periodically received material from a building demolition. A sample site along the path to the landfill could be very significantly impacted by such an event. Aside from these considerations, the determination of the effect an emission has at a given site involves the use of source inventories, assemblage analysis, industrial tracers, and source samples. Together these are the foundation of the receptor model.

Source Inventories

Source inventories are available in all monitored areas though they are not always complete or up to date. This information along with the location of the sample site can be very beneficial in evaluating the impact of different sources at the site. Many sources emit similar materials. Fused glass spheres from the melting of small quartz, calcite, or clay particles may be from an accidential fire, an incinerator, a coal fired power plant, or a hog fuel boiler. There is nothing unique about a fused mineral that would differentiate these sources. A source inventory showing the presence or absence of these possible sources would simplify the identification of the source.

A knowledge of the types of sources in an area also helps to establish the optical categories that are least likely to overlap. The absence of a source inventory can and has significantly limited results when it was discovered that two or more sources that could have been resolved, have been lumped into one category because the source types were not known. Source inventories are occasionally incomplete with regard to identifying the type of source. When only company names are present, the analyst is often not aware of the nature of that company's business. Some additional information would be needed in this instance.

Assemblage Analysis

Another property of most sources is that along with a major particle type, a variety of other types of particles are emitted. Often the presence of these other types of particles will permit a simple type of mass balance to help distribute the contents of a category with a number of different sources. Quartz would be an example. In an environment that contains cement plants, glass manufacturing, rock crushing, and roads, the apportionment of the quartz category can be very complex. Assemblage analysis can be very beneficial in such an area. The cement industry would contribute calcite, quartz, and blast furnace slag as a raw product assemblage, clinker, and lime spheres from roasting, and alite, belilite, celite, and crushed clinker from the finishing grading, and packaging or shipping facility. Each of these industries has its own typical assemblage for different parts of its processes. By recognizing these assemblages, part of the quartz category could be distributed among those sources for which assemblages were present.

Assemblage analysis can also indicate unknown sources and indicate how particles from those sources are reaching the sampling site. The shipping of landfill materials past a sampling site can produce an assemblage characteristic of a source not in the local area. The size distribution and mix of particles in the assemblage would indicate bulk transport into the area.

Assemblage analysis is a very useful technique, but it may require more information about the sources and more source samples than are often available.

Industrial Tracers

Many industries have a particle type or assemblage that is unique to that industry or for that industry in a given area. These particles become a tracer for that industry or source. Pre-hydrated cement minerals, zinc tetrads, lead oxide prisms, gypsum crystals, etc., are just a few examples. There are many such tracers, too many to list here, but they all share two attributes: they are characteristic of one source and they represent only a fraction of the emissions from that source.

Source Sampling

Source sampling for light microscopy is a little different than that for any other type of analysis. Dust collected with cellophane tape from flat surfaces around an industry can be very useful in terms of characterizing fugitive emissions and assemblages. Samples from baghouses or surfaces within the plant can indicate specific assemblages characteristic of specific operations. Soil samples indicate both local geology as well as indicating the amount of modification from deposited industrial particles. Tape lifts from the surface of the ground indicate the composition of the top layer of dust, that which may be lifted by the wind on a windy day. Particles filtered from the air during periods of specific wind direction indicate typical air transport assemblages.

The collection of source samples for the light microscope is very simple in terms of sampling equipment required. A roll of tape is often sufficient. The ability of the microscopist to screen types of particles that will not become airborne from the sample being examined will result in useful information from what would be useless for an elemental analysis aproach. These samples from the source also indicate the variety of emissions from the source. The most critical part of source sample collection for the light microscope is explicit record keeping.

Conclusion

Light microscopy can be a very valuable asset in the evaluation of a suspended particle problem. It has its greatest potential when applied by local personnel to local airshed problems, but can still be very effectively used by outside laboratories with a skilled microscopist. The light microscope has capabilities beyond the range of any other analytical instrument used in the analysis of airborne particles, but it also requires a very skilled analyst with years of industrial experience or intimate familiarity with the particle types in a given airshed.

If you are contracting for this service you should be paying for a skilled analyst. The following progression should be evident in a quantitative analysis: a qualitative analysis, the establisment of optical categories, sample preparation for quantitative analysis, quantitative analysis with particle count, size, density and percentage by weight calculation for all categories, quality assurance data, and finally, discussion of sources.

If only an estimate of source impact is requested you are paying for opinion that is only as valuable as the microscopist is skilled. Skill can not be determined by years of exposure to a microscope. Photographic documentation with an explanation of the assumptions underlying the estimate should help indicate the knowledge of the analyst. The three papers listed below contain additional information on the quantitative analysis of airborne particles.

1. Crutcher, E.R. and L.S. Nishimura (1978), "Estimation of Error in Quantitative Microscopic Analysis of Compounds", Proceedings of the Fourth International Symposium on Contamination Control, Washington, D.C., pp. 150-158.

2. Crutcher, E.R. (1979), "A Standardized Approach to Airborne Particulate Analysis Using Quantitative Light Microscopy", Proceedings of the Institute of Environmental Sciences, Seattle, Washington.

3. Fitzsimmons, D. and E.R. Crutcher (1979), "A Formalized Mathematical Evaluation of Error in Quantitative Optical Microscropy", Proceedings of the Institute of Environmental Sciences, Seattle, Washington.

All of these references are available through the Institute of Environmental Sciences, 940 East Northwest Highway, Mt. Prospect, Il. 60056.

Appendix

When a beam of light strikes a material, be it gas, liquid or solid, it interacts with the electrons that occupy the volume filled by that material. The interaction takes two forms, either energy is absorbed by the electrons from the light (heating) or the electron is accelerated by the beam and becomes an emitter (diffractor) of light with the same frequency. Only a small fraction of the incident beam actually interacts with the electrons in the material but the change in the total beam provides detailed information on the distribution of electrons, their bonding energies, and their number per unit volume of material.

All chemical compounds are uniquely characterized by their refractive indices. The ability to uniquely identify all compounds is limited by the ability to measure their refractive indices. In order to fully utilize optical microscopy and understand its limitations it is essential that the concept of the refractive index be understood.

The Concept of Light in Light Microscopy

Light is a three-dimensional transverse electron-magnetic wave with wavelengths between 280 and 1200 nanometers. If fluorescent and infrared microscopy are excluded, the active band of wavelengths is reduced to the range between 380 and 700 nanometers. As with any wave it has a velocity, c, a wavelength, , and a frequency, f, with the following relationship:

$$c = f\lambda$$

The speed of light is a fixed value at 3×10^{10} centimeters per second but it also has a phase velocity that may be more or less than that value. The phase velocity is the speed of a specific part of the wave which may be increased or decreased by interference effects. The significance of phase velocity will become apparent in the discussion of matter-light interactions. A light wave consists of a traveling electrical field in one plane with a magnetic field in a plane normal to the first plane. Both planes are parallel to the traveling direction of the wave. Both field amplitude vectors are at 90 degrees to the traveling direction of the wave. The electric field direction carries all of the force in the wave and can be modeled by an electric field vector whose amplitude varies sinusoidally. In all of the models used to discuss light it will be the electrical vector that will underlie the model. The source of these electrical field waves is accelerating, oscillating electrons.

The Concept of Matter in Light Microscopy

Matter for the most part is empty space with number of small electrons orbiting insignificantly smaller atomic or molecular nuclei. Every system has a specific number of electrons within predetermined orbital bands. Within these bands are orbits of varying energy levels. The electrons closest to the center are strongly fixed and have natural vibrational frequencies that are very high, far higher than those of light. In some systems the electrons are very loosely held. The outermost electrons may even form a cloud with individual electrons free to move from one nucleus to another with little effort. These are the metals or highly conductive materials. Matter is bounded by a surface or a concentration gradient that is different than the bulk material in terms of the distribution of electrons. The interface between two different materials is very useful in comparing the electron densities and energies of the two materials, the particle and the media in which it is mounted.

Light-Matter Interaction and Refractive index

The index of refraction can be expressed in a variety of ways, each adding to the understanding of this basic property of matter and electro-magnetic energy. One of the simplest equations relates the dielectric constant, ε, the magnetic permeability, μ, and the refractive index, n, of a material:

$$n = \sqrt{\varepsilon\mu}$$

The most commonly used equation is:

$$n = c/v$$

Where c = speed of the wave in vacuum
v = phase velocity of the wave in the material

For the purpose of understanding the optical properties of matter in terms of the refractive index, the following equation is useful:

$$n^2 = 1 + (4\pi N q^2 / M)\,[1 / (f_o^2 - f^2)]$$

where N = number of outer electrons per unit volume (resonances per molecule per cc)
q = charge on the electron
M = mass of the electron
f_o = electron harmonic or natural resonance frequency
f = frequency of light used

This equation is a simplification based on the assumption that the frequency of light used is far from any important resonance frequency. From this equation it can be seen that any increase in the number of outer shell electrons, assuming the number of molecules per cubic centimeter is constant, will increase the refractive index. As the atomic number of the elements in a compound increases so does the refractive index for the compound.

Another relationship that can be seen from the refractive index equation is the relationship between the wavelength of light used and the refractive index. Most colorless objects have the resonance frequencies in the ultra violet (1 x 10 cycles per second) end of the spectrum. If blue light (6.6 x 10 cycles per second) is used to determine the refractive index, the term $(f_o^2 - f^2)$ is smaller than if red light (5.0 x 10 cycles per second) is used.

If the $(f_o^2 - f^2)$ term is made smaller, then the refractive index becomes larger. This property is known as dispersion. All colorless compounds have higher refractive indices in the blue end of the spectrum than in the red end.

The refractive index is a complex number consisting of both a refracting coefficient, the real term, and an absorption coefficient, the imaginary term. Both can be measured using optical microscopy but generally only the real term is quantified. The variability of these terms with wavelength produces colored particles if the absorption frequency is in the visible band of light. Color is therefore a special topic under refractive index.

As a final note, the terms "N" and "f_o" in the above equation are not the same in all directions through most materials. This is the origin of anisotropy. For more information on this subject, consult a text on optical crystallography.

FORMULATION AND APPLICATION OF A HYBRID SOURCE RECEPTOR MODEL

Dr. Robert J. Yamartino
Environmental Research & Technology, Inc.
696 Virginia Road
Concord, Massachusetts

An approach, developed initially to solve for the set of area-source strengths, Q_j, that would yield the minimum residual of observed and modeled concentrations, is extended to include buoyant points sources and is applied to hourly SO_2 concentrations measured at a 31 station network. In the original formulation (Yamartino and Lamich, 1979) the residual minimization condition led to a set of equations $A_{ij}Q_j = B_i$, where

$$A_{ij} \equiv R_{ikt}R_{jkt}/(\Delta C_{kt})^2, \quad B_i \equiv C_{kt}R_{ikt}/(\Delta C_{kt})^2$$

C_{kt} is the concentration at station k at time t, ΔC_{kt} is the corresponding measurement error, and R_{jkt} is the modeled coupling coefficient between source j and receptor k at time t. The equation set, solved with a non-negative least squares algorithm, produced $Q_j \geq 0$ that could be associated with fixed locations in x-y space. For buoyant point sources, the Q_j cannot be associated with fixed locations in 3-d space because the plume rise, and hence the effective source location, is a function of source characteristics (e.g., heat output) and meteorology (e.g., wind speed). The approach taken was to input the known major point source locations and stack parameters and to compute the coupling coefficients based on this information, as traditionally done in source-based modeling. The receptor model equations could then be solved as before. The region modeled was 30 km x 30 km. Stack parameters were input for 27 known major point sources and each 3 km x 3 km region was considered a potential area source. The resulting set of 127 equations was rather expensive to solve, so the area sources were assigned to land-use categories: rural, suburban, and urban. This reduced the number of unknown area-source strengths from 100 to 3 without sacrificing the basic 9 km^2 spatial resolution, greatly reduced program execution time, and greatly improved the stability and reasonableness of the results.

FORMULATION AND APPLICATION OF A HYBRID SOURCE RECEPTOR MODEL

Introduction

Most of the developments in receptor modeling have been motivated[1] by efforts to relate chemical analyses of aerosols at receptors to the chemical signatures associated with specific source emissions. The essential aspect of these receptor models is the multispecies nature of the problem, and dispersion meteorology considerations have received cursory, if any, attention. The approach to be described in this paper grew out of a desire to analyze single-species, gaseous concentrations observed at receptor networks. Objectives of such an analysis include:

- an independent means of checking source inventories and determining source culpabilities
- the ability to locate inadvertent leaks of hazardous effluents into the atmosphere, and
- as an enforcement tool for air pollution regulatory agencies.

The single species nature of this problem allows primary attention to be focused on inclusion of the varying source-receptor coupling coefficient, though the multi-species dimension of the problem is easily reinserted into the formalism.

The "hybrid" nature of the approach actually is forced upon the formalism because

a) even in the case of purely ground level sources, relative source-receptor coordinates enter into the problem, and

b) in the case of buoyant point sources, source related variables (e.g. stack height, heat output) in concert with meteorological variables (e.g. stability, wind speed), affect where in space the receptor "sees" the source.

However, to a large extent the formalism can be easily reworked on a case-by-case basis so that "hybrid" can indicate a situation where one inputs a variety of known quantities into the formalism in order to determine another set of unknowns.

In this paper the approach is applied to SO_2 concentration and meteorological data from West Berlin, Germany in order to address the questions:

a) what fraction of observed concentration variances are explained by sources within W. Berlin, and

b) are predicted W. Berlin source emission strengths consistent with the known emissions inventory?

Formalism Development

Air quality dispersion models are most often concerned with the determination of pollutant concentrations at a receptor given known source strengths and locations. The concentration, C_{kt}, observed at the kth receptor during the tth time interval can be modeled as

$$C'_{kt} = \sum_j R_{jkt} Q_j \quad , \tag{1}$$

where C′ denote modeled concentrations, Q_j is the time-independent strength of the jth source and R_{jkt} is the transport coupling coefficient between the jth source and k^{th} receptor for the meteorological conditions existent during the tth time interval. For simple, short range transport cases, this coupling coefficient might be computed from the time-independent, Gaussian plume formalism as:

$$R_{jkt} = \frac{1}{\pi u \sigma_y \sigma_z} \exp\left[\frac{-1}{2}(y_j - y_k)^2/\sigma_y^2\right] \exp\left[\frac{-1}{2}(z_j - z_k)^2/\sigma_z^2\right], \tag{1a}$$

where $\sigma_y = \sigma_y(|x_j - x_k|)$, $\sigma_z = \sigma_z(|x_j - x_k|)$ and x/y are measures in the downwind/crosswind coordinate frame. In more complex situations, the R_{jkt} might be computed by a sophisticated numerical dispersion model.

The solution of the simple inverse problem $Q_j = \sum_k R^{-1}_{jkt} C_{kt}$ using measured concentrations, is not of particular interest as a unique solution might exist only if the number of receptors, K, equalled the number of source candidates, J. A more reasonable problem would consist of determining the set of Q_j values leading to the minimization of the quantity χ^2, defined as

$$\chi^2 = \sum_{t,k} \frac{\left(C_{kt} - \sum_{i=1}^{J} R_{ikt} Q_i\right)^2}{(\Delta C_{kt})^2} \tag{2}$$

where ΔC_{kt} is the uncertainty in C_{kt}, i is a dummy index, and a summation over both t and k is required. Choice of ΔC_{kt} that reflect all sources of uncertainty, including fluctuations from the ensemble mean predicted by the model, are necessary if the χ^2 is to be appropriately normalized for subsequent evaluation in the statistical terms of confidence level and chisquare per degree of freedom. However, this degree of knowledge concerning ΔC_{kt} is seldom available. In our applications we have used a ΔC_{kt} that has reflected measurement errors only (i.e., either a constant ΔC_{kt} or a value proportional to observed concentration, as $\Delta C_{kt} = 0.1\ C_{kt}$). In such cases the normalization serves a useful relative weighting role, enabling one to discuss results in terms of fractional variance explained, but one is clearly barred from attaching confidence levels to such results.

Minimization of equation (2) will generally lead to a unique set of J source strengths provided the number of measurements M (nominally KT, where T is the number of time periods of data available) exceeds J, and secondly, that the sources Q_j of interest actually couple to the data in hand (i.e., $R_{jkt} = 0$ for all j and some k,t). Given these conditions, one may write down the set of J equations generated by the relations $\partial\chi^2/\partial Q_j = 0$. These equations are of the form

$$\sum_j A_{ij}\, Q_j = B_i \tag{3}$$

where

$$A_{ij} \equiv \sum_{t,k} \frac{R_{ikt}\, R_{jkt}}{(\Delta C_{kt})^2} \tag{4a}$$

and

$$B_i \equiv \sum_{t,k} \frac{C_{kt}\, R_{ikt}}{(\Delta C_{kt})^2} \tag{4b}$$

and upon obtaining the inverse A^{-1} of the positive definite matrix A, one arrives at the solution

$$Q_j = \sum_i A^{-1}_{ji}\, B_i \tag{5}$$

This approach, with no constraint on the Q_j, is quite acceptable provided the number of candidate source locations J is small and source candidate locations are well separated; however, if nothing is known about the source locations and, instead, an array of point or area sources is conjectured to exist on an X-Y grid of size n x m, then one must content with a matrix A of order (n x m) by (n x m), which, in turn demands a minimum of n x m concentration measurements. In addition, the solution Q_j may be quite unstable, with adjacent grid point Q_j values oscillating wildly positive and negative in a region known to be source free. Fortunately the availability of a non-negative least squares algorithm, developed by Lawson and Hanson[2], enables one to solve equation (3) subject to the physically reasonable requirement that all source strengths Q_j be positive (i.e, $Q_j \geq 0$).

It should be stressed that the Q_j are assumed to be time-independent. Thus, one must select time periods for which this assumption can be assumed to be reasonable or else assume that the Q_j represent maximal values and attempt to consider the effect of reduced emissions as an additional known fractional coefficient in the R_{jkt} terms.

In the case of multiple species where the fraction f_{sj} defines the mass fraction of species s emissions from source j, the problem can either be solved completely separately for each species s, yielding the quantities $f_{sj}Q_j$, or, if the f_{sj} are known, all species can be folded into the computation of a single χ^2. In this latter case, equations (1),(2), and (4) become

$$C_{skt} = \sum_j R_{jkt} f_{sj} Q_j \tag{1'}$$

$$\chi^2 = \sum_{s,k,t} \frac{\left(C_{skt} - \sum_{i=1}^{J} R_{ikt} f_{si} Q_i\right)^2}{(\Delta C_{skt})^2} \tag{2'}$$

$$A_{ij} = \sum_{s,k,t} \frac{R_{ikt} f_{si} R_{jkt} f_{sj}}{(\Delta C_{skt})^2} \tag{4a'}$$

$$B_i = \sum_{s,k,t} \frac{C_{skt} R_{ikt} f_{si}}{(\Delta C_{skt})^2} \tag{4b'}$$

while equations (3) and (5) remain unchanged.

As described above, the f_{sj} are envisioned as time-independent constants; however, complications such as gas-to-particle conversion could be built into the f_{sj} by making them functions of travel time or distance. This would, at worst, result in the f_{sj} becoming f_{sjkt}, but the normalization condition would remain simply $\sum_s f_{sjkt} = 1$, and equations (1')-(4') would remain unchanged.

Proper treatment of the measurement uncertainties ΔC_{skt} becomes even more important in multi-species applications as different species may have relative errors, $\Delta C_{skt}/C_{skt}$, differing by orders of magnitude. Failure to include the proper errors will yield a χ^2 that is truly a sum of "apples and oranges".

Isotropic/Anisotropic Separation

In the previous section the C_{kt} under consideration represented the total concentration measured at a given receptor k during time period t. Ideally one would prefer that the C_{kt} were background subtracted concentrations, but accurate extraction of the background is seldom possible. One is then led to consider the separation of the network measured concentrations into an isotropic or spatially averaged concentration

$$C_t^I = \frac{1}{K} \sum_k C_{kt}$$

and the anisotropic concentrations

$$C_{kt}^A = C_{kt} - C_t^I$$

Accompanying this separation must of course be a similar decomposition of the coupling coefficients R_{jkt} into

$$R^I_{jt} = \frac{1}{K} \sum_k R_{jkt}$$

and

$$R^A_{jkt} \equiv R_{jkt} - R^I_{jt}$$

The problem as defined by equation (2) then can be re-expressed as the two problems

$$\chi^2_A = \sum_{t,k} (C^A_{kt} - \sum_j Q_j R^A_{jkt})^2 / (\Delta C_{kt})^2$$

and

$$\chi^2_I = \sum_t (C^I_t - \sum_j Q_j R^I_{jt})^2 / (\Delta C_t)^2$$

where

$$(\Delta C_t)^2 = \frac{1}{K} \sum_k (\Delta C_{kt})^2$$

each of which may be explored in exactly the same manner as equation (2).

The utility of this isotropic/anisotropic separation becomes evident when one considers the behavior of the attendant sensitivity function, which arises from consideration of the expansion of χ^2 about its minimum value χ^2_{min}. To lowest order one obtains

$$\delta\chi^2 \equiv \chi^2 - \chi^2_{min} \simeq \sum_{i,j} A_{ij}\, \delta Q_i\, \delta Q_j$$

which, in the case of a single source Q_j, reduces to

$$\delta\chi^2 \equiv A_{jj} (\delta Q_J)^2 \quad .$$

Using the conventional definition of significance, $\delta\chi^2 = 1$, one may define the sensitivity or reciprocal source error as

$$S = \frac{1}{\delta Q_j} = \sqrt{A_{jj}} \quad .$$

As S grows with the number of measurements M it is convenient to define the normalized sensitivity

$$\hat{S} = S/\sqrt{M} \quad ,$$

which is a function only of the network geometry and mean dispersion conditions. Figure 1 shows a plot of these normalized sensitivities*, for 5 receptors equally spaced along the circumference of a circle of radius R, as a function of distance r from the center of the network. All three sensitivity functions show a rather mild variation with r for $r<R$ followed by rather rapid drop-offs for $r>R$; however, the attractive property of the anistropic component is its relatively high sensitivity to sources inside the network radius R, followed by a sensitivity for $r>R$ which falls more rapidly (e.g., in this case $1/r^3$ vrs $1/r^2$) than for either the total or isotropic components. This behavior of the anisotropic component, coupled with its insensitivity to background (i.e., isotropic) concentrations, make it a reasonable formulation for many applications.

The preceding consideration of the network sensitivity also leads to some rather practical results. The fact that the sensitivity of a network increases only as $\sim \sqrt{M} \sim \sqrt{KT}$ provides one with a tool for optimizing the sensitivity of an experiment subject to the cost constraints associated with the number of receptors and duration of the experiment. In addition, a related analysis indicates that the spatial resolution of a network increases only as $(KT)^{1/4}$ so that if network geometry is not optimized initially for the desired resolution, one is not likely to acquire it easily through increased statistics.

Model Application

Data Base

In a previous application of this approach to the determination of ground-level CO sources at an airbase, Yamartino and Lamich[3] found an excellent match between the solution and the known major CO emission sites when a full year of data from five, on-base, hourly concentration monitors was utilized. In this application, half-hour average SO_2 concentrations from a 31-station network in West Berlin, Germany are considered. As seen in Figure 2, the monitors are quite regularly spaced (about 4km apart) and provide nearly complete coverage of the 30km x 30km computational region. In addition, the coordinates (plus stack heights and heat outputs) of the city's 27 major point sources, emitting in excess of 1.5 gm/sec, were input in order to determine their emission strengths on the basis of the SO_2 concentration observations.

Meteorological data were obtained from the monitoring station operated by the Free University of Berlin and consisted of wind speeds and directions at 10 m and 60 m elevation, surface temperature, relative humidity, and stability class. Mixing height data was not available.

*Additional assumptions for these computations include use of a bivariate Gaussian form for R_{ikt} where $2\sigma_z = \sigma_y = r \sin(\sigma_\theta)$, $\sigma_\theta = 15°$, and a uniform distribution of wind directions.

The initial test runs considered only data obtained during August 1975. The further demand that at least 20 of the 31 SO_2 monitors be functioning reduced the sample to 178 hours (5285 receptor-hours) during the 8-hour nighttime period and 372 hours (11085 receptor-hours) during the 16-hour daytime period).

Computation of R_{jkt}

The coupling coefficients were computed using a segmented, Gaussian plume type model similar to that expressed by equation (1a). The plume trajectory was allowed to change direction each half-hour and was followed backward in time, in $\sim$ 5 minute time steps for up to six hours, in order to find the emission time that was currently impacting the receptor. Wind speed at emission height and stability class at the time of emission were then used to compute initial plume dilution and buoyant rise (for point sources only). Dispersion rates and transport speed along the trajectory were also permitted to change each half-hour based on changes in stability class and wind speed respectively.

Area source candidates were obtained by subdividing the 30km x 30km region into 3km x 3km squares (i.e., Δx = 3km), yielding 100 possible candidates. The coupling coefficients were computed using the pseudo-upwind point source approximation with initial dispersion coefficients of $\sigma_y = \Delta x/2.4$ and σ_z = 10m.

In order to focus computational effort on the point sources, a land-use map of Berlin was employed to assign the area source candidates to three categories; rural (including forests and lakes), suburban, and innercity. This procedure, introduced by Prahm et al[4], reduced the number of unknown area source strengths from 100 to only 3, without sacrificing the basic 3km x 3km resolution of the initial area source grid.

Model Results

The methodology, expressed by equations (2)-(5), was applied to the observed total, anisotropic, and isotropic concentration fields. Computed quantities included the three area source and 27 point source emission strength estimates, their associated errors, and initial (i.e., with all $Q_j = 0$) and final χ^2 values. Area source strength predictions were found to be in quantitative agreement with estimated summer values of rural, suburban, and innercity emissions. Estimated point source emissions using the nighttime data spanned the known inventory range of 1.5 gm/s to $\sim$ 600gm/s but there was poor correlation between predictions and inventory estimates on a stack-by-stack basis. This shortcoming is attributed to low coupling of nighttime elevated plumes to ground-level receptors and to the fact that computed sensitivities (i.e., $\sqrt{A_{jj}}$) to nighttime point source emissions were overshadowed by area source sensitivities by factors of 10 to 500. For the daytime data this sensitivity ratio dropped four-fold, yielding total estimated point source emissions within 25% of inventory values; however, comparison on a stack-by-stack basis continued to show poor correlation Statistical properties of these preliminary runs suggest that the variance is most reduced by the anisotropic method. This is as expected since the assumed sources all lie within the receptor network and the anisotropic method has the highest sensitivity ratio for sources inside versus outside the receptor region. This also corresponds with the theoretical results shown in Figure 1.

Conclusions

In the previous application(3) to a two-dimensional (i.e. x and y) array of candidate, ground-level, area sources, the solution of the matrix problem $\sum_j A_{ij} Q_j = B_i$, subject to the constraint $Q_j > 0$, led to a set of source strengths which agreed in magnitude and spatial location with the known CO emissons inventory. In the three-dimensional application (i.e., elevated point sources also included) considered in this paper, preliminary results indicate continued good resolution of ground-level area sources but rather poor determination of point source strengths. While better results are found for daytime conditions because of the higher point-source/area-source coupling ratio to ground-level receptors, even more promising results for point sources might be obtained if one were to:

- consider using elevated receptors (to avoid domination by ground level area sources),
- weight more heavily the data from ground-level receptors outside of regions dominated by ground-level sources, or
- choose species emitted exclusively by point souces.

Aside from some rather obvious areas for improvement, such as using a more sophisticated dispersion model for improved R_{jkt} determination, trying newer non-negative least squares algorithms with improved error analyses, and a more thorough pre-investigation of a network's sensitivity and resolution functions, significant increases in predictive power may be achieved through the consideration of a multi-species data base.

Acknowledgement

This work was supported by internal R&D funds from ERT Inc. and IVU/GEOMET, West Berlin, Germany. I also wish to thank Rainer Stern of the Free University of Berlin for his assistance in using and interpreting the West Berlin monitoring data.

References

1. Watson, J.G., 1981. "The state of the art of receptor models relating ambient suspended particulate matter to sources". OAQPS Report EPA-600/2-81-039.

2. Lawson, C.L. and R.J. Hanson, 1974. Solving Least Squares Problems, Prentice-Hall Inc., Englewood Cliffs, N.J.

3. Yamartino, R.J. and D.L. Lamich, 1979. "The formulation and application of a source finding algorithm". Fourth Symposium on Turbulence, Diffusion and Air Pollution, Jan. 15-18, Reno, NV. Published by the American Meteorological Society, Boston, Mass.

4. Prahm, L.P., K. Conradsen, and L.B. Nielsen. "Regional source quantification model for sulphur oxides in Europe". Atmospheric Environment, 14, 1027-1054.

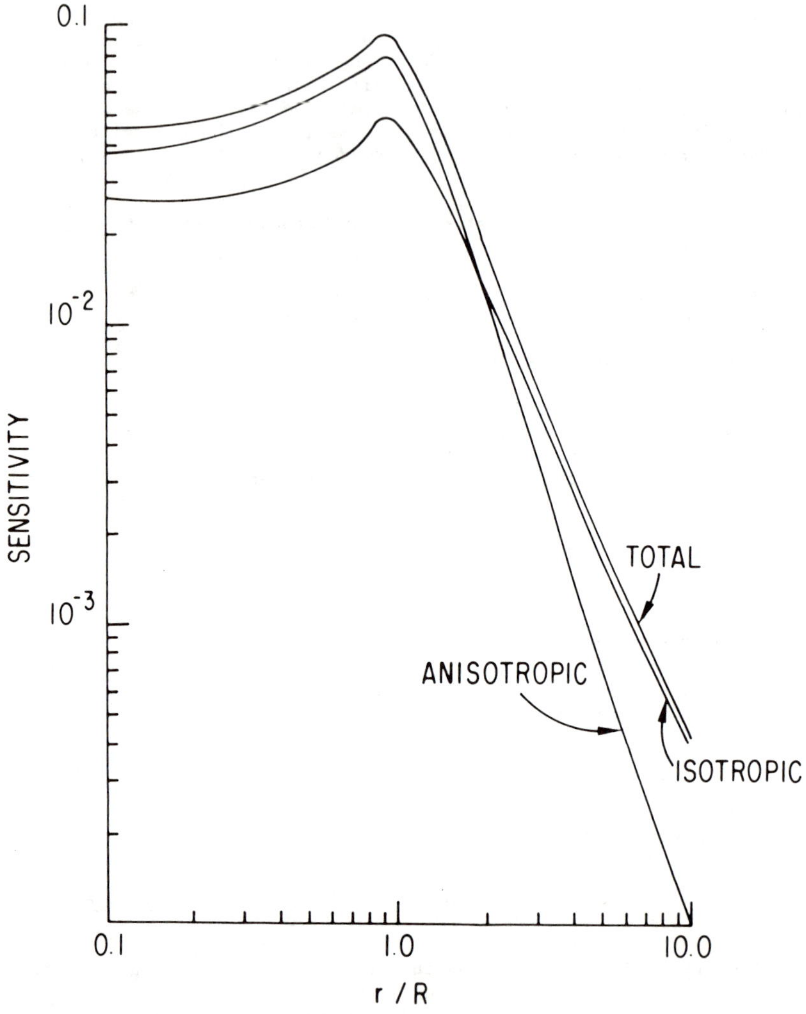

Figure 1 Normalized Sensitivities, S, as a Function of Distance from the Center of a Circular Network of Receptors. (Additional Details are Given in the Text)

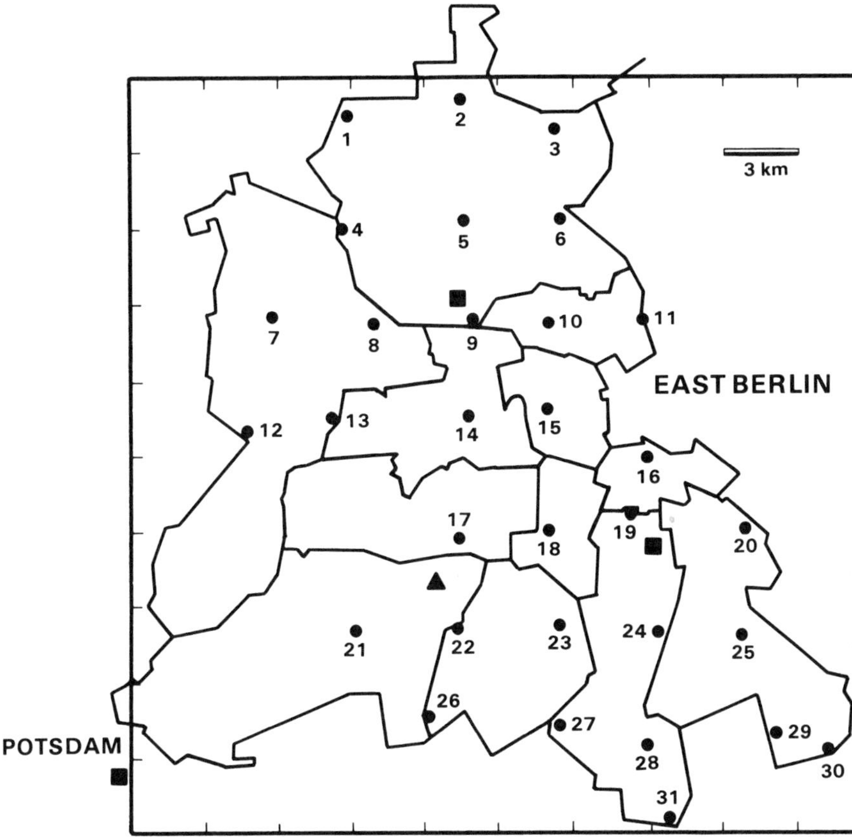

● SO_2 MONITOR
▲ METEOROLOGICAL STATION
■ AIRPORT METEOROLOGICAL STATIONS (Not used in this study)

Figure 2 West Berlin SO_2 Monitoring Network. Boundaries Indicate Political Subdivisions Within West Berlin

8210047

RECEPTOR-ORIENTED SOURCE-RECEPTOR ANALYSIS

Douglas W. Cooper
Environmental Health Sciences Department
Harvard School of Public Health
665 Huntington Avenue
Boston, MA 02115

Our approach is to view the receptor as having a response to sources that is a function of source location and averaging time. Given information about meteorological parameters that control transport, one can determine which sources could contribute significantly to exposure and can thus estimate their relative contributions. This should be quite useful in making siting decisions and in other aspects of control strategy. These response function coefficients can also be used to infer source strengths, given information about concentrations and species found at the receptor. A five-year (1967-1971) joint frequency distribution of wind direction, wind speed, and atmospheric stability class for Boston was used to determine the response characteristics for sites having this meteorology. The response function is the concentration per unit emission rate, as a function of distance and direction from the receptor. The computations were simplified by using the "forward-backward" method, taking advantage of the symmetry of the Gaussian dispersion model. The radial dependence of the response function of a receptor in Boston for this five-year period was found to be inversely proportional to distance to the 1.43 power, meaning that doubling the source-receptor separation would decrease the source impact by a factor of 2.7. The directional dependence was such that the response to locations in the most sensitive direction (WNW) was three times that to the response in the least sensitive direction (SSE). The transport relationships were used to form a set of linear equations, the solutions to which give source strengths, by location, in the vicinity of the sampler. Examples of applying this technique are presented. This approach should be especially useful in augmenting chemical element balance techniques in those instances (such as soil sources and coal fly ash sources) where the chemical element make-ups of different sources are quite similar.

RECEPTOR-ORIENTED SOURCE-RECEPTOR ANALYSIS

Introduction

In the past, the major focus of attention regarding the relationships between air pollution sources and air pollution receptors was often the calculation of concentrations and doses as functions of position (x,y,z) and time (t) for various sources of emissions. Currently, attention has shifted to trying to use measurements made at receptor sites to infer the sources of the exposure, using knowledge of the chemical composition of the exposure (as determined by analyses of sampled material) and of the emissions. Knowledge of the relative contributions of the sources would be valuable in formulating optimal control policies and in conducting epidemiological studies to relate disease to source types and emission rates.

Gordon (1980) discussed source-receptor models at length, and he offered this analysis of the shortcomings of these models as currently used: "For many elements, the composition patterns of soil and particles emitted from coal-fired plants are so similar that it is difficult to resolve the contributions." "The compositions of components are not well enough known." "Even if one has very accurate composition data for in-stack particles, they may not reflect the composition of airborne particles from the source at some distance away. In-stack measurements will include neither fugitive emissions from the source nor, in high-temperature sources, all of the 'condensibles' that may initially be in the vapor state but will later condense." "While some of the [chemical element balances] appear to provide a reasonably accurate idea of the contributions to TSP from major *classes* of sources, most are not definitive enough to identify contributions of individual sources within a class." "In its normal form, factor analysis has several weaknesses which limit its usefulness: Because of the extensive normalization of raw data, one cannot extract the concentrations of species of a component. The results indicate only the fraction of a species' variation that is explained by the component. Since the method operates on variations rather than absolute values, it is not effective when applied to data sets that have little variation in time or space. If a very strong, but almost constant component were present among other weaker, but more variable components, the former might be undetected by factor analysis. Factor analysis is apparently not able to resolve components that have similar concentrations for most elements." In some limited cases, the meteorological history of a site and knowledge of the source types and emission rates have been used to determine trajectories of pollutant types to aid in identifying the major contributors to dose (see Gordon, 1980).

As noted by Yocum et al. (1981): "An analytic method which identifies the physical and chemical properties of a sample of suspended particulate matter on a particle-by-particle basis is an extremely powerful tool for helping to understand the ancestry of particulate matter. However, such information cannot specify the route followed by each particle from its primary source to the sampler. A particle may be identified as iron oxide from a metallurgical process, but such an analysis does not tell us if it came from an electric furnace, basic oxygen furnace, or foundry cupola. Neither does the analysis indicate if the particle arrived directly from the primary process of generation, or if it had settled out and been reentrained one or more times before collection. In developing control strategies for traditional and nontraditional particulate matter, it is important to know, not only the character of the particles, but also the routes followed to the sampler."

The new approach we present here is to view the receptor as having a response to the sources that is a function of source position and of time. The receptor will experience a concentration increment from each source, which increment depends upon the location of the source and the conditions affecting transport. Given information about transport conditions, one can determine which sources could contribute significantly to exposure at the receptor and estimate their relative contributions. The value of this information for formulating control strategies is evident. The response function coefficients could also be used to infer source strengths, given information about the concentrations at the receptor. We will show that these response coefficients can be obtained relatively conveniently from meteorological and topographical information. Further, the receptor-oriented perspective is revealing in itself.

Theoretical Bases

In general, one is trying to determine the response coefficients a_{ij} which link the measured concentrations C_i of pollutant type i with the rates of emissions x_j from sources j. Under certain assumptions, the relationship between C_i and x_j has the following linear form:

$$C_i = \Sigma a_{ij} x_j \tag{1}$$

in which the a_{ij} coefficients are transfer coefficients or (the term we shall use) response coefficients. Almost all receptor-oriented analyses at present use information about C_i and x_j to infer the a_{ij}. (See the review articles by Gordon, 1980 and by Cooper and Watson, 1980.) In the past, much use was made of dispersion models; these give a_{ij} by solving for the transport of the material from each source due to atmospheric dispersion. For each major source in a region, the pattern of its contributions to pollution was calculated, and then these contributions were summed to form an overall pattern. If many sources were modeled, one could approximate the concentration expected at site i using this approach, calculating the concentration contributions for each source separately, then adding them together. This focused on where the pollution was going rather than on where it was coming from.

In what follows, we will obtain the a_{ij} coefficients by starting from the receptor rather than from the sources, using information about the conditions that affect transport. This helps identify the causes of the pollution experienced at a receptor site.

Gaussian Dispersion: Ground Level Sources and Receptors

The Gaussian dispersion model for the concentration (g/m^3) at ground level $\chi(x,0,0)$ due to a source at ground level having source strength (mass emission rate, g/s) of $Q(0,0,0)$ is:

$$\chi = Q(\pi \sigma_y \sigma_z u)^{-1} exp[-y^2/2\sigma_y^2] \quad (2)$$

in which σ_y is the horizontal dispersion parameter; σ_z is the vertical dispersion parameter (m); u is the mean wind speed (m/s); and y is the horizontal distance from the mean wind direction (plume axis) (m). This equation assumes "reflection," rather than absorption, of the concentration at the ground surface (accomplished in the analysis by using a virtual source as far below ground as the real source is above ground).

If the positions of the source and the receptor were interchanged and the wind velocity reversed, the computed concentration (and the real concentration) would be the same.

The group χ/Q gives the concentration at the receptor per unit source strength and is the response function a_{ij} for this situation. Presentation of the χ/Q relationship is simplified by using $\chi u/Q$. Figure 1 (IAEA,1969) shows values of this group for various types of weather (stability classes A through F) versus distance from the source. Due to symmetry, these are also the response coefficients versus distance from the receptor to the sources.

Over the range of x from 10 to 10^4 meters, the approximation that $\chi u/Q$ is 0.1 x^{-2} (for x measured in meters) comes within a factor of three of the values for C and D stabilities shown in the figure. Thus the response of the receptor to ground-level sources decreases approximately as the inverse square of their distances from the receptor, for any single stability class. (A source ten times further away in the same direction as a second source would be expected to contribute to the concentration experienced by the receptor only one-hundredth as much as the first source per unit emission rate.) The relative importance of quite local sources suggests itself immediately.

Though the response to individual sources decreases rapidly with distance from the receptor, the area to which the receptor responds increases almost as rapidly. A policy which reduced all sources in an area by a factor of two, for example, could have substantial effect even if instituted in locations so far away that the reduction due to eliminating a single source might well be negligible.

The inverse-square approximation for the relationship between response (χ/Q) and distance upwind arose from some simplifying assumptions about the dependence of the dispersion parameters on distance downwind. The more widely used analysis is that of Turner (1969), based on the work of Pasquill and of Gifford. Figure 2 (Turner, 1969) shows equal-concentration contours ("isopleths"), which are equivalent to equal-response contours, for a ground-level source and a ground-level receptor under the most common atmospheric conditions, stability category D. Figure 3 is for the next most common condition, stability category C. For D stability, the profiles for values of $\chi u/Q$ greater than 10^{-3},10^{-4},10^{-5} are seen to be limited to distances of 0.4, 1.2, and 5.2 km respectively. They are even closer to the receptor for C stability. Such figures could be overlaid on a map of sources (with their source strengths) for a quick first approximation of the relative importance for that receptor of local sources under such transport conditions. (The use of averaging times different from those assumed for the calculations underlying Figures 2 and 3 would change the profiles, however, so consideration must be given to matching the calculation's parameter values to the averaging time.)

Gaussian Dispersion: Elevated Sources

Turner (1969) summarized the approach we take ("forward-backward" or "reversed-plume"), as follows: "Sometimes, especially for multiple sources, it is convenient to consider the receptor as being the origin of the diffusion coordinate system. The source-receptor geometry can then be worked out merely by drawing or visualizing an x-axis oriented upwind from the receptor and determining the crosswind distances of each source in relation to this x-axis. As pointed out by Gifford (1959), the concentration at (0,0,0) from a source at (x,y,H) on a coordinate system with the x-axis oriented upwind is the same as the concentration at (x,y,0) from a source at (0,0,H) on a coordinate system with the x-axis downwind (Figure [4])."

Therefore, one could calculate a series of isopleths for a source located at the origin and having several values of height above ground level (H); these would constitute iso-response contours for the sampler at the origin and at ground level for sources having these several values of H. If one had to use as many values of H as there are sources, one would lose the advantage of this approach, so judicious selection of H values is needed.

Figure 5 (IAEA, 1968) shows ground-level values of $\chi u / Q$ for a source height of 10 m. (The figure uses "h" instead of "H".) This figure shows that, depending upon stability, sources closer than 10^2 to 10^3 m can be substantially less effective in contributing to the concentration at the receptor than if they too were at ground level. At 2000 (2 km), for D stability, the elevated source (H=10 m) has $\chi u / Q = 4.3 \times 10^{-5}$ and the same source on the ground would have a value of $\chi u / Q$ of just under 5×10^{-5}. Thus, ground-level iso-response contours tend to overstate the contribution of relatively high sources nearby (for x within 100 H or so) but should be useful approximations for sources which are not as elevated or farther away (in other words, for sources farther than about 100 H or lower than about $x/100$).

Values of $\chi u / Q$ may be obtained from a variety of computer programs or from direct or interpolated use of the figures given by Turner (1969) showing profiles for all six stability classes at various heights, including H=0.

Because we wanted to use the reversed-plume or forward-backward approach cited above by Turner, we tried it for a few conditions for the EPA computer program PTMTP (Khanna, 1976), which is designed to give hourly concentration values for up to 25 sources and up to 30 receptors. The conditions chosen were: Q=1000 g/s; U=5 m/s; stack gas temperature = 400°K and ambient temperature = 300°K; volume flow rate = 1 m^3/s; mixing height = 200 m (deliberately chosen to be atypically low.) For both source and receptor at ground level, the predicted concentrations were the same to two or three significant digits regardless of whether the 24 locations were assumed to have sources or receptors and whether the origin was assumed to have a source or a receptor. Similar agreement (two or three significant digits) was found for computer runs for situations with the source at (x,y,z)=(0,0,100 m) and receptors at (x,y,0) or for the sources at (x,y,100 m) and the receptor at (0,0,0) both for an unusually low mixing height (L=200 m) or an unusually high mixing height (L=3000 m). (The figures presented by Holzworth, 1972, show annual mean morning mixing heights in the contiguous United States to range between 300 and 900 m and annual mean afternoon mixing heights to range between about 800 and 2800 m. We were not sure that the method of handling of the mixing height by the computer program would preserve the symmetry of the approach, but it seemed to.)

Applications: Iso-Response Contours

In this section, we obtain an annual mean response function for a location, and study its use with information about sources or sampled material.

Seasonal weather patterns make the annual mean a useful measure for trying to interpret long-term relationships between the receptor and sources in its locale. One approach to obtaining such an annual mean would be to take the hourly records for an entire year, calculate the response functions, and average these over the entire year. The number of calculations could be reduced by sampling (perhaps stratified sampling) from the over eight thousand hourly sets for the year. Although the use of hourly data (or at least data taken several times per day) will be explored below, here we take a different approach, using the joint frequency distribution for the most important meteorological variables for transport.

The National Climatic Center in Ashville, NC has records of the joint probability distributions of stability, wind speed, and wind direction for many locations in the United States and for many years. The joint probability distribution is tabulated as F_{ijk}, where the first subscript runs from 1 to 6 and refers to stability classes A through F, the second runs from 1 to 6 and refers to wind speed categories, and the third runs from 1 to 16 and refers to 22.5° sectors for wind direction. The values are normalized so that

$$\sum_{i=1}^{6}\sum_{j=1}^{6}\sum_{k=1}^{16} F_{ijk}=1 \qquad (3)$$

Making some assumption about mixing height (such as choosing 1000 m), one can then obtain response functions by using the values for wind speed, wind direction, and stability in the dispersion computer codes. These response functions then can be added, with weights proportional to the joint frequency distribution values. The annual mean response function would then show annual mean response as a function of distance and as a function of direction (the sixteen sectors).

Table 1 gives the values for five-year (1967-1971) mean response function for Boston for the sixteen sector directions for the distances 0.5,1,2,4,8 km from the receptor. These were supplied through the courtesy of the Six Cities Study group here at HSPH.

Table 2 summarizes Table 1 in a set of sixteen regression equations for response function versus distance, obtained from least-squares regression of log (χ/Q) versus log$x$ where x is the distance (0.5,1,2,4,8 km). The coefficients of determination (R^2) were all 0.999, indicating the fraction of the variance of the log (χ/Q) values attributable to the regression equation. The regression equations are quite useful for interpolating between the various distances given in Table 1.

This ground-level five-year mean response function would be a first approximation to the response function of any site in Boston over this five-year period or perhaps any period of a comparable duration. Of course, this approximation has all the shortcomings of the simplified dispersion model from which it comes, aggravated by the aggregation over wind speeds and directions, aided by the degree to which the averaging process gives values for the mean which are less uncertain than any of the individual values.

The regression results show that the radial component of the spatial dependence can be well approximated by $r^{-1.43}$. Doubling the source-receptor distance would typically decrease the concentration contribution of the source by 2.7. This reduction is somewhat less than the factor of 1/4 that one would predict for any one stability class. (Our analysis averages over all stability classes.)

The radial and angular dependencies of the response function are summarized in Figure 6, a contour map of the χ/Q values (superimposed on the numerical values calculated). The directional dependence is such that the Boston receptor is most sensitive to sources to the WNW, due to the prevailing winds, for any given radial distance. From Table 1 it can be seen that a source located WNW of a receptor (at, say 1 km away) would produce three times the mean concentration contribution as a source located SSE from the receptor (1.327/0.449 = 3.0). Thus, for Bostonians, development involving new sources of air pollution have greater impact if located WNW of Boston than if located in other directions, other things being equal.

Another method besides the contour map for summarizing the response function of the receptor $f(r,\vartheta)=\chi(0,0,0)/Q(r,\vartheta)$ is to use the multiplicative ("separable") form:

$$f(r,\vartheta)=R(r)\Theta(\vartheta) \tag{4}$$

We have already found $R(r)\approx 1/r^{1.43}$ and can obtain $\Theta(\vartheta)$ at any value of r which is convenient. We choose r = 1 km and look for relations between ϑ and the values 10^{a} for the multiplicative model

$$f(r,\vartheta)=10^{a}/r^{b} \tag{5}$$

in which a is now a function of ϑ.

To find a pattern for the exponents a, one can take their Fourier transform, which indicates the spectrum of terms of the form $\cos(n\vartheta)$ and $\sin(n\vartheta)$ which approximates $\Theta(\vartheta)$. It is conceptually clearer to perform the analysis on their deviations from the mean. We performed this analysis, using a program for the "fast Fourier transform," and found that the function

$$f(\vartheta)=\frac{a_0}{2}+\sum_{n=1}^{7}(a_n\cos(n2\Pi N/16)+b_n\sin(n2\Pi N/16) \qquad (6)$$

should be a good approximation where for N = 1 to 16, the directions are S, SSW, ..., SE, SSE. Table 3 lists the values predicted by this equation for the sixteen directions (at r=1) versus the values of χ/Q (from Table 1) for these same positions.

Thus, one could use the function

$$f(r,\vartheta)=\chi/Q=(1/r^{1.43})(14.72/16+\Theta'(\vartheta))x\,10^{-2} \qquad (7)$$

where $\Theta'(\vartheta)=\sum_{n=1}^{7}[a_n\cos(n2\Pi N/16)+b_n\sin(n2\Pi N/16)]$ to estimate the response function anywhere in the domain for which it was determined $(0<r\leq 8km, 0\leq\vartheta\leq 2\Pi)$. Figure 7 shows eighty such determinations for r = 1 km.

Application: Transport Matrix Analysis

The chemical element balance (CEB) method infers the relative impact of an area's source types on the basis of the quantitative elemental analysis of ambient particulate samples. The elemental composition of a source type's emissions can be similarly determined by analysis or estimated by other means. Most of the important source types usually emit one or two elements in high quantity that are almost unique to that source, thus serving as its tracer. Knowing these two sets of information, the receptor concentration of each element and the relative elemental concentration for each source type, the relative importance of each source type, or source-strength, can be determined. This is represented mathematically as:

$$C_i=\sum_j b_{ij}m_j \qquad (8)$$

where:

b_{ij} = the concentration of the ith element from the jth source type,

C_i = the concentration of element i in the sample, and

m_j = the "strength" of source j (Gordon, 1980).

Factor analysis also requires elemental analysis of several filter samples at a receptor. Rather than associating absolute elemental concentrations with known sources, however, factor analysis tries to explain the variance observed for each element with a set of hypothetical factors. These statistically derived factors are later compared with the elemental compositions of known sources to explain which source types have contributed most to the sample. Factor analysis, however, requires no prior assumptions about important sources and it can disclose important "unexplained" factors.

The method we have developed, transport matrix analysis (TMA), identifies the relative impact of sources by location only. TMA takes advantages of air pollution dispersion modeling to identify the locations of major contributions to a collected sample. Mathematically similar to the other receptor models, TMA derives a receptor's meteorological and directional response (transfer coefficient) to sources:

$$C_i = \sum_j a_{ij} x_j \tag{9}$$

where:

C_i = measured concentration of pollutant (or at position i or time i) i

x_j = emission rate from source at location j

a_{ij} = response (transfer) coefficient for pollutant i from source at location j.

Receptor modeling is a useful tool for pollution control planning. These models identify the most important sources or source types and can be used to determine which sources to control preferentially. TMA explores the meteorological effects unique to an area, which could shed light on siting issues and control priorities. Additionally, TMA could potentially be used in conjunction with CEB to determine locations of the most important sources of a specific particulate type.

TMA Research Description

The preliminary application of TMA to small data sets has yielded mixed results. To date, we have investigated 2 cases, both total suspendable particulate (TSP) high-volume air sampler (hi-vol) sample sets collected in Watertown, Massachusetts as part of Harvard's Six Cities Study. The first set we analyzed was collected between January and October 1980, the second was collected from January to June 1979.

General Method

We used TMA to try to derive area source strengths around the Watertown sampler(s). The EPA dispersion model PTMTP was run to calculate a response coefficient matrix (a_{ij}) for each day used in the studies. Each matrix contains a response coefficient for each of 24 sectors arranged in a 20 km x 20 km square grid centered on the sampler(s) shown in Figure 8. These coefficients are the concentrations to be expected from 1 kg/s emission rate. For this purpose, source and receptor information had to be consistent for all PTMTP runs. We chose: source emission rate of 1000 g/s; volume flow rate of 0.39 m^3/s; stack gas temperature = 301°K, source height = 10 m, receptor height = 0 m. (These were chosen to give effective stack heights near 10 m.)

It would be preferable to use known stack heights for the sources, but this would involve much more data collection. The stack height uncertainty contributes to predicted concentration uncertainty in much the same way that uncertainty about distance from the plume centerline would. The directional data are in 360/16 = 22.5° increments; even a 10° error in direction would correspond to a 180-m error in source location at 1 km and correspondingly greater error at further distances, exceeding most plausible stack height errors caused by assuming a single stack height. The directional errors would tend not to introduce bias, but would increase uncertainty. To prevent bias, the stack height chosen should similarly be close to the weighted-mean stack height, if possible.

Other meteorological inputs were held constant since they have little effect on the model solutions: ambient temperature = 300°K; mixing height = 1000 m. For each day included in the studies, meteorological data measured at Logan Airport, Boston, were obtained from the National Climatic Center (NCC). Each 3-hour-average wind speed and wind direction reported by NCC was used as input, and one daytime and one nighttime stability class was summarized for each day. Thus we obtained the transport coefficients.

The transport coefficients for each sector were then entered as independent variables in a multiple regression analysis where the hi-vol sampler TSP concentration was the dependent variable (C_i). The resulting regression coefficients (x_j) are the source location strengths for each sector modeled.

Case 1 (January-October, 1980)

TSP concentrations have been measured every other day (approximately) at a single Watertown site. Thirty-two samples were chosen, 8 each from January, April, July and October, between days 1 and 18 of each month.

Once the daily response (transport) matrices were derived for the sample days, several different regression analyses were performed to try to explain the measured concentrations and estimate sector source location strengths. Two groups of 16 days each (Group A: days 11-18; Group B: days 2-10, 4 days from each month) were analyzed separately; in each case the sampler concentration was regressed on 8 source sector response coefficients. The sectors chosen formed a 10 x 10 km grid about the sampler as shown in Figure 9.

The transport coefficients for Group A yielded only moderately effective explanation of the observed concentrations, with a coefficient of determination of R^2 = 54%. The Group B data performed better with an R^2 = 75%. However, when data for the 2 groups were combined in regression, the degree of explanation was worse, R^2 = 29%.

The combined data (32 days) provided enough information to model a 20 x 20 km grid of 24 source locations about the receptor as shown in Figure 8. A favorable R^2 of 85% was obtained; however, almost half of the statistically significant regression coefficients (source strengths) obtained were negative. (Statistical significance was chosen as $|\text{t-ratio}| \geq 1$, where t-ratio = regression coefficient/standard error). The large number of negative, physically implausible coefficients is disturbing and not easily explained.

We investigated a directional explanation by modeling 8 radial sectors about the monitor, where the transport coefficient was as shown in Figure 10 representative of an entire wedge-shaped sector with a radius of 10 km. This model performed only marginally better than the 3 x 3 grid, with A and B data combined, R^2 = 34%.

There are several plausible reasons why TMA performed poorly with this data set:

1. Only a small number of days were modeled; random variation in sample data may have been significant.

2. Only a small grid of sources was modeled, and each source represented a large (5 x 5 km) area. Therefore, any real sources very near the monitor or far from modeled source points might not have received adequate representation.

Background

We hypothesized that the unexplained quantity of TSP concentration is a background effect which is independent of our modeled sources. If two TSP monitors were located near one another and operating identically and simultaneously, one could derive the source strengths from the differences between respective response coefficients and samplers. This calculation effectively removes the background factor for each day (B):

$$C_1 = \sum_j a_{1j} x_j + B \tag{10}$$

$$C_2 = \sum_j a_{2j} x_j + B \tag{11}$$

$$C_1 - C_2 = \sum_j (a_{1j} - a_{2j}) x_j \tag{12}$$

where:

C_1, C_2 = concentration at monitor 1 and 2

a_{1j}, a_{2j} = response coefficient for monitor 1 or 2 to source location j

x_j = source strength of location j

B = constant background factor

Case 2 (January-June, 1979)

Case 2 is based on 21 days from January to June, 1979 when 2 hi-vol TSP samplers were simultaneously operated, 0.6 km apart, in Watertown. Each sampler was analyzed separately first, and the difference analyzed later. The results for the individually treated set of sampler data were considerably better than those obtained for Case 1. A similar 3 x 3 grid analysis was performed, as well as a regression for 16 of the 24 sectors, arranged as shown in Figure 11 about the samplers.

Each of the 8 sectors not included (of the 24) was often highly correlated with one of the 8 sectors defining the 10 x 10 km grid about the receptors on the same radius from the sampler. Results are shown in Table 4. In this table, R^2 values are given for the cases indicated: the model based on sampler 1 ("Receptor 1") and either the 3x3 grid (Figure 9) or the 5x5 grid (Figure 8) or the model based on sampler 2 or the model based on the differences between the two samplers and their two transport matrices.

Figure 12 indicates calculated source strengths for sectors which gave significant and positive coefficients in the regression models for both samplers (note: the regression coefficients are the source location strengths as discussed above). For our purposes, significance is defined as coefficients with a t-ratio ≥ 1. Disturbingly, some of the calculated source strengths are negative and statistically significant by our criterion; however, in only one sector was a negative source term derived for both samplers. Three other sectors had consistently positive, but inconsistently significant source strengths.

Source resolution by analysis of the differences between monitors and respective response coefficients was disappointing. The 16-sector R^2 = 78% was reasonably high; however, a relatively large number (4) of significant, negative source strengths were derived. Of the 5 significant, positive sources resolved, only one coincided with those identified by analysis of the individual samplers. Plausible explanations for the apparent failure of the difference analysis are:

1. A systematic bias may exist between the hi-vol samplers used.

2. Too little data may have been used for accurate, stable analysis.

3. The grids may be too coarse to represent the true source distributions accurately.

4. The errors in dispersion modeling may have been too large for accurate source location determination.

5. Neglecting the grid in which the sampler is located may introduce excessive error, given the strong distance dependence of the transport relations.

6. It may be necessary to screen the data sets beforehand, to take advantage of periods when the wind shifts very little (Gordon, 1980).

7. The "background" contribution may well not be independent of direction, especially where gas-to-particle conversion takes place, and such conversion is not part of the model.

Discussion

Subsequent to the presentation of this paper, we learned that the transport matrix analysis approach had been published earlier by Yamartino and Lamich (1979), who used it successfully to identify sources of gaseous emissions at an airfield.

In general, methods which try to determine sources from measurements at receptors may obtain incorrect, sometimes obviously wrong, answers due to errors in the model, coarseness of the linear equations approximation to the model, and/or errors in the data. The prediction of concentrations from source strengths due to dispersion modeling is rarely achievable to better than a factor of two accuracy. The error in the linear equations approximation can be estimated from quadrature formulas given in numerical analysis texts; the errors typically decrease in proportion to the square of the number of intervals used in any dimension. We used few such intervals because of the need to have our limited data set contain more observations than grid locations estimated.

The errors of measurement for high-volume air samplers are likely to be much less than the errors in the dispersion model, but these data errors can be made less important by the collection of much larger data sets.

The shortcomings of these results are fairly evident. It would have been preferable to be able to compare the few positive and statistically significant emission rates with source inventories in the area, if our resources had allowed this. A pattern of agreement would have been encouraging but by no means definitive, however. A pattern of disagreement would have had to be explained by the same factors discussed above.

Conclusions

The receptor-oriented response function provides a convenient way to discriminate betwen sites for potential new pollution sources. The transport matrix analysis approach to source identification did not perform as well as it did for an earlier investigation (Yamartino and Lamich, 1979).

Acknowledgments

This work was funded by a grant made to the Harvard School of Public Health by the Mellon Foundation, and that support is appreciated, as was the cooperation of my colleagues, Professor John D. Spengler (and associates in the Six Cities Study) and Arthur D. Schatz and David Sylvain, graduate students here.

References

J. A. Cooper, J.G. Watson, "Receptor oriented models of air particulate source apportionment," *J. Air. Poll. Control Assoc.*, *30* (10): 1116-25 (1980).

F. A. Gifford, "Computation of pollution from several sources," *Int. J. Air Poll.*, *2:* 109-110 (1959).

G. E. Gordon, "Receptor models," *Environ. Sci. Technol.*, *14* (7): 792-800 (1980).

G. C. Holzworth, "Mixing heights, wind speeds, and potential for urban air pollution throughout the contiguous United States," Environmental Protection Agency, Research Triangle Park, N.C. (1972).

International Atomic Energy Agency, "Application of meteorology to safety at nuclear plants," International Atomic Energy Agency, Vienna (1968).

S. B. Khanna, "Handbook for UNAMAP," Abcor, Inc., Walden Research Division, Cambridge, MA (March 1976).

D. B. Turner, "Workbook of atmospheric dispersion estimates, " Public Health Service, U.S.D.H.E.W., Cincinnati, OH (1969).

R. J. Yamartino and D. J. Lamich, "The formulation and application of a source finding algorithm," Fourth Symposium on Turbulence, Diffusion, and Air Pollution, Reno, NV, January 1979.

J.G. Yocum, E. T. Brookman, R. C. Westman, O. P. Ambardar, "Determining the contributions of traditional and non-traditional sources of particulate matter," *J. Air Pollut. Control Assoc.*, *31:* 17-23 (1981).

Table 1. Values of χ/Q at Receptors, 10^{-3} s/m^3.

Direction	Radius (km) 0.5	1.0	2.0	4.0	8.0
S	3.054	1.257	0.477	0.168	0.060
SSW	2.205	0.908	0.342	0.121	0.043
SW	2.131	0.858	0.328	0.115	0.041
WSW	3.178	1.295	0.483	0.169	0.061
W	3.244	1.317	0.494	0.173	0.061
WNW	3.236	1.327	0.497	0.175	0.063
NW	3.015	1.212	0.462	0.161	0.058
NNW	2.043	0.838	0.314	0.111	0.040
N	2.959	1.216	0.461	0.162	0.058
NNE	1.330	1.549	0.207	0.073	0.026
NE	1.330	0.538	0.206	0.072	0.026
ENE	1.324	0.541	0.202	0.071	0.025
E	2.382	0.966	0.362	0.126	0.045
ESE	2.108	0.853	0.316	0.110	0.039
SE	1.503	0.592	0.222	0.077	0.027
SSE	1.096	0.449	0.168	0.059	0.021

Table 2. Values of a and b for $\chi/Q \approx 10^a/r^b$ for Boston 1967 through 1971 by Direction of Source with Respect to Receptor (r in km).

Direction To Source	Exponents a (10^{-3} s/m^3)	b
S	3.082	1.42
SSW	2.940	1.42
SW	2.921	1.43
WSW	3.094	1.44
W	3.103	1.44
WNW	3.104	1.43
NW	3.070	1.43
NNW	2.905	1.43
N	3.068	1.42
NNE	2.721	1.42
NE	2.718	1.43
ENE	2.715	1.43
E	2.968	1.44
ESE	2.913	1.44
SE	2.760	1.45
SSE	2.634	1.43

Table 3. Comparison of Approximate Response Function $f(r,\vartheta)$ with Calculated χ/Q Values (from Table 1) at r = 1 km.

Direction To Source	χ/Q Calculated ($\times 10^3$)	$f(r, \vartheta)$ Approximation ($\times 10^3$)	Ratio $f/(\chi/Q)$
S	1.257	1.182	0.94
SSW	0.908	0.983	1.08
SW	0.858	0.783	0.91
WSW	1.295	1.370	1.06
W	1.317	1.242	0.94
WNW	1.327	1.402	1.06
NW	1.212	1.137	0.94
NNW	0.838	0.913	1.09
N	1.216	1.141	0.94
NNE	0.549	0.624	1.14
NE	0.538	0.463	0.86
ENE	0.541	0.616	1.14
E	0.966	0.891	0.92
ESE	0.853	0.928	1.09
SE	0.592	0.517	0.87
SSE	0.449	0.524	1.17

Table 4. R^2 Values.

	8-sector (3 x 3 grid)	16-sector
Receptor 1	65%	94%
Receptor 2	52%	96%
Difference	19%	78%

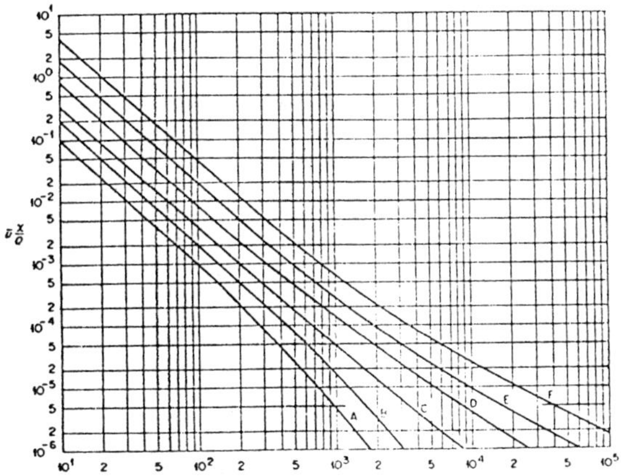

Figure 1. Values of $\bar{u}\chi/Q$ as a function of downwind distance, x, meters, for various types of weather; source at surface.

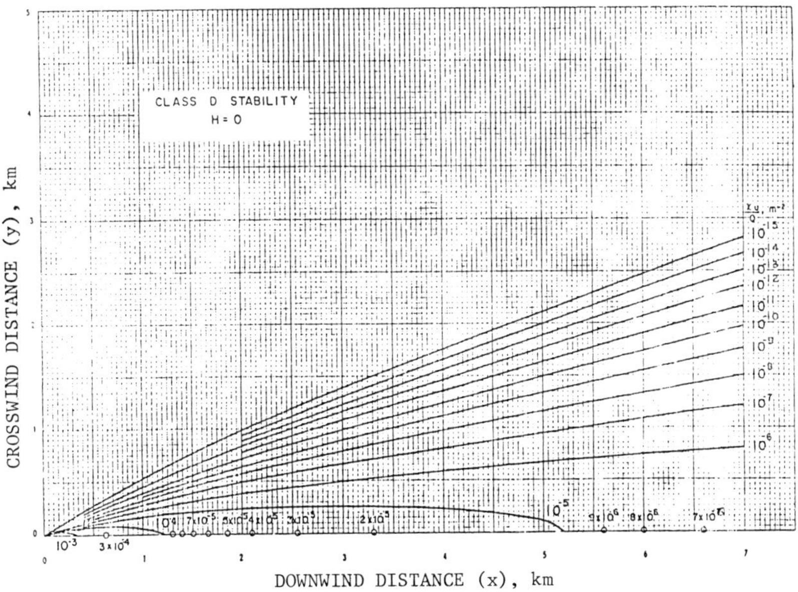

Figure 2. Isopleths of $\chi\bar{u}/Q$ for a ground-level source, D stability.

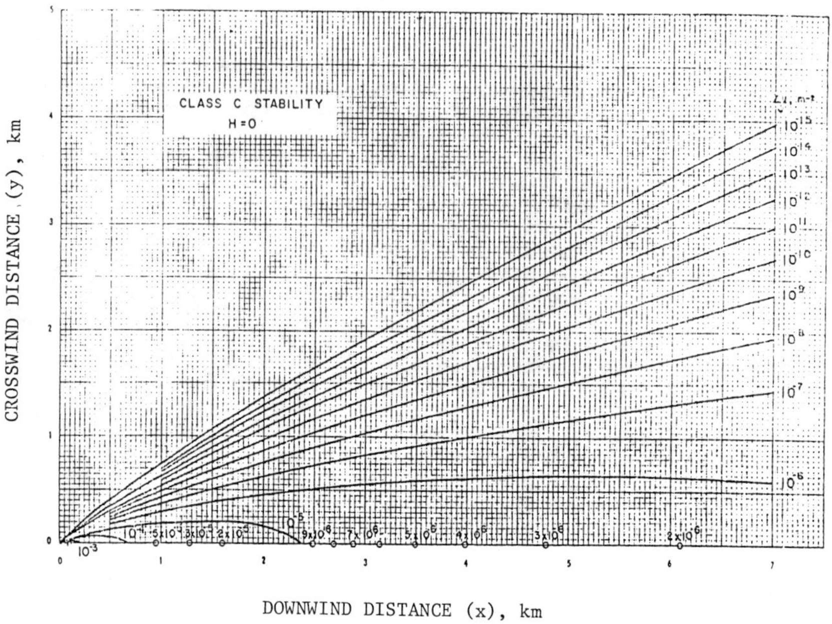

Figure 3. Isopleths of $\chi \bar{u}/Q$ for a ground-level source, C stability.

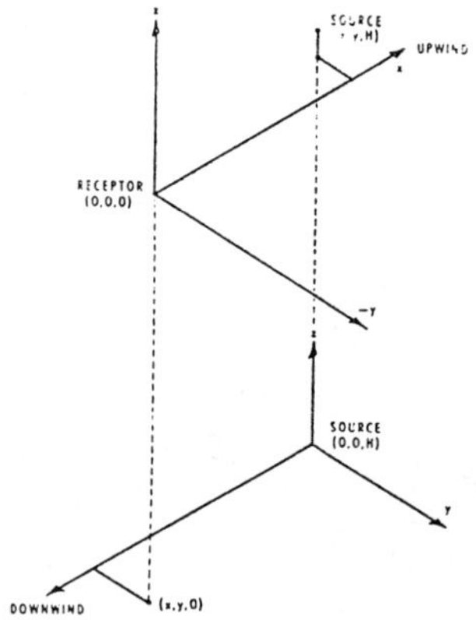

Figure 4. Comparison of source-oriented and receptor-oriented coordinate systems.

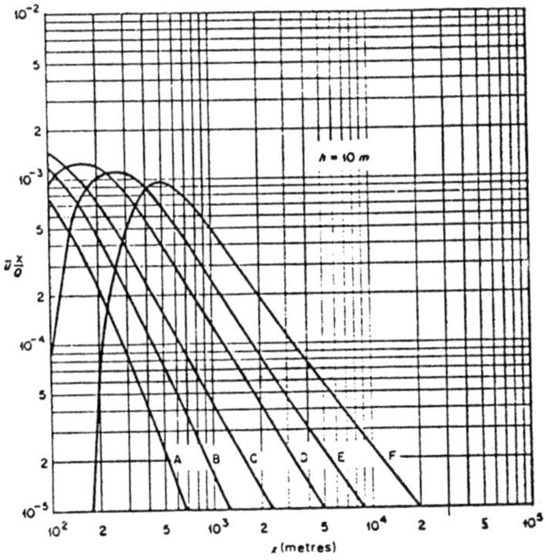

Figure 5. Values of $\overline{u}\chi/Q$ as a function of downwind distance, x, meters, for various types of weather; h = 10 m.

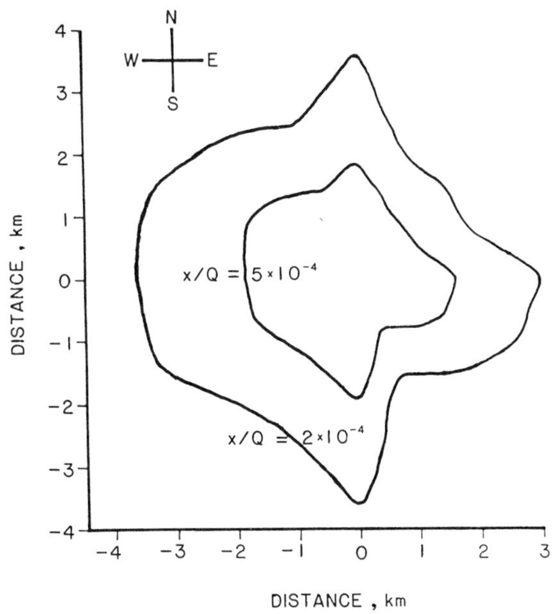

Figure 6. Response function for five-year-average meterology in Boston, MA region.

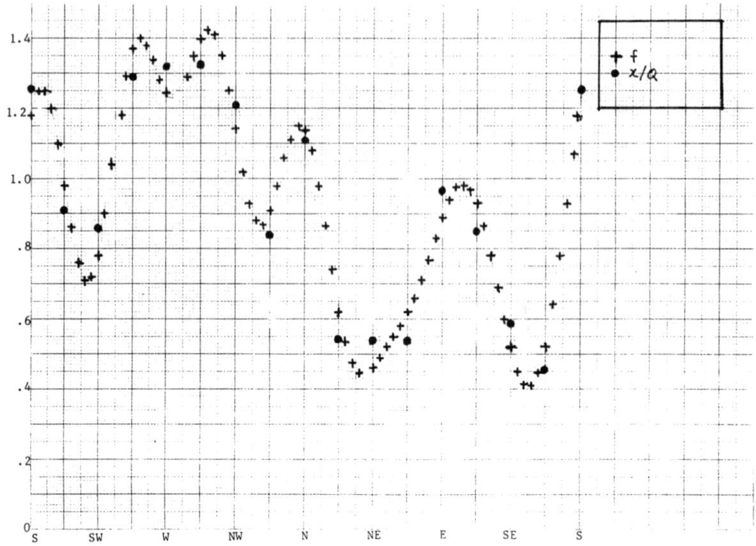

Figure 7. 1000(χ/Q) for Boston 1967-71 at r = 1 km and interpolating function. f.

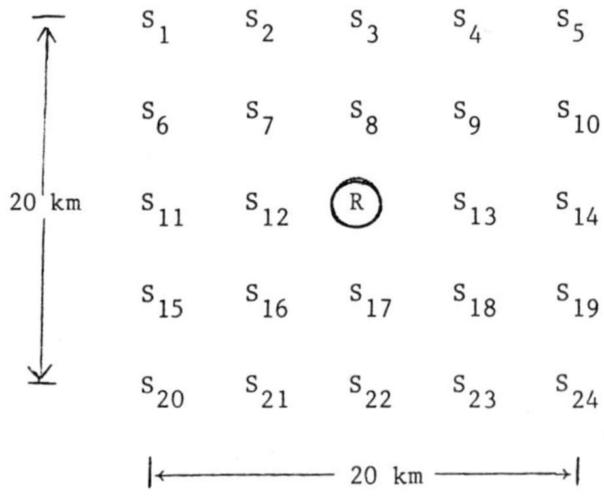

Figure 8. Transport matrix analysis grid, 5x5, for 20x20 km area.

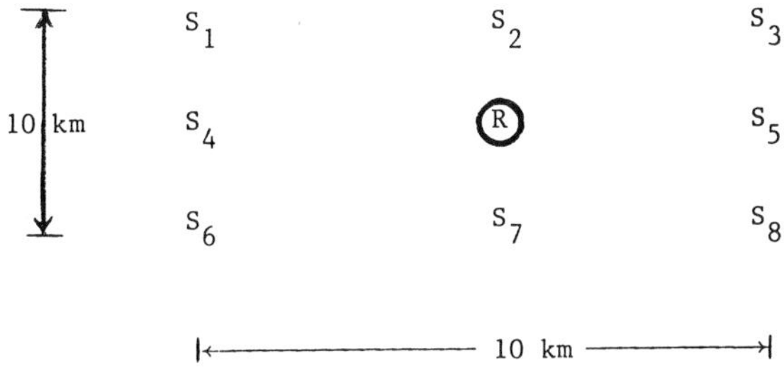

Figure 9. Transport matrix analysis grid, 3x3, for 10x10 km area.

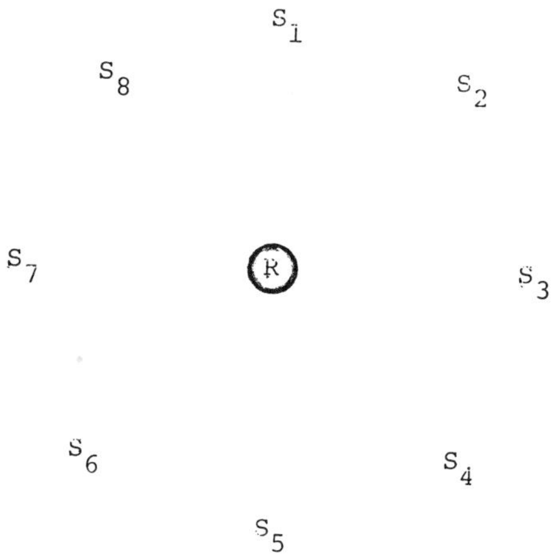

Figure 10. Transport matrix analysis grid, directional.

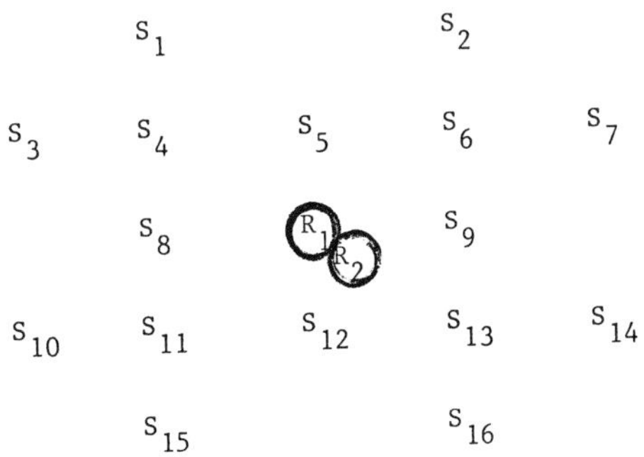

Figure 11. Transport matrix analysis grid, differences between samples.

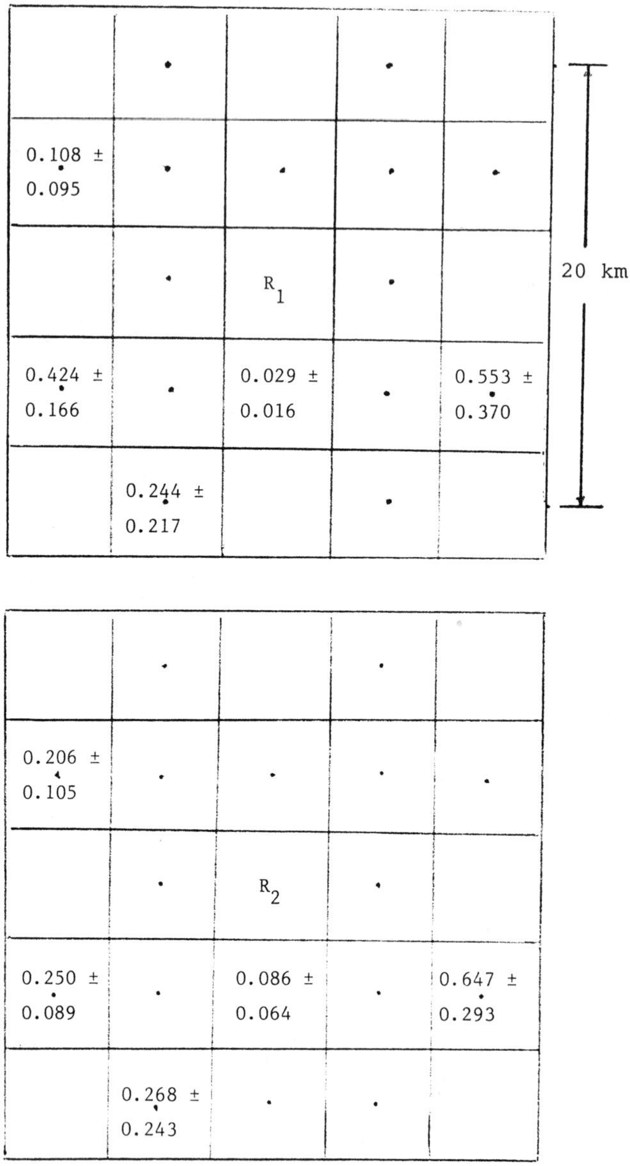

Figure 12. Source-strength values (mg/s) and standard errors, 20x20 km, for samplers 1 and 2, Case 2.

TIME SERIES ANALYSIS FOR IDENTIFICATION OF PARTICLE EMISSION SOURCES

Chang Sik Hwang ⟶
and Philip K. Hopke
Institute of Environmental Studies and
Department of Civil Engineering
University of Illinois
Urbana, Illinois

Spectral and autoregressive analyses were made on a suite of 2640 concentration measurements of 5 elements, S, Cu, Zn, Br, and Pb, determined in fine airborne particle samples collected at RAPS sampling stations 103 and 105 every 2 hours over the course of 3 weeks in July and August 1975. Sulfur was found to be related to a long term transport, copper and zinc related to the point sources, and bromine and lead primarily related to automobiles. However, the Br values at site 105 did not show the distinct frequency pattern for automobiles possibly caused by the presence of missing data points.

1. Introduction

There is concern over possible deleterious effects associated with exposure to anthropogenic airborne particulate matter. Large scale efforts are being made to reduce the potential hazard to public health and welfare by identifying the sources of airborne particulate matter and, where possible, applying appropriate control technologies. Because of the complexities of urban ecosystem, quantitatively apportioning the aerosol particles between the various sources is a very difficult problem. Modern receptor methodologies for determining the contribution of the various sources to the total level of suspended particulate matter (TSP) require the collection and elemental analysis of large sets of ambient aerosol samples. It is then necessary to examine the voluminous amounts of data in order to provide recommendations that can be used by local officials to mandate the necessary controls. The volume of data necessitates the use of statistical analysis techniques to resolve the contributions of the various sources.

One such method is time series analysis. The variation in the measured elemental composition in a sequence of samples from a single site may be employed to determine associations between certain elements. Elements that are strongly correlated in time are assumed to originate from the same source or source-type. This assumption is not valid when two sources are co-located and the variation in their contributions to a receptor site are governed primarily by the meteorology of dispersion rather than disparate operating schedules. There are a number of airborne particle sources that have significant diurnal variations such as automobile traffic, electric power plants, smelters and steel plants depending on the production schedule. Therefore, for those data sets where short duration sampling and analysis of airborne particles has been performed, it may be possible to determine the types and amounts of source emissions by various types of time series analysis. Only a single prior report has appeared in the literature (1) where such methods have been employed for aerosol source resolutions. They employed a Box-Jenkins model to obtain a smoothed average sequence for a particular element and examined the deviations or "residual shock" to the system. This approach although it gave some indication of the causes of time specific events did not permit much understanding of the sources and variations in bulk of the aerosol mass. Recently, Murray *et al.* (2) have used the Box-Jenkins technique to examine the correlation between visibility and sulfate. In the last three decades, spectral analysis has been widely used by many scientists, particularly in electrical engineering, physics, meteorology and marine science because of its usefulness in spectrum and frequency analysis. The application of spectral analysis has been well studied by Chatfield (3) and Jenkins (4). In order to explore the utility of time series techniques to environmental sample source resolution, the source resolution of five elements S, Cu, Zn, Br, and Pb measured in fine particles (diameter less than 2.5 microns) at RAPS site 103 and 105 in St. Louis, MO in July and August 1975 will be examined.

2. Description of Time Series Model

For any given sequence of N observed concentrations of a particular element, each observation X(t), is considered to be composed of a part that is dependent on other values in the sequence and an independent part that is purely random. The autocovariance function is given by

$$\gamma(\tau) = \frac{1}{N} \sum_{t=1}^{N-\tau} [(X(t) - \overline{X})(X(t + \tau) - \overline{X})] \tag{1}$$

where X(t) is the concentration for time period t, $\overline{X}$ is the mean value and τ is the lag. The autocorrelation function between X(t) and $X(t + \tau)$ is given by

$$\rho(\tau) = \gamma(\tau)/\gamma(0) \tag{2}$$

where $\gamma(0)$ is simply the variance of X(t). The function, $\rho(\tau)$, can be interpreted as a correlation coefficient between values within a series differing in time by lag τ. If there were no periodic processes, $\rho(\tau)$ would be zero for all $\tau > 0$. The objectives are to separate the original time series into the periodic and random series by the following set of sequential steps.

2.1 Sample Mean of the Series X_t

Let $X_1, X_2, \ldots, X_n$ denote n discrete values of random variable X_t. If the sample mean, $\overline{X}$ is subtracted from the series, X_t, new series X_{t1} will be

$$X_{t1} = X_t - \overline{X} \tag{3}$$

2.2 Spectral Analysis

One way of analyzing the dependent part of time series is based on the assumption that it is made up of sine and cosine waves with different amplitudes and frequencies. After removing the periodic components, the second residual, X_{t2} is given as

$$X_{t2} = X_{t1} - \sum_{i=1}^{M} (A_i \sin \omega_i t + B_i \cos \omega_i t) \tag{4}$$

where A_i and B_i are amplitudes of sine and cosine functions, respectively, and ω is frequency. The spectral intensity, $I(\omega)$ is expressed

$$I(\omega) = [(\sum_{t=1}^{N} X_{t2} \sin \omega t)^2 + (\sum_{t=1}^{N} X_{t2} \cos \omega t)^2/N\pi \tag{5}$$

The spectral intensity, $I(\omega)$, gives a measure of the importance of the particlar frequencies within the data set. By plotting $I(\omega)$ against ω, large values of $I(\omega)$ are seen as peaks in this frequency plot. The largest of these peaks can then be used to choose the most significant frequencies in the data. The significant periodic components are eliminated until the autocorrelation function of X_{t2} will satisfy the condition such that

$$\rho(\tau) \geq 2/N \qquad (6)$$

where $\rho(\tau)$ is autocorrelation function, τ is a number of lags, and N is a total number of data points.

2.3 Autoregressive Process

Even if the periodic properties are removed, the series may still contain the history of past values. This component also should be removed to get a final random residual series, so called white noise. Suppose that X_{t2} depends on the p past values. Then the final residual series, Z_t is written

$$Z_t = X_{t2} - \sum_{j=1}^{p} \alpha_j X_{tz-j} \qquad (7)$$

where p and α_j are the order and the coefficient of autoregressive process, respectively. The pth order of autoregressive process is simply abbreviated as AR(p). The order of regressive process was determined iteratively by examination of the significance of the coefficients and at the test for randomness of the final residual series. The Kolmogorov-Smirnov test of goodness of fit is used to check the randomness of Z_t. For the details, see R. J. Miller, Jr. (5), L. H. Miller (6), Wine (7). Thus, the overall expression of the time series model is given by

$$X_t = \overline{X} + \sum_{i=1}^{M} (A_i \sin \omega_i t + B_i \cos \omega_i t) + \sum_{j=1}^{p} \alpha_j X_{t-j} + Z_t \qquad (8)$$

The mean value includes a sum of background concentration and an average contribution from all sources. Periodic phenomena is presumably related to the emissions from sources. For instance, the variation in the amount of particles from automobiles or a factory with time depends on the schedule or operation and/or production. Autoregressive processes should include the meteorology over time and region. The transport of particles depends on the synoptic conditions, especially wind direction. The final term in the equation is purely random and should include the non-periodic source emissions as well as any random errors in sampling and analysis. Since the mean concentration, amplitudes, frequencies, coefficients, and the order of autoregressive process can be calculated, the prediction of the particle concentrations is possible within a certain range of error.

3. Data

Two data sets each containing 264 samples taken for 2 hour intervals during the three-week period; July 12 to August 2, 1975, at RAPS (Regional Air Pollution Study of St. Louis, MO) sites 103 and 105 have been examined in this study. The location of two sites are shown in figure 1. Five elemental concentrations, S, Cu, Zn, Br, and Pb, were selected for fine particle (<2.5 m) samples. Since a short period set of data is used, systematic changes in trend and seasonal variation are not expected. Two data points are missing at site 103. Approximate values were obtained by linear interpolation between the two adjacent data points. Values of Cu and Zn that are below the detection limits were replaced with values equal to one-half the detection limit. A total of sixteen data points are missing at site 105 with gaps of four and ten consecutive data points. These data gaps make it difficult to analyze the time series at site 105 since a continuous, complete sequence is required. The missing values of Cu, Zn, Br, and Pb were replaced with the values from other days in the data set having a similar wind pattern and time of day. This problem requires further study. For the unavailable values of S, the S values at the same time at site 103 are employed. The other missing values are calculated by interpolation.

4. Results and Discussion

Figure 2a, b, c, d, and e show the concentration of S, Br, Pb, Cu, and Zn as a function of time at site 103, respectively. Cu and Zn have a similar time-concentration pattern and the correlation coefficient between these two elements is 0.81 at site 103. However, the large difference between the highest and lowest values made it difficult to satisfy equation (6). Thus a logarithm of the concentration values is employed to minimize the range for Cu and Zn concentrations. Br and Pb also show the profiles similar to each other and their correlation coefficient is 0.83. The correlation coefficients between the elements for site 103 are given in Table 1.

The elemental concentrations at site 105 are shown in figure 3a-e. The S series looks very similar to that observed at site 103 indicating that there are not strong local sources of particulate sulfate and that sulfate concentrations are governed by regional processes. For Cu and Zn, both sites show intermittant high values and many values are again below detection limits. A logarithmic transformation was again required to complete the spectral analysis. Most of the peaks in the site 103 Cu and Zn data are in the early part of the period while high values at site 105 are seen later in this time interval. Pb and Br are again quite complex. The correlation coefficients between the elements at site 105 are given in Table 2.

Spectral analysis can be a useful exploratory diagnostic tool in the analysis of many types of time series. The periodogram for S, Br, and Pb are shown in figure 4a, b, and c. Figure 4d and e show the intensity obtained from the log-concentration site 103 data set of Cu and Zn. The spectral frequencies and order of autoregressive process for each of the 5 elements at each of the two sites are listed in Table 3.

For S, frequencies 1, 2, and 5 (period, 22, 11, and 4.4 days) are dominant and the order of the autoregressive process, AR is 4 (8 hours) at both sites. The correlation between original and residual concentration is 0.43 at site 103. This weak correlation indicates that sulfur is not likely to be related to point sources. The long period, high order of AR process, and weak correlation with residuals indicate that S is affected by long-range transport of particles. In contrast, the correlation coefficients of the original values with the residuals, R, for Cu and Zn are 0.75 and 0.80 at site 103, respectively. This result indicates that Cu and Zn are strongly correlated with point sources. The dominant frequency of Cu is 2 and AR value is 1 at site 103. Four critical frequencies, 1, 2, 5, and 3, and AR (2) are found at site 105. Cu at site 105 has more periodic components and time dependent characteristics. Zn has frequency, 2 and 22, and AR (3) at site 103 and frequency, 22, 18, 28, 1, and 5, and AR (1) at site 105. Therefore, a small frequency (2) or long period (11 days) is dominant at site 103 and high frequency (22, 18, and 28) or short period (about one day) is dominant at site 105. Thus, it is expected that Zn concentration at site 103 depends more on past values than those at site 105. AR (3) at site 103 and AR (1) at site 105 are well matched with such an explanation. The feature of short period of Zn at site 105 may be a clue to finding sources related to Zn near site 105. A copper products plant that produces fine Zn particles is located to the east of site 105 (8).

The correlations of the original data and the residuals for Br and Pb are 0.65 and 0.66 at site 103, respectively. They show intermediate values compared to S, Cu, and Zn. The significance of this range of correlations between raw data and random residuals is not clear. Pilotte (9,10) found that Pb is time correlated most strongly with Br from the study of lead in the atmosphere of St. Louis, MO. Br and Pb are well correlated with each other at site 103 and the correlation coefficient between original Br and Pb concentrations is 83%. Br has frequency, 22 and 44 (period, 24 and 12 hours) and AR (2), and Pb has frequency, 19 and 44 (27.8 and 12 hours) and AR (2) at site 103. Thus, Br and Pb have roughly two kinds of period at site 103. One is a day and the other is a half day. These periods are well matched with daily peak traffic hours. However, they may be confounded by periodic changes in the mixing height that occurs at about the same time of day. The model suggests the second-order autoregressive process for Br and Pb at both sites. At site 103 it appears that Br and Pb are mostly from automobile traffic. However, since Br and Pb have several minor

Spectral analysis can be a useful exploratory diagnostic tool in the analysis of many types of time series. The periodogram for S, Br, and Pb are shown in figure 4a, b, and c. Figure 4d and e show the intensity obtained from the log-concentration site 103 data set of Cu and Zn. The spectral frequencies and order of autoregressive process for each of the 5 elements at each of the two sites are listed in Table 3.

For S, frequencies 1, 2, and 5 (period, 22, 11, and 4.4 days) are dominant and the order of the autoregressive process, AR is 4 (8 hours) at both sites. The correlation between original and residual concentration is 0.43 at site 103. This weak correlation indicates that sulfur is not likely to be related to point sources. The long period, high order of AR process, and weak correlation with residuals indicate that S is affected by long-range transport of particles. In contrast, the correlation coefficients of the original values with the residuals, R, for Cu and Zn are 0.75 and 0.80 at site 103, respectively. This result indicates that Cu and Zn are strongly correlated with point sources. The dominant frequency of Cu is 2 and AR value is 1 at site 103. Four critical frequencies, 1, 2, 5, and 3, and AR (2) are found at site 105. Cu at site 105 has more periodic components and time dependent characteristics. Zn has frequency, 2 and 22, and AR (3) at site 103 and frequency, 22, 18, 28, 1, and 5, and AR (1) at site 105. Therefore, a small frequency (2) or long period (11 days) is dominant at site 103 and high frequency (22, 18, and 28) or short period (about one day) is dominant at site 105. Thus, it is expected that Zn concentration at site 103 depends more on past values than those at site 105. AR (3) at site 103 and AR (1) at site 105 are well matched with such an explanation. The feature of short period of Zn at site 105 may be a clue to finding sources related to Zn near site 105. A copper products plant that produces fine Zn particles is located to the east of site 105 (8).

The correlations of the original data and the residuals for Br and Pb are 0.65 and 0.66 at site 103, respectively. They show intermediate values compared to S, Cu, and Zn. The significance of this range of correlations between raw data and random residuals is not clear. Pilotte (9,10) found that Pb is time correlated most strongly with Br from the study of lead in the atmosphere of St. Louis, MO. Br and Pb are well correlated with each other at site 103 and the correlation coefficient between original Br and Pb concentrations is 0.83. Br has frequency, 22 and 44 (period, 24 and 12 hours) and AR (2), and Pb has frequency, 19 and 44 (27.8 and 12 hours) and AR (2) at site 103. Thus, Br and Pb have roughly two kinds of period at site 103. One is a day and the other is a half day. These periods are well matched with daily peak traffic hours. However, they may be confounded by periodic changes in the mixing height that occurs at about the same time of day. The model suggests the second-order autoregressive process for Br and Pb at both sites. At site 103 it appears that Br and Pb are mostly from automobile traffic. However, since Br and Pb have several minor frequencies, the possibility of other source impacts should be checked. Assuming that all of the bromine comes from automobiles, the amount of lead attributable to automobiles can be estimated. Using the linear-correlation relationship between Br and Pb at site 103, the following equation has been found.

$$(Pb)_{auto} = 2.95\ (Br)_0 + 200 \tag{9}$$

Here $(Pb)_{auto}$ is the lead concentration contributed from the automobile and $(Br)_0$ is the original bromine concentration. To find the lead contribution from other sources like a copper-products plant or an incinerator, the lead contribution from automobile, $(Pb)_{auto}$, is subtracted from the original lead concentration. The residual concentrations of lead, $(Pb)_r$ show peak values at data points with wind direction and speed of 185/3.8, 186/3.8, 184/3.4, and 189/2.8 ms^{-1}, respectively. Presumably southerly winds contributed to the transport of lead coming from a copper products plant and/or an incinerator. Actually, a copper-products plant is located to the south of site 103. To confirm the contribution of lead from the copper-products plant, the correlations of $(Pb)_r$ and the original values of Cu and Zn, $(Cu)_0$, and $(Zn)_0$, are examined. The correlation coefficient between $(Pb)_r$ and $(Cu)_0$ is 0.65 and for $(Pb)_r$ and $(Zn)_0$ is 0.68. It is known that fine particles of copper and zinc come from the copper-products plant to the south (8) and the highest values of $(Pb)_r$ are associated with southerly winds. The moderate correlations of $(Pb)_r$ with the original concentrations of Cu and Zn indicate that a portion of the lead comes from the East St. Louis area of Illinois. From other results (8), it appears that the most likely source is the copper products plant.

The correlation coefficient between Br and Pb concentrations at site 105 is 0.73, about 0.10 less than the value of site 103, indicating that other processes are important to one or both of these elements. Br has frequencies of 3 and 4 (period 7.3 and 5.5 days) and Pb has frequency, 19, 22, and 5 (period, 27.8, 24, and 106 hours) at site 105. Br has a very long period (5 to 7 days) and Pb also has a relatively long period (1 and 4 days). Neither of them shows a half-day period. Such a difference in frequencies may be indicative of contributions from sources of Br and Pb besides automobiles. Possible lead sources include a copper products plant, an incinerator, and a lead smelter to the south of St. Louis. The operating schedules of these sources are presumably different from peak traffic hours.

In order to examine these data for other sources, the same procedures have been applied to the lead data from site 105. The linear correlation computation gives

$$(Pb)_{auto} = 3.05\ (Br)_0 + 300 \tag{10}$$

The peak values of the lead residuals, $(Pb)_r$, show wind directions and speeds of 186/2.4, 176/2.7, 116/1.1, 140/0.9, and 113/1.9 ms^{-1}. Peak values occur for southerly and east southeasterly wind directions. Apparently there are two areas with sources that contribute lead to site 105. Since the incinerator is to the north of site 105, it is not a significant source of lead to this site during this sampling period. The copper products plant is to the east southeast of this location and three directions (113,116, and 140) generally point toward the plant. The lead plant is located to the south of site 105 near site 124. It may also contribute to the lead at site 105. Wind directions of 186 and 176 point toward that plant. The correlation between $(Pb)_r$ and $(Cu)_0$ is 0.46, and between $(Pb)_r$ and $(Zn)_0$ is 0.26. These correlation coefficients are much lower than the corresponding values at site 103. Thus, Pb at site 105 may be the result of complex mixing of these sources leading to poorer relationships than seen at site 103.

It is difficult to find a physically realistic mechanism to account for lack of a short period for Br. The long period for Br may represent reduced traffic volume on weekends. We are concerned that the choice of values used to fill the data gaps may affect the frequency determination. Therefore, further study of the effect of different choices for these missing values will be made. A significant problem will exist if this choice is critical since it is often possible for a sampler to fail and data from several consecutive periods to be missing. Thus, the ability to successfully fill data gaps will be important to the utility of these methods.

5. Conclusions

The time series model has produced these results.

1. Sulfur is related to long-term transport which depends on the past values up to 8 hours.

2. Copper and zinc are related to point sources. Cu and Zn at site 105 have more complex sources compared to site 103. The source of Zn is expected to be near site 105 because of the short periodicity of Zn values at site 105.

3. Br and Pb at site 103 were found to strongly suggest automobiles as the principal source of the observed lead values. The Br values at site 105 do not show the expected frequency pattern for automobiles as they do at site 103. This result may have been obtained from the choice of values made to fill in missing data points and further studies are in progress.

The residual analysis method developed by Box (11) may be a useful approach to investigate the sources of particles. Forecasting the future values of an observed time series is an important problem in the study of air pollution. Techniques of linear prediction for climatology with periodic statistics were recently discussed by Hasselmann (12). Since all the necessary parameters in the spectral analysis and autoregressive method are known, the prediction of particulate concentrations may be possible within limits. Determination of sources associated with pollutants and forecasting of future values will be useful for ambient quality control strategy development. The time series analysis will be a useful technique to solve such problems.

6. Acknowledgement

This work was supported by the U.S. Department of Energy under contract DE AC02 80 EV10403. The assistance in initiating these studies of Gregory Pugh and Charles Smithe is greately appreciated.

7. References

1. Zinsmeister, A.R. and T.C. Redman. A Time Series Analysis of Aerosol Composition Measurements. Atmos. Environ. 14:210-215, 1980.

2. Murray, L.C. and R.J. Farber. Time Series Analysis of Historical Visibility Data Base. Atmos. Environ., 16:1-10, 1982.

3. Chatfield, C. The Analysis of Time Series, 2nd Ed., Chapman and Hall Ltd., London, England, 1980.

4. Jenkins, G.M. and D.G. Watts. Spectral Analysis and Its Application, Holden-Day, Inc., San Francisco, California, 1968.

5. Miller, Jr., R.J. Simultaneous Statistical Inference, 2nd Ed., Springer-Verlag New York Inc., New York, NY, 1981.

6. Miller, L.H. Table of Percentage Points of Kolmogorov Statistics. Journal of American Statistical Association, 51:113-121, 1956.

7. Wine, R.L. Beginning Statistics, Winthrop Publishers, Inc., Cambridge, Massachusettes, 1976.

8. Chang, S.N., P.K. Hopke, S.G. Rheingrover, and G.E. Gordon, Target Transformation Factor Analysis of Wind-Trajectory Selected Samples, Air Pollution Control Association Paper No. 82-23.1, 1982.

9. Pilotte, J.O. Aerosol Composition Relationships in Time Sequence Samples of an Urban Atmosphere. M.S. thesis, Department of Oceanography, Florida State University, FL, 1977.

10. Pilotte, J.O., J.W. Winchester, and J.W. Nelson. Components of Lead in the Atmosphere of St. Louis., Missouri. J. Appl. Met., 17:627-635, 1978.

11. Box, G.E.P., and G.M. Jenkins. Time Series Analysis: Forecasting and Control , rev. ed., Holden-Day, San Francisco, 1976.

12. Hasselmann, K. and T.T. Barnett. Techniques of Linear Prediction for System with Periodic Statistics. J. Atmos. Sci., 38:2276-2283, 1981.

Table 1. Correlation Coefficients Between The Original Elemental Concentrations For Site 103

	S	Cu	Zn	Br	Pb
S	1.00	0.04	0.03	0.11	0.14
Cu		1.00	0.81	0.04	0.39
Zn			1.00	0.03	0.41
Br				1.00	0.83
Pb					1.00

Table 2. Correlation Coefficients Between The Original Elemental Concentrations For Site 105

	S	Cu	Zn	Br	Pb
S	1.00	0.08	0.28	0.20	0.04
Cu		1.00	0.28	0.06	0.38
Zn			1.00	0.05	0.23
Br				1.00	0.73
Pb					1.00

Table 3. Frequencies of Spectrum and Order of Autoregressive Process for Elemental Concentrations at Sites 103 and 105

Site	Element	Frequencies (in order of decreasing spectral intensities)	Order
	S	1,2,5,6,3,7,10,16,8,17,13	4
	Cu*	2,4,21,24,18,22	1
103	Zn*	2,22,7,18,13,16,17,29,10	3
	Br	22,44,19,6,9,10,20,16	3
	Pb	19,44,22,20,7,6,1	2
	S	1,2,5,3,6,7,10,16,17,21,8,19,13	4
	Cu*	1,2,5,3,22,17,11,7,46,26,28,33, 6,40,16,25,19,9,27,34	2
105	Zn*	22,18,28,1,5,27,3,24,33,4,13, 23,17,42,19,2,26,16,6,38	1
	Br	3,4,5,19,27,8,7,66,2,16,17,6,11,36,44,23	2
	Pb	19,22,5,27,7,17,14,23,66	2

* Log concentration was used for Cu and Zn.
Period is given by (264/frequency) x 2 hours.

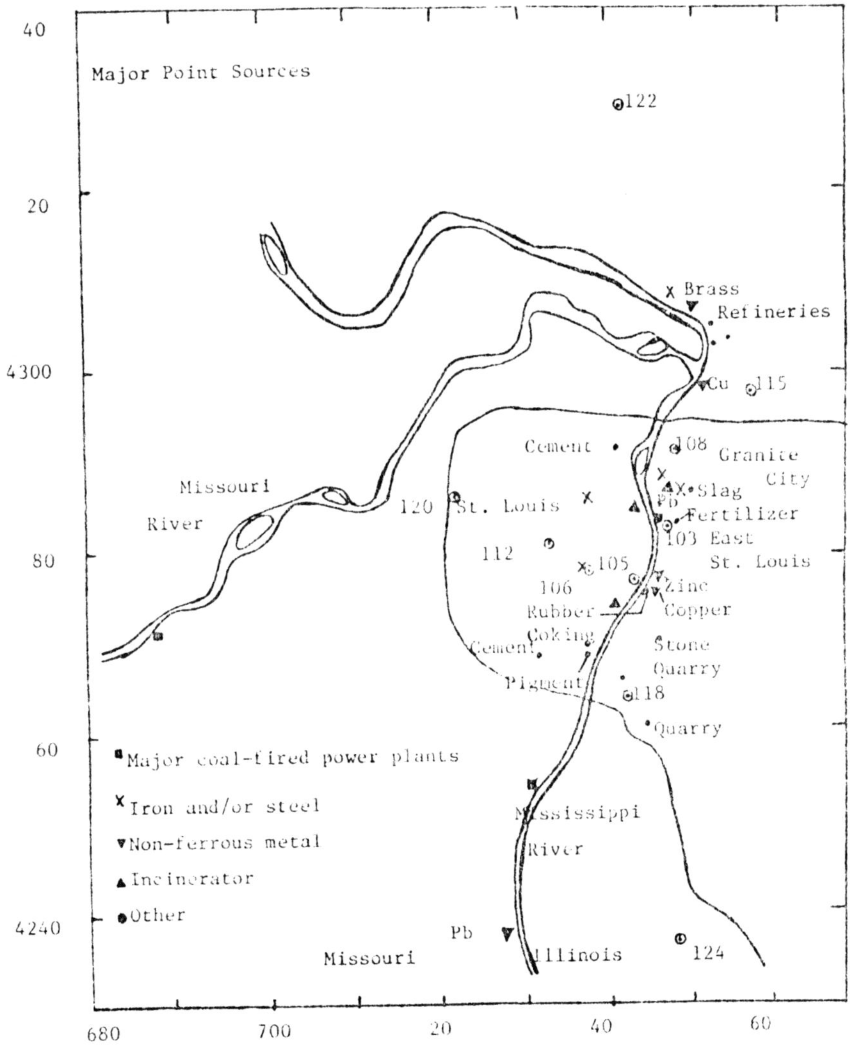

Figure 1. Location of the Regional Air Monitoring Systems sites and major emission sources in the vicinity of St. Louis, Missouri.

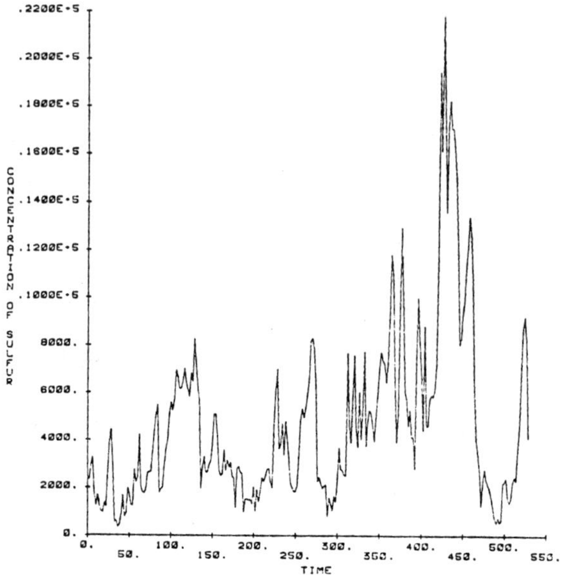

Figure 2a. Concentration of Sulfur (ng/m^3) at site 103 as a function of time (hours) after 0:00 hours on July 12, 1975.

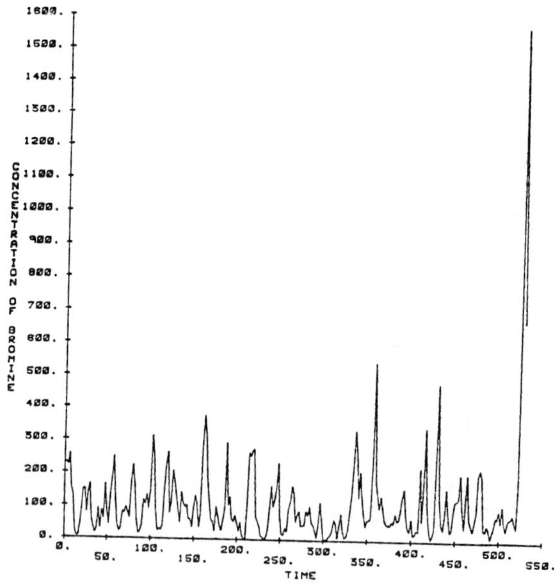

Figure 2b. Concentration of Bromine (ng/m^3) at site 103 as a function of time (hours) after 0:00 hours on July 12, 1975.

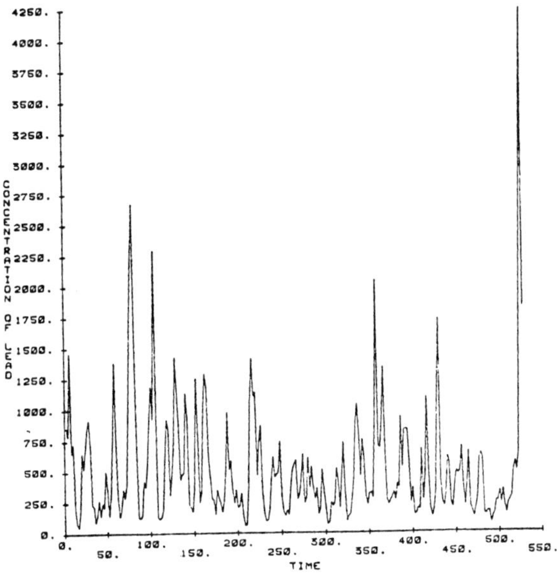

Figure 2c. Concentration of Lead (ng/m^3) at site 103 as a function of time (hours) after 0:00 hours on July 12, 1975.

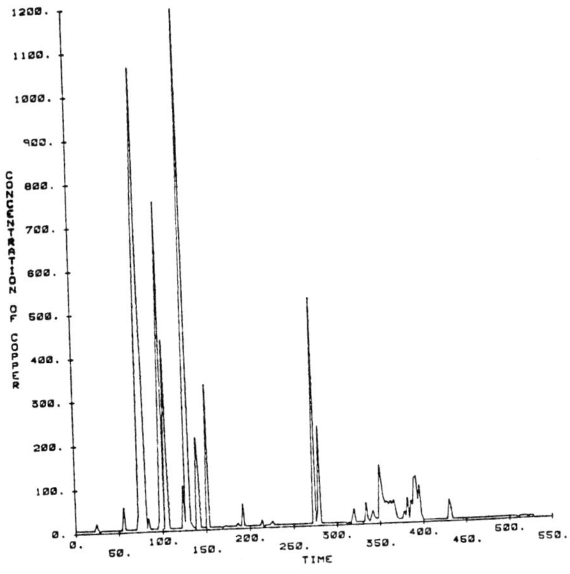

Figure 2d. Concentration of Copper (ng/m^3) at site 103 as a function of time (hours) after 0:00 hours on July 12, 1975.

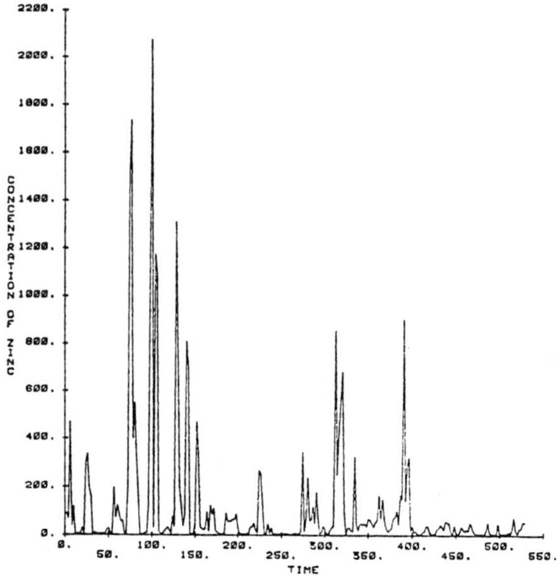

Figure 2e. Concentration of Zinc (ng/m^3) at site 103 as a function of time (hours) after 0:00 hours on July 12, 1975.

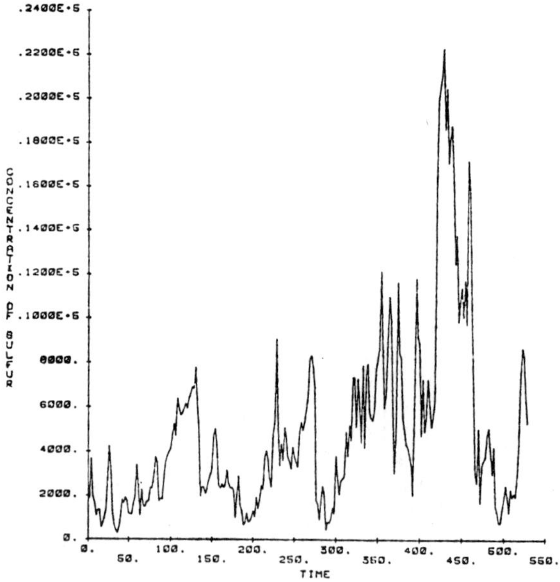

Figure 3a. Concentration of Sulfur (ng/m^3) at site 105 as a function of time (hours) after 0:00 hours on July 12, 1975.

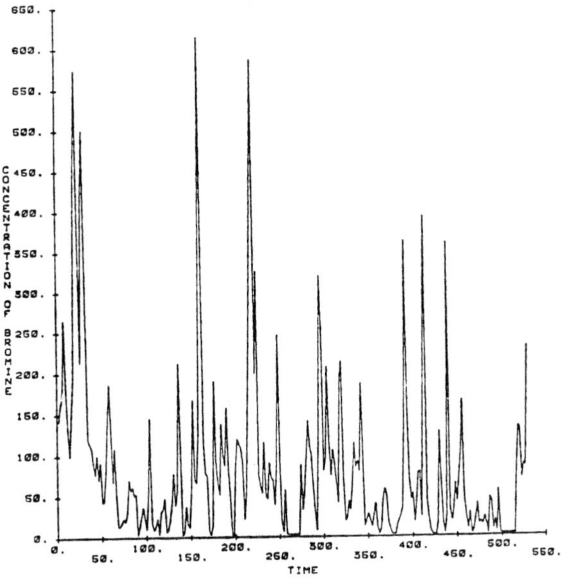

Figure 3b. Concentration of Bromine (ng/m^3) at site 105 as a function of time (hours) after 0:00 hours on July 12, 1975.

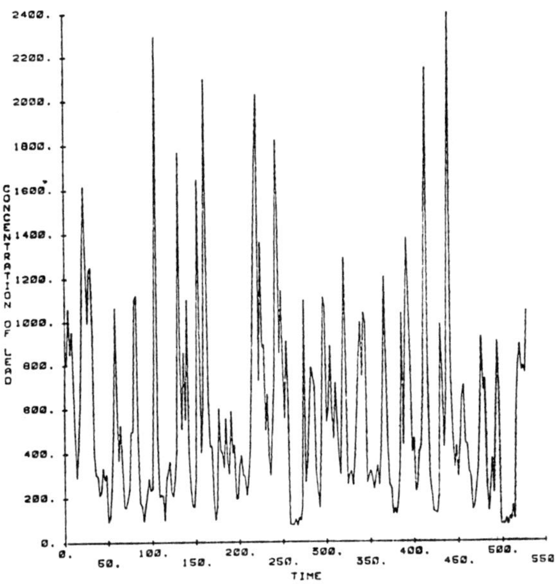

Figure 3c. Concentration of Lead (ng/m^3) at site 105 as a function of time (hours) after 0:00 hours on July 12, 1975.

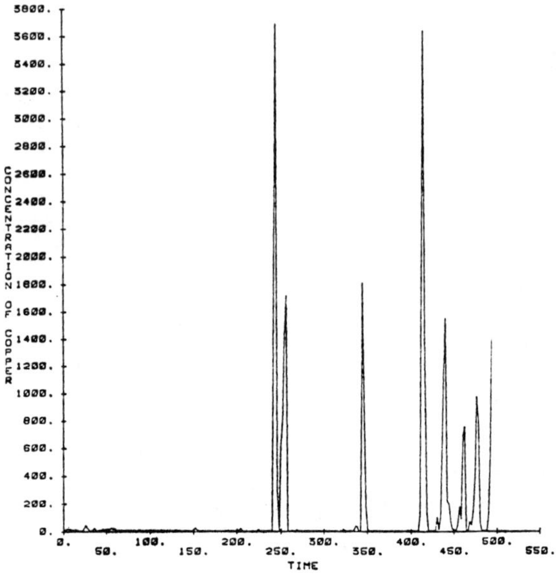

Figure 3d. Concentration of Copper (ng/m^3) at site 105 as a function of time (hours) after 0:00 hours on July 12, 1975.

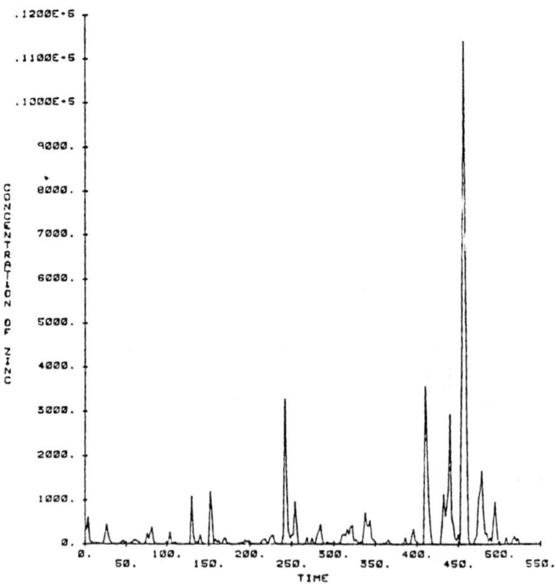

Figure 3e. Concentration of Zinc (ng/m^3) at site 105 as a function of time (hours) after 0:00 hours on July 12, 1975.

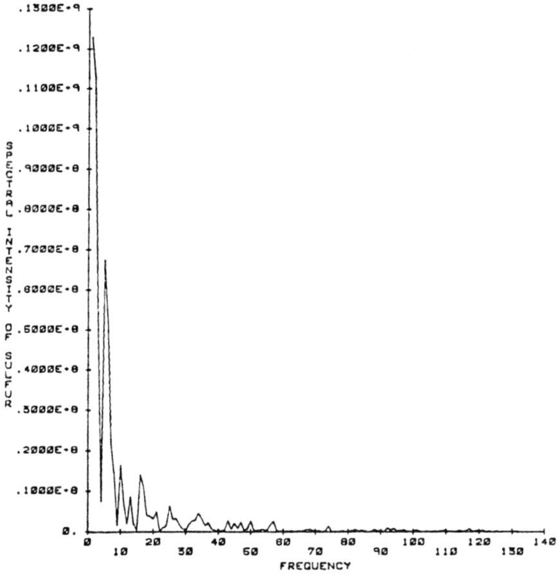

Figure 4a. Spectral intensity, I(ω), as a function of ω for Sulfur concentrations at site 103.

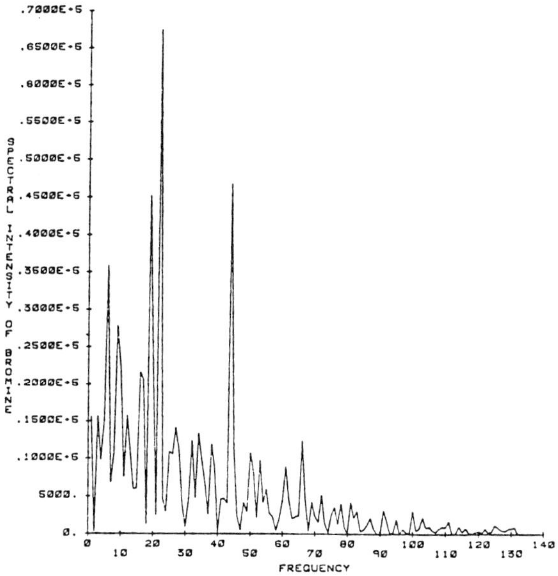

Figure 4b. Spectral intensity, I(ω), as a function of ω for Bromine concentrations at site 103.

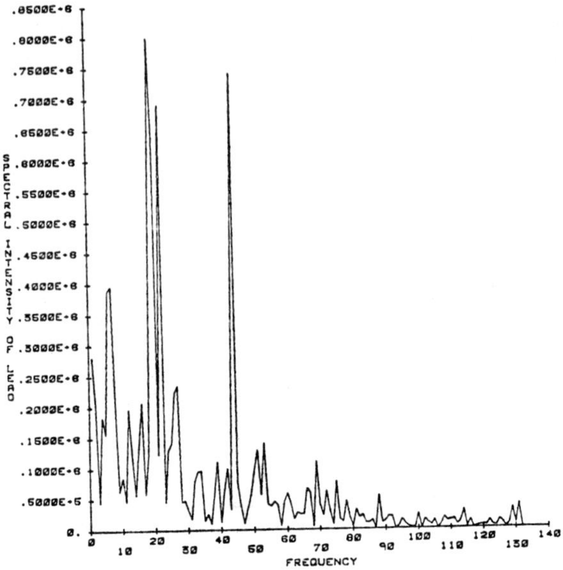

Figure 4c. Spectral intensity, I(ω), as a function of ω for Lead concentrations at site 103.

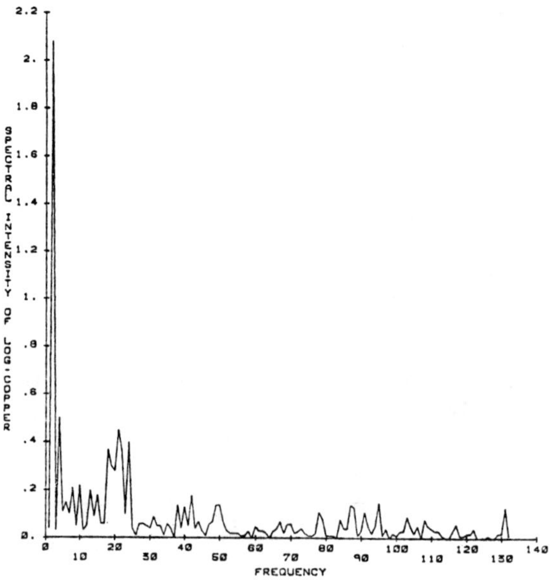

Figure 4d. Spectral intensity, I(ω), as a function of ω for logarithm of the copper concentrations at site 103.

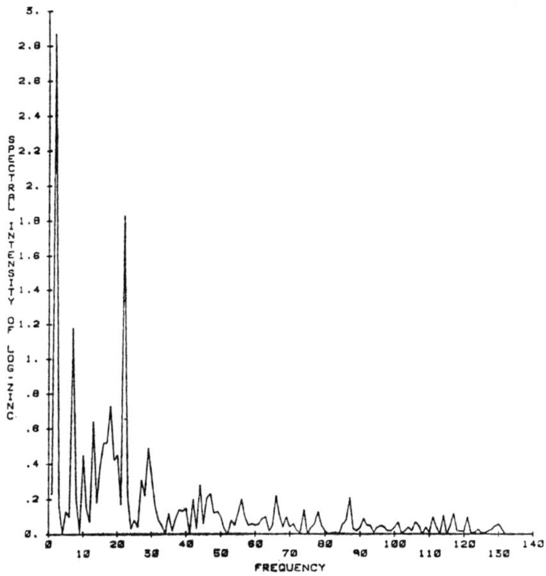

Figure 4e. Spectral intensity, I(ω), as a function of ω for logarithm of the Zinc concentrations at site 103.

APPLICATION OF RECEPTOR MODELING TO VOLATILE ORGANIC COMPOUNDS

Ivar Tombach,

AeroVironment Inc.
Pasadena, California

Receptor modeling has been used primarily with particulate matter, since the variety of elements in particulate matter conveniently provides conservative signatures for a number of common air pollutant sources. In principle, similar methods are equally applicable to gaseous organic compounds (VOC), but receptor modeling has not seen widespread use for such compounds.

Factors contributing to this limited use could include the relatively high costs of VOC speciation analyses and concern with how to handle the chemical transformations that some VOC undergo in the atmosphere. But, high analysis costs and chemical reactivity are also concerns of recent complex particulate matter receptor modeling efforts involving carbon, sulfate and nitrate. Furthermore, as we illustrate by two examples in this paper, effective VOC studies often do not need to deal quantitatively with the reactivity issue.

We discuss here the application of receptor modeling methods to VOC. General procedures for such applications are reviewed, and both the strengths and practical limitations of VOC receptor modeling methods are presented. To illustrate the applicability of the method, two examples of actual VOC source apportionment studies are discussed. One, using a method similar to the chemical mass balance (CMB) approach, allocated VOC emissions in Southern California to major source categories. The second, using an inverse dispersion modeling approach, allocated VOC emissions to various processes in an oil refinery complex. Special techniques, such as those utilizing the differing reaction rates of several species as an aid to source identification, can be used to enhance the effectiveness of VOC receptor modeling approaches.

APPLICATION OF RECEPTOR MODELING TO VOLATILE ORGANIC COMPOUNDS

Background

Receptor modeling has become a powerful method for identifying and quantifying the contributions of various sources or source types to some types of ambient air pollution. The first receptor modeling methods were developed for particulate matter and subsequently have been applied primarily to particulate matter. One of the receptor modeling approaches, involving microscopic identification of individual particles, is obviously limited to particulate matter, but the chemical mass balance (CMB), multivariate, and source/receptor hybrid approaches can be used with either gaseous or particulate pollutants.

Receptor modeling generally involves the assumption that the mass of an individual element or species is conserved during the transport of material from source to receptor. For nonradioactive elements this is unarguably true, excluding the removal of larger particles by fallout, and therefore the most fundamental approaches have used the elemental composition of the aerosol as its signature. More recently, efforts have been made to extend the approach to other chemical species, with particular interest in the sulfate, nitrate, and ammonium ions. These species are generally not conservative, but rather are created in the atmosphere by gas-to-particle conversion and by particle interactions, and therefore the receptor modeling approach has to deal with the production and removal of a species as it is transported from the source to a receptor. Friedlander[1] considers the receptor modeling of reactive (decaying) species, and describes a specific example in which the polycyclic aromatic hydrocarbon contributions from refinery emissions were determined.

Extension of receptor modeling methods to volatile organic compounds (VOC) involves exactly the same considerations of species identification, conservation of mass, and reactivity as for the more complex particulate problems. In many cases (two examples of which will be given below), however, the VOC analyses can be as simple as the simplest particulate analyses and therefore the application of receptor modeling to VOC can be as useful as it has been for particulate matter.

The Problem

The typical VOC emissions apportionment problem starts out much the same as one for particulate matter. A set of chemical analyses of VOC in ambient air has been performed and information is available as to the species signature of each sample. Such analyses will generally have been performed by gas chromatographic (GC) separation of the component species, followed by flame ionization detection (FID) or by further separation and detection by mass spectrometry (MS).

Given this data set, the objective of the receptor modeling analysis is to determine what fraction of the total VOC, or of each VOC species, can be attributed to a specific type of source. As the examples below will show, the sources of interest could include automobiles, gasoline, natural emissions, and natural gas, or specific processes (e.g., refining, fuel handling, etc.) in a plant complex.

To carry out the receptor modeling, knowledge is needed of the types of primary sources (ones that directly emit the species of interest) that contribute to the ambient air VOC concentrations, and of the chemical composition of the VOC emissions of those sources. Atmospheric reactions that create or deplete the species of interest need to be identified and, if those species are actually to be modeled, the reaction rates need to be quantified.

The subsequent process of receptor modeling is then the same as for particulate matter, except that the species being studied are compounds such as acetylene, isopentane, and propane rather than the elements or ionic species familiar in particulate matter studies. There are, however, a few special considerations:

(1) Many VOC are reactive. Generally one uses only those compounds that are relatively inert over the transport time-scale of interest because precise reaction rates in the atmosphere can be very difficult to quantify. On the other hand, the ratio of concentrations of a reactive and a nonreactive compound emitted by the same source can be used to infer the age of the air sample and hence can help deduce the location and type of the source.

Especially useful nonreactive species are: acetylene, which is almost exclusively an auto exhaust product in most urban situations[2]; butane and pentane, which are present in gasoline vapor; and propane and ethane, which are present in natural gas and liquefied petroleum gas (LPG). Mayrsohn and Crabtree[3] and Mayrsohn et al.[4] discuss relative abundances of these species in various source categories.

The most useful reactive species are ethene and propene, both of which are formed primarily in auto exhaust but which react in the ambient air at a substantially faster rate than acetylene. Stephens and Burleson[5] have used this characteristic to estimate the duration of transport of air samples they collected in the eastern part of the Los Angeles basin and Watanabe and Stephens[6] have reported similar analyses for several locations in the United States and Japan.

(2) Sampling and analysis of hydrocarbons can be expensive, and for those heavier than C_{10}, can be very difficult as well. Generally, one uses only C_{10} and lighter, and often C_6 and lighter, to describe source signatures. If enough specificity can be achieved by the GC-FID method, which will generally be the case if only the lighter ($\leq C_6$) hydrocarbons are measured, then analysis costs will be between $100-$200 per analysis, considerably less than the $350-500 analysis cost for the more detailed GC-MS procedure.

The discussion below describes two studies which have used receptor modeling methods to apportion VOC emissions. These studies have been selected for their illustrative value, and are not the only ones of their types. The first study, which assigned VOC emissions to source categories in the Los Angeles area, has been described in the published literature[3] and in California Air Resources Board reports. The second study, performed by AeroVironment and involving the participation of this author, has been described in a conference paper and in contractor reports.

Example 1: Determining Hydrocarbon Sources in Los Angeles

Approach

The objective of this study[3] was to determine the major contribution to ambient VOC levels in the Los Angeles Basin. Measurements of the concentrations of seven relatively inert nonmethane hydrocarbon compounds were made at several sampling sites, using the GC-FID method. Samples were collected during the summer months at three times of the day: late night, early morning peak traffic hours, and in the early afternoon.

Six categories of sources were assumed to represent most of the VOC emissions in the Los Angeles area: auto exhaust, gasoline, gasoline vapor, commercial natural gas, geogenic natural gas, and liquefied petroleum gas. Samples of each of these sources were analyzed for the same seven hydrocarbon species as were the ambient samples.

The chemical mass balance (CMB) method was used to derive the fractional contribution of each source to the ambient concentrations. The mass fractions were determined by minimizing (in a least squares sense) the residuals between the measured ambient chemical signatures and the computed composite source contribution signatures. The equations resulting from this optimization process were simply the matrix form of the mass balance equations multiplied by the transpose of the source signature matrix, viz:

$$\mathbf{C}^T\mathbf{C}\mathbf{B} = \mathbf{C}^T\mathbf{A} \quad , \tag{1}$$

where **A** is the column matrix of atmospheric hydrocarbon compound concentrations (normalized to isopentane or propane), **B** is the (unknown) column matrix of mass fractions of VOC attributable to each source category, and **C** is the matrix of hydrocarbon compound concentrations (columns) for all source categories (rows), and the superscript **T** represents the transpose.

By multiplying Equation (1) by $(\mathbf{C}^T\mathbf{C})^{-1}$, the inverse of the product of $\mathbf{C}^T$ and **C**, we obtain

$$(\mathbf{C}^T\mathbf{C})^{-1}\,\mathbf{C}^T\mathbf{C}\,\mathbf{B} = (\mathbf{C}^T\mathbf{C})^{-1}\,\mathbf{C}^T\,\mathbf{A} \quad . \tag{2}$$

And, since $(\mathbf{C}^T\mathbf{C})^{-1}\,\mathbf{C}^T\mathbf{C} = \mathbf{I}$, the identity matrix, this gives the unknown coefficients in **B** as

$$\mathbf{B} = (\mathbf{C}^T\mathbf{C})^{-1}\,\mathbf{C}^T\mathbf{A} \quad . \tag{3}$$

Thus, the solution is a simple exercise in matrix manipulation; Mayrsohn and Crabtree[3] give a numerical example in their paper.

In actual calculations, some of the elements of **B** may turn out negative, which is not a physically meaningful result. When this occurred, Mayrsohn and Crabtree deleted a source category from the analysis and repeated the calculation. Logic was used to limit the possible numbers of source combinations to 12. For example, since acetylene was found in every ambient sample, and auto exhaust was the only source that contained acetylene, then auto exhaust must always be included in the analysis. By similar reasoning with other species, they concluded that auto exhaust, either gasoline or gasoline vapor, and either natural gas or liquefied petroleum gas had to be included in each source combination. The optimum distribution was then one of the 12 distributions that (a) had all non-negative coefficients and (b) had a minimum residual $\mathbf{D}^T\mathbf{D}$, where $\mathbf{D} = \mathbf{CB} - \mathbf{A}$.

Findings

The contribution, by weight, of each source to the ambient nonmethane hydrocarbon concentrations in the Los Angeles Basin was, on the average, as follows:

Auto exhaust	50%
Gasoline and gasoline vapor	30-35%
Commercial and geogenic natural gas	20%

The distribution varied somewhat with the time of day.

A subsequent study[4] using the same approach considered more receptor sites, and found the following average distributions:

Auto exhaust	53%
Gasoline	12%
Gasoline vapor	10%
Commercial natural gas	5%
Geogenic natural gas	19%
Liquefied petroleum gas	1%

The improved resolution resulting from the larger scope of this study is apparent. A major revelation of this second study was the high contribution of geogenic natural gas, emanating directly from underground petroleum-bearing formations in the petroleum-producing areas of the basin.

Example 2: Determining Hydrocarbon Emissions From a Refinery

Approach

Refinery hydrocarbon emissions are primarily fugitive and emanate from an almost limitless number of sources. The objective of this study was to quantify the total VOC emissions from the various types of operations that occur at a refinery. The traditional approach for such determinations is to measure leakage from typical valves and various process elements, and then to estimate total emissions by multiplying these emission factors by the number of components. As a supplement to the traditional approach, receptor modeling was also used in this study.

The distribution of nonmethane hydrocarbon (NMHC) concentrations in the ambient air around the refinery was determined using fifteen samplers downwind and one upwind. Atmospheric dispersion coefficients for a dispersion model were obtained by releasing SF_6 tracer gas within the refinery and collecting samples at the same sixteen locations. These samples were then analyzed for both NMHC and SF_6.

To define the emission sources, five source types were considered: Q_1 -- process areas and emulsion plant; Q_2 -- asphalt plant; Q_3 -- loading-unloading facilities; Q_4 -- fixed roof storage tanks; and Q_5 -- floating roof storage tanks. The refinery was subdivided into 31 areas (see Figure 1) and each area was assigned to one of the source types. For each source type

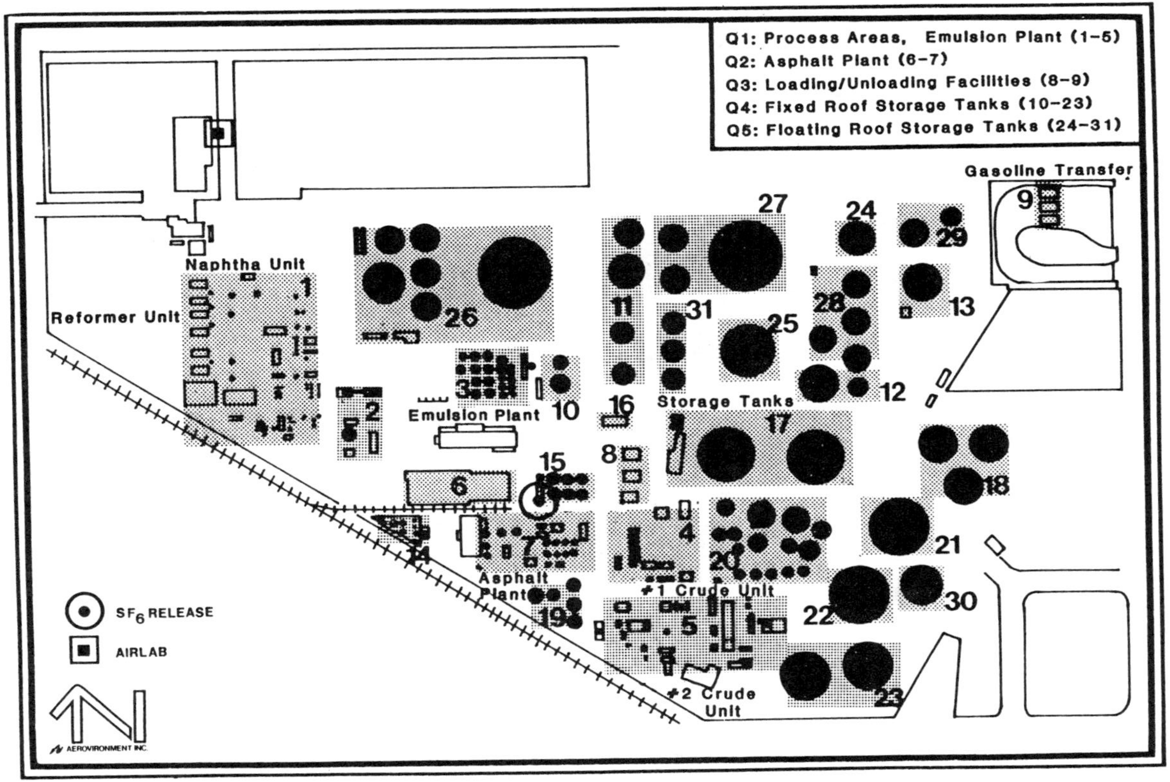

Figure 1. Grouping of refinery sources by type.

the emission rate per unit area of ground was assumed constant. Then, for modeling purposes, each source area was represented by a grid of equal-strength point sources.

A dispersion model (AVQUAL[8]) was then applied to calculate the effect of all sources on each receptor, according to the algorithm

$$\chi_i = \sum_{j=1}^{5} \sum_{k=1}^{n_j} q_j f_{ijk} , \qquad (4)$$

where χ_i = concentration of NMHC at receptor i

q_j = strength of source type j, per unit area, so $Q_j = n_j q_j$

f_{ijk} = meteorological dispersion term, defining transport from the k-th unit area source of type j, to a receptor i

and n_j = total number of unit area sources for each type of source.

The factor f_{ijk} was computed by the dispersion model for each combination ijk, using the SF_6 tracer gas data to quantify the diffusion and transport portion of the dispersion algorithm. In other words, for a single source and receptor, one can write an idealized generalized dispersion model as

$$\chi = q \, g(x) \, h(met) , \qquad (5)$$

where g(x) is only a function of the relative positions of the source and receptor and h(met) is only a function of meteorology (wind speed and turbulence). The factor f is then given by

$$f_{ijk} = g(x) \, h(met) \qquad (6)$$

Here, h is determined from Equation (5), where the tracer measurements define χ, Q, and the g(x) applicable to the tracer study. The coefficients j and k determine each g(x) in Equation (6).

Since there were only five source categories, q_j, only five equations of the form of Equation (4) are needed to fully define the problem. Data from the other ten downwind receptors provided redundancy, resulting in an over-determined set of equations (fifteen equations for five unknowns). These equations were solved using a linear programming approach with a non-negative Q constraint. This technique was very sensitive to the "noise" in the receptor concentration data, however. To obtain stable answers, therefore, it was necessary to smooth the measured ambient concentration profiles before using the ambient hydrocarbon data.

As a check on the whole process, the modeled SF_6 concentrations were compared with the measured ones. An example of the agreement obtained is shown in Figure 2.

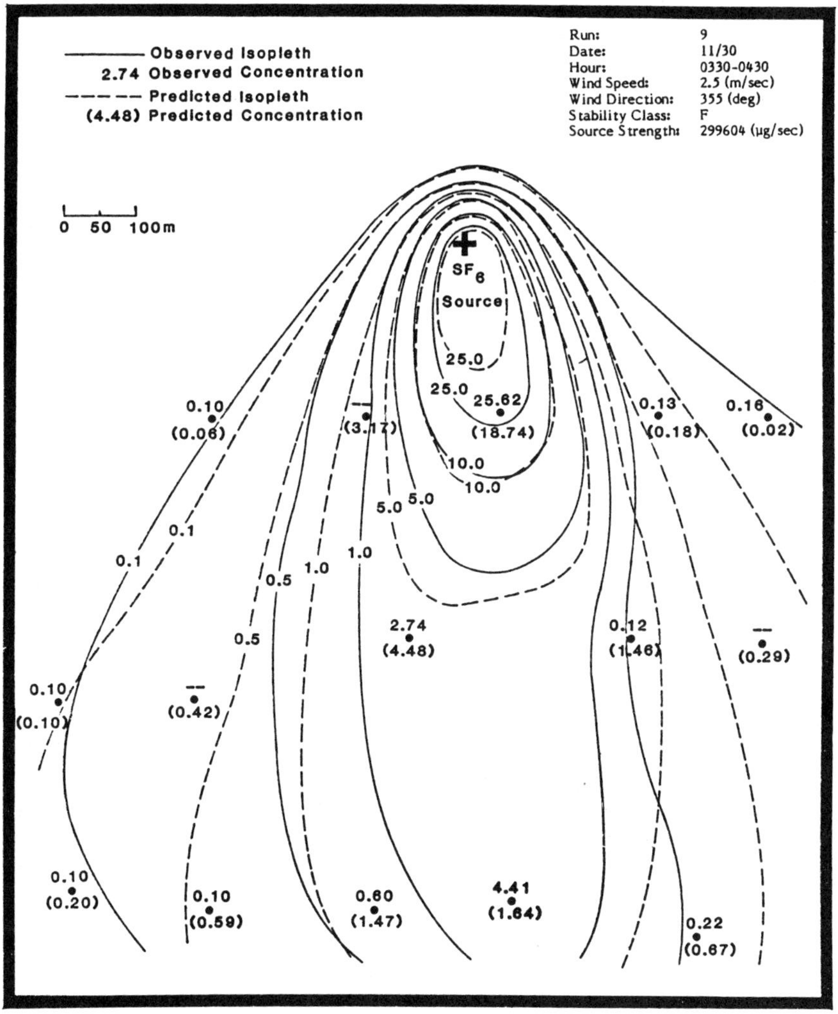

Figure 2. Example of comparison of model with measurements.

Findings

Average NMHC emission fractions for each source type were estimated by the receptor modeling approach to be:

Q_1	(process areas and emulsion plant)	58%
Q_2	(asphalt plant)	0%
Q_3	(loading-unloading facilities)	0%
Q_4	(fixed roof storage tanks)	4%
Q_5	(floating roof storage tanks)	38%

These agree qualitatively with proportions derived by traditional measurements at the same refinery. However, the total emission rate (ΣQ_i) was computed to be 12 kg/hr, which is about one order of magnitude less than the value derived by the traditional method.

The approach used underestimates the effects of stack emissions and other elevated emissions unless those emissions are mixed downward rapidly enough to reach the samplers. Although these tests were performed under stable atmospheric conditions near the surface, it probably would have been better if the atmosphere had been less stable, perhaps with an inversion aloft to limit plume rise. In this case, the estimate of total emissions might have been greater.

Concluding Comments

As the preceding discussion and examples have shown, receptor modeling methods can be effectively used for VOC source attribution studies. The mathematical approaches are the same for VOC as for particulate matter, while the sample analysis methods are quite different.

A thorough understanding of the experimental methods used to derive source signatures and ambient composition is necessary for the correct application of any receptor modeling approach. Such an understanding is particularly critical with VOC studies, since VOC can change chemical form considerably during sampling, storage, and analysis, which could lead to erroneous conclusions. In addition, many VOC are considerably lighter than air and, therefore, this buoyancy causes them to diffuse preferentially upward, which can lead to errors in understanding, especially in stable atmospheric conditions. (Gravitational settling of particles is of similar concern in particulate studies.)

Although traditional receptor modeling approaches require that only inert compounds be considered, relative changes in concentrations between inert and reactive VOC species emitted from the same source type can be a useful tool for the identification of sources.

VOC analyses, especially of the GC-MS type, can be forbiddingly expensive if many samples are to be analyzed. An excellent understanding of the problem at hand and of organic chemistry is needed to design an effective experimental and receptor modeling program. In VOC work one can rarely afford to add a few extra analyses "just in case."

But, for the analyst who understands these constraints and who has a good feel for the chemical processes involved, receptor modeling of VOC can be a very useful tool.

References

1. S.K. Friedlander, "New developments in receptor modeling theory," in Atmospheric Aerosol: Source/Air Quality Relationships, E.S. Macias and P.K. Hopke, editors, American Chemical Society, Washington, D.C., 1981 (pp. 1-19).

2. R.A. Whitby and E.R. Altwicker, "Acetylene in the atmosphere: Sources, representative ambient concentrations and ratios to other hydrocarbons," Atmos. Environ., 12:1289-1296 (1978).

3. H. Mayrsohn and J.H. Crabtree, "Source reconciliation of atmospheric hydrocarbons," Atmos. Environ., 10:137-143 (1976).

4. H. Mayrsohn, J.H. Crabtree, M. Kuramoto, R.D. Sothern, and S.H. Mano, "Source reconciliation of atmospheric hydrocarbons," 1974, Atmos. Environ., 11:189-192 (1977).

5. E.R. Stephens and F.R. Burleson, "Distribution of light hydrocarbons in ambient air," J. Air Poll. Cont. Assoc., 19: 929-936 (1969).

6. I. Watanabe and E.R. Stephens, "Relationships between individual species of ambient C_2-C_5 hydrocarbons. J. Japan Soc. Air Pollut., 15:342-347 (1980).

7. A.H. Huang and S.J. Head, "Estimation of fugitive hydrocarbon emissions from an oil refinery by inverse modeling," Third Symposium on Fugitive Emissions: Measurement and Control, U.S. Environmental Protection Agency (1978). (Also technical paper TP8162, AeroVironment Inc., Pasadena, CA.)

8. P.B.S. Lissaman, "A simple unsteady concentration model explicitly incorporating ground roughness and heat flux," Paper 73-129, 66th Annual Meeting, Air Pollution Control Association (1973). (Also technical paper TP311, AeroVironment Inc., Pasadena, CA.)

RECEPTOR SOURCE APPORTIONMENT MODELS FOR TWO POLYCYCLIC AROMATIC HYDROCARBONS

Joan M. Daisey
New York University Medical Center
Institute of Environmental Medicine
550 First Avenue
New York, New York

The feasibility of developing receptor models for polycyclic aromatic hydrocarbons (PAH) has been investigated using benzo(a)pyrene (BAP) and chrysene (CHR) as prototype compounds. Factor analysis was used to identify elemental source tracers and then a multiple linear regression equation was developed to relate concentrations of individual PAH to Pb and V, tracers for traffic-related and oil combustion emissions in New York City. Statistically significant equations were obtained for both compounds, despite the somewhat limited number of cases available for modeling. Oil-burning accounted for 114 ± 22% of airborne BAP during the heating season; 62 ± 27% of airborne particulate chrysene was traffic-related while 19 ± 22% was related to oil-burning. The multiple regression coefficient for Pb in the CHR model was in good agreement with the estimated ratio of CHR/Pb in airborne particles sampled in tunnels. Further validation of the models was not possible due to the lack of appropriate source emissions measurements.

RECEPTOR SOURCE APPORTIONMENT MODELS FOR TWO POLYCYCLIC AROMATIC HYDROCARBONS

Introduction

The polycyclic aromatic hydrocarbon (PAH) fraction of airborne particles is the most carcinogenic in animal bioassays and is, therefore, of major interest from the standpoint of environmental health[1]. Many of the PAH, including benzo(a)pyrene and chrysene, have been classified by the EPA as priority pollutants. Although these compounds are known to originate from the combustion of carbonaceous matter such as oil, coal or refuse, methods to determine the relative contributions of various types of combustion sources at a given location have not been developed. In principle, it should be possible to identify major contributing sources and to estimate their contributions at a given location using receptor source apportionment modeling techniques. In practice, limited source composition data, volatilization of lower molecular weight compounds during sampling and atmospheric reactions occurring between source and receptor may hinder the development of these types of receptor models.

A preliminary investigation was undertaken to determine the feasibility of developing receptor models for PAH compounds and to identify some of the problems which might arise in model development. The FA/MRRM (factor analysis/multiple regression receptor model) approach developed by Kneip, Kleinman and co-workers[2-4] was used in the present analysis. Factor analysis was used to identify independent source tracers and then a multiple regression equation was developed relating the concentrations of the individual PAH to the source tracers. This approach was selected because of the limited source composition data for trace metals and PAH determined simultaneously. In addition, changes which may occur between the source and the receptor are automatically incorporated into the coefficients of the source tracers with the method.

Two compounds were selected for model development. Benzo(a)pyrene (BAP) was selected because it has been widely used as an index of the PAH fraction and it is concentrated in the particulate phase ($\geq$ 90%)[5, 6]. Chrysene (CHR) was selected because it is distributed between the gaseous and particulate phases and the distribution changes with temperature[6]; thus, it presents a more complex problem for source apportionment via receptor modeling.

Experimental

Weekly samples of total (TSP) and respirable ($\leq$ 3.5 μm, A.D.) (RSP) suspended particulate matter were collected at the New York University Medical Center Roof Station, which is located on the east side of Manhattan and has been described[7] in a previous report. The RSP samples were collected on pre-ignited filters (Gelman AE) from August 1, 1978 to August 30, 1979. Half of each RSP sample was extracted sequentially with cyclohexane, dichloromethane and acetone, respectively, as described previously[7]. Samples were stored at -35°C prior to extraction, but were extracted within a few days of collection. Samples were protected from light at all times. Extracts were stored in a freezer until PAH analysis. The cyclohexane extracts were analyzed for PAH by a thin-layer/gas chromatographic method[8]. Concentrations of BAP were below the limit of detection (0.1 ng/m^3) from 8/1-9/28/78 and from 4/11-8/30/79, leaving for this compound data only associated with the heating season.

A portion of each RSP and TSP sample was analyzed for selected trace metals: Pb, V, Cu and Mn. These metals had been previously identified as source tracers for major sources in New York City[3, 4]. Portions of the TSP and a second RSP sample, collected on Spectrograde filters, were analyzed for water-soluble sulfate[3, 4]. Daily values of the 7 A.M. mixing height and windspeed aloft, measured at Fort Totten, N.Y., were obtained from the National Weather Service. Weekly averages were computed for use in the modeling. Weekly averages of daily average temperature measures in Central Park (in Manhattan) were also calculated.

The Statistical Package for the Social Sciences (SPSS), version 8[9, 10] was used for statistical analyses and model development. Listwise deletion of missing data was employed so that the analyses were based only on cases which included a complete set of data for all variables. Factor analysis (principal factoring, R-type, with Varimax rotation) had been used previously on this data set to identify source types of tracers[7]. In general, V, Pb, $SO_4^=$ and Cu were found on separate factors. Manganese in the RSP was associated with Pb. This appeared to be due to the use of methylcyclopentadienyl tricarbonyl as an octane booster in gasoline during this period[11].

Stepwise multiple linear regression models were then developed. These were of the form:

$$[PAH] = k_1T_1 + k_2T_2 + \ldots k_iT_i + R \quad (1)$$

where [PAH] is the ambient concentration of the PAH species of interest,
T_i = the ambient concentration of the source tracer,
k_i = the regression coefficient of source tracer, T_i,
R = the residual, i.e., the concentration of PAH not associated with the modeled sources.

Respirable Pb and V, tracers for traffic-related and oil-burning sources, respectively[3, 4], were used in the models since these elements were most highly correlated with CHR and BAP.

The constants k_i relate source emissions to ambient concentrations according to the equation:

$$k_i = \frac{\alpha_{ij}[PAH]_j}{T_j} \quad (2)$$

where $[PAH]_j$ = the concentration of PAH in emissions from source j,
T_j = the concentration of the tracer element in emissions from source j, and
α_{ij} = the coefficient of fractionation which describes changes in the ratio $[PAH]_j/T_j$ which may occur between the source and receptor as a result of physical and chemical processes.

All ambient concentrations of PAH and source tracers were normalized to constant dispersion conditions[12] using the dispersion factor proposed by Holzworth[13], e.g.,

$$[PAH]_N = \frac{[PAH]\ h\cdot s}{4200m^2/sec} \quad (3)$$

where $[PAH]_N$ is the dispersion normalized concentration, h is the 7 A.M. mixing height, s is windspeed aloft and 4200 m^2/sec is an average product of h·s for New York City. This technique has been used previously[7].

The distributions of the ambient concentrations of both BAP and CHR were examined prior to the regression analyses. These are shown in Figure 1. The 22 values of BAP were approximately normally distributed, but the CHR appeared to follow a bimodal distribution. The lower concentrations were generally associated with the summer months. This suggested that shifts in the vapor-particle distribution might be the source of the bimodal distribution. Vapor phase concentrations of CHR were then estimated using the equation of Yamasaki, et al.[6], which is based on the Langmuir model for a vapor adsorbed on a solid;

$$\log\ (CHR)_v = \frac{-5826}{T} + 24.89 + \log \frac{(CHR)_p}{(TSP)} \tag{4}$$

where $(CHR)_v$ and $(CHR)_p$ are the concentrations of CHR in the vapor and particle phases, respectively, T is temperature in degrees Kelvin and TSP is the concentration of total suspended particulate matter. The distribution of the concentrations of total CHR (vapor + particle) is shown in Figure 2 and the data are obviously more normally distributed than the original particle phase concentrations. A source apportionment model for total chrysene (CHR_T) was also developed.

Results

The results of the regression analyses are summarized in Table I. All three equations were statistically significant at the $p \leq 0.02$ level. The values of R^2 were relatively low compared to models of this type developed for other variables[2-4, 7], but fewer data were available for the PAH compound models, particularly for BAP. The reduced set of data for the BAP was inadequate to obtain a coefficient for Pb. Comparing the models for CHR_P and CHR_T, the coefficients for Pb did not differ significantly between the two models but the standard error of the Pb coefficient was greater for the CHR_T model. The coefficients of V did differ slightly, but were not statistically different. This and the greater residual for the CHR_T model presumably are associated with the additional uncertainties introduced into the data from the calculated vapor concentrations.

Discussion

The average source contributions calculated from the models are presented in Table II. For BAP, the oil-burning sources in New York City during this winter period accounted for 114 ± 22% of the total concentration. Based on the model, traffic-related sources contributed, at most, only a few percent of the BAP. This is consistent with the reduced BAP emissions associated with catalyst equipped cars[14] and the location of the sampling site on the roof of a 15 story building.

Traffic-related sources accounted for 1.42 and 1.65 ng/m^3 of the particulate and total chrysene concentrations while oil-burning was a lesser contributor. The model for particulate chrysene accounted for about 80% of the chrysene. The model based on the more normally distributed data for the total chrysene accounted for only 58% of the total. Although, in principal, this is a more physically appropriate model, the uncertainties in the calculated vapor concentrations led to a less satisfactory model in terms of both R^2 and the residual CHR. It should be noted, however, that the equation used to calculate vapor phase concentrations was determined for TSP rather than RSP samples and was based on measurements made in a different city.

In Table III, the average percent deviation of the concentration predicted by each of the particulate models from the actual is presented as a function of concentration. The greatest deviations were observed for low concentrations, which are associated with the greatest analytical uncertainties. The relative deviations between actual and predicted concentrations for these compounds, in general, are of the same order of magnitude as those observed for many trace elements in receptor models[15].

In relationship between the coefficients of the models and the ratios of BAP and CHR to the source tracer in source emissions, can provide evidence of the validity of the developed models, as shown in eqn. (2). However, in only a few instances have both PAH and trace element compositions been measured simultaneously. Kebbekus and co-workers[16] recently measured the concentrations of certain PAH compounds and of Pb for 4 samples collected in the Lincoln Tunnel (between Weehawken, N.J. and Manhattan). The ratio of BAP/Pb averaged 0.00145 ± 0.0056; chrysene was not determined. This ratio cannot be compared directly to the models, since there were insufficient data to fit the Pb coefficient in the BAP model. Grimmer and co-workers[17] reported concentrations of PAH for 4 samples collected in a tunnel with car traffic in Essen, Germany. The BAP/CHR ratio was 0.65 ± 0.14. If the data for the two tunnels are combined, the CHR/Pb ratio can be estimated as 0.0022, which is in very good agreement with the Pb coefficient of the model for particulate chrysene.

Grimmer and co-workers[17] have also determined concentrations of PAH in an area of Essen dominated by oil-burning. The ratio of BAP/CHR for the area was 0.3. The oil-burning contributions for New York City during the winter months can be calculated from the models as 1.61 ± 0.35 and 0.63 ± 0.72 ng/m^3, respectively, for BAP and CHR, giving a BAP/CHR ratio of 2.6 ± 1.2. The agreement in this instance is poor; however, the differences in types of oil (No. 2 vs. No. 4 and No. 6) and oil-burning equipment may affect such ratios. Ideally, comparisons should be with source emission measurements made in New York City.

Conclusions

This preliminary study has demonstrated the feasibility of developing FA/MR models for PAH compounds. The Pb coefficient of the CHR model is in quite good agreement with available source emission data and supports the validity of the CHR model. The poor agreement between actual and predicted concentrations at low PAH concentrations suggests a need for improved analytical methods. Corrections for shifts in the vapor-particle distribution of CHR due to seasonal temperature variations have been found to give an approximately normal distribution for concentrations of total CHR, although these corrections did not give any improvement in the model.

The greatest need for the development of such models is the need for appropriate source emissions composition data. Source emission samples must be collected at close to ambient conditions, i.e., temperature and dilution, in order to insure that vapors have condensed and adsorbed on the particles. Source samples should be analyzed for trace elements as well as PAH compounds. Since much more is known about the sources and stability of the trace elements, these species can be very useful for model development and verification.

Acknowledgement

The suggestions of Dr. Paul Lioy during the preparation of this manuscript are gratefully acknowledged. This work was supported by Grant No. RP-1222 from the Electric Power Research Institute, by the American Petroleum Institute and by Cooperative Agreement No. CR810300-01-0 from the U.S. Environmental Protection Agency and is part of a Center Program supported by the National Institute of Environmental Health Sciences, Grant No. ES 00262 and the National Cancer Institute, Grant No. CA 13343. The information in this document has not been subjected to the Environmental Protection Agency's required peer and administrative review and, therefore, does not necessarily reflect the view of the Agency and no official endorsement should be inferred.

References

1. D. Hoffmann and E. L. Wynder, "Organic particulate pollutants - chemical analyses and bioassays for carcinogenicity," In Air Pollution, A. C. Stern, Ed., 3rd ed., 2: pp 361-455, Academic Press, Inc., New York, N.Y. 1977.

2. T. J. Kneip, M. T. Kleinman and M. Eisenbud, "Relative contribution of emission sources to the total airborne particulates in New York City," Proceedings of the Third IUAPPA International Clean Air Congress, Dusseldorf, Germany, pp C96-C97 (1973).

3. M. T. Kleinman, B. S. Pasternack, M. Eisenbud and T. J. Kneip, "Identifying and estimating the relative importance of sources of airborne particulates," Environ. Sci. & Technol., 14: 62-65 (1980).

4. T. J. Kneip, R. P. Mallon and M. T. Kleinman, "Multiple regression models for source apportionment," Atmos. Environ., In Press.

5. W. Cautreels and K. VanCauwenberghe, " Experiments on the distribution of organic pollutants between airborne particulate matter and the corresponding gas phase," Atmos. Environ., 12: 1133-1141 (1978).

6. H. Yamasaki, K. Kuwata and H. Miyamoto, "Effects of ambient temperature on aspects of airborne polycyclic aromatic hydrocarbons," Environ. Sci. & Technol., 16: 189-194 (1982).

7. J. M. Daisey and T. J. Kneip, "Atmospheric particulate organic matter: Multivariate models for identifying sources and estimating their contributions to the ambient aerosol," In Atmospheric Aerosol: Source/Air Quality Relationships, E. S. Mascias and P. K. Hopke, Eds., ACS Symposium Series No. 167, pp 197-221, American Chemical Society, Washington, D.C. (1981).

8. J. M. Daisey and M. A. Leyko, "Thin-layer gas chromatographic method for the determination of polycyclic aromatic and aliphatic hydrocarbons in airborne particulate matter," Anal. Chem., 51: 24-26 (1979).

9. N. H. Nie, C. H. Hull, J. G. Jenkins, K. Steinbrenner and D. H. Beat SPSS Statistical Package for the Social Sciences, 2nd ed., McGraw-Hill Book Co., New York, N.Y. (1975).

10. C. H. Hull and N. H. Nie, Eds., SPSS Update, New Procedures and Facilities for Releases 7 and 8, McGraw-Hill Book Co., New York, N.Y. (1979).

11. P. J. Lioy, R. P. Mallon and T. J. Kneip, "Long-term trends in total suspended particulates, vanadium, manganese, and lead at near street level and elevated sites in New York City," J. Air Pollut. Control Assoc., 30: 153-156 (1980).

12. M. T. Kleinman, T. J. Kneip and M. Eisenbud, "Seasonal patterns of airborne particulate concentrations in New York City," Atmos. Environ., 10: 9-11 (1976).

13. G. C. Holzworth, "Mixing depths, wind speeds and air pollution potential for selected localities in the U.S.," J. Appl. Meteorol., 6: 1039 (1967).

14. R. L. Williams and S. J. Swarin, "Benzo(a)pyrene emissions from gasoline and diesel automobilities," SAE Technical Paper No. 790419, Society of Automotive Engineers, Inc., Warrendale, PA 15096 (1979).

15. D. J. Alpert and P. K. Hopke, "A determination of the sources of airborne particles collected during the Regional Air Pollution Study," Atmos. Environ., 15: 675-687 (1982).

16. B. Kebbekus, et al., New Jersey Institute of Technology, Newark, N.J. Personal Communication.

17. G. Grimmer, K.-W. Naujack and D. Schneider, "Changes in PAH-profiles in different areas of a city during the years," In Polynuclear Aromatic Hydrocarbons: Chemistry and Biological Effects, Fourth International Symposium, A. Bjorseth and A. J. Dennis, Eds., pp 107-125, Batelle Press, Columbus, Ohio (1980).

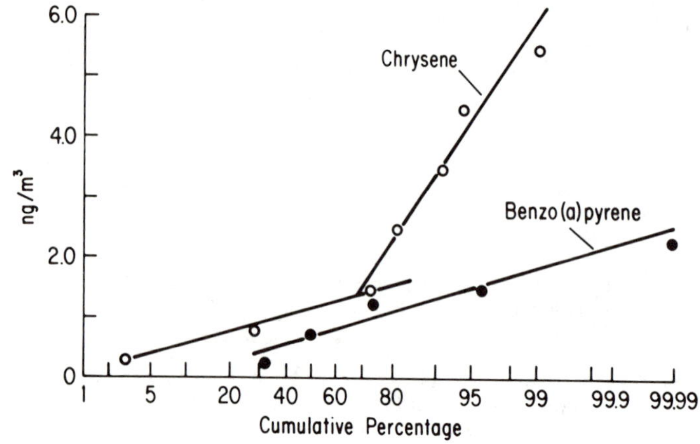

Figure 1. Distributions of particulate benzo(a)pyrene and chrysene ambient concentrations

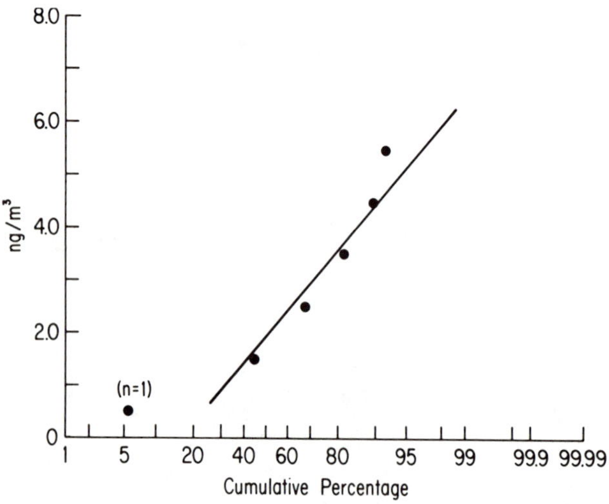

Figure 2. Distribution of total chrysene (particle + vapor) concentration.

TABLE I. STEPWISE LINEAR REGRESSION MODELS FOR BAP AND CHR

Dependent Variable	n[a]	Coefficients of Source Tracers ± S.E. PB	V	Residual	R^2	F	P
BAP	22	-	0.017 ±0.003	-0.23±0.41	0.31	4.4	0.02
CHR_P	39	0.0016±0.0007	0.0059±0.0068	0.43±0.45	0.41	12.5	<0.001
CHR_T	35	0.0017±0.0010	0.0025±0.0106	1.38±0.62	0.27	6.15	0.005

a. n - number of cases.

TABLE II. AVERAGE SOURCE CONTRIBUTIONS CALCULATED FROM REGRESSION MODELS

PAH	Sampling Period	Source Contributions, ng/m^3 Traffic-related	Oil-Burning	Residual
BAP	10/12/78-4/11/79	-	1.84±0.36 (114±22%)	-0.23±0.41 (-14±25%)
CHR_P	8/ 1/78-8/30/79	1.42±0.62 (63±27%)	0.43±0.49 (19±22%)	0.42±0.45 (19±20%)
CHR_T	8/ 1/78-8/30/79	1.65±0.97 (51±30%)	0.18±0.77 (6±24%)	1.38±0.62 (43±19%)

TABLE III. PERCENT DEVIATION OF PREDICTED FROM ACTUAL VALUE AS A FUNCTION OF CONCENTRATION

Dispersion Normalized Concentration (ng/m^3)	Average % Deviation of BAP[a.] (n)[b.]	Average % Deviation of CHR_P (n)[b.]
0 -0.49	81 (6)	198 (7)
0.5-0.99	82 (5)	81 (6)
1.0-1.99	54 (2)	83 (7)
2.0-2.99	31 (6)	32 (7)
3.0-3.99	22 (3)	38 (3)
4.0-4.99	-	30 (6)
>5.0	-	53 (3)

a. 100% x (actual - predicted)/actual.
b. n - number of cases.

Receptor Models Applied to Contemporary Air Pollution Problems

A Panel Discussion

Theo. J. Kneip, Chairman
Institute for Environmental Medicine
NYU Medical Center
Tuxedo, NY

Jacques Guertin
Electric Power Research Institute
Palo Alto, CA

Thompson G. Pace
OAQPS
USEPA
Research Triangle Park, NC

John G. Watson
Desert Research Institute
Reno, NV

The keynote of the conference, both in the formal sessions and in the discussion, was the assertion of the need for a more organized approach to source apportionment efforts. Much of the panel discussion focused on the need to define the objectives in each study, to state clearly the assumptions involved in the conceptual model, and to discuss the limitations to potential applications of the conceptual model owing to deviations from model assumptions and limitations of the actual data collected. The latter step is very critical since resource availability may restrict data collection to less than the optimum set. In many cases the conceptual models may be used despite the fact that the data sets may deviate from the assumptions, but it is important to evaluate the extent of those deviations. Such deviations have rarely, if ever, been discussed but they should be. We most particularly owe this direct and open communication of our approaches to the non-experts, who may be easily confused by unstated and unevaluated deviations from assumptions.

The need to plan the sampling and analytical systems in relation to the objectives and the conceptual model is implicit in this general discussion of models. Careful development of the operating protocols will reveal potential deviations from critical assumptions in advance, and avoid wasted expenditures. In the development of the working protocols, strict attention must be given to sampling and analytical compatibility. For example, in the choice of sampling substrates, we must take into account a wealth of information available as to the requirements of a number of analytical methods. We must be aware of ongoing analytical and sampling developments, since they may change sensitivities, reduce interferences or otherwise affect the ability to collect data. Development of new or improved substrates is also of importance, as current limitations in data collection may be overcome by such efforts.

A general sampling and analytical protocol can be described at present for any given conceptual model; however, a careful discussion of the assumptions and limitations in each case will help to focus future efforts to further develop and to improve such protocols. In these efforts, there is a need for differentiation between the conceptual model, the analytical tools and the interpretive approach. We should not confuse the potential user of the receptor modeling appproach by mixing measurement and model in our nomenclature, i.e., there is no ´atomic absorption regression model´ or ´optical microscopy receptor model.´

Development of protocols and overall program plans requires careful use of data already available. Compositional data from source or plume sampling and air quality data from local or regional studies should be carefully evaluated before undertaking new programs. Data obtained in any given program must also be examined by graphical techniques or other methods that allow the investigator to personally detect errors, recognize relationships and guide the further evaluations. The computer is simply a more sophisticated tool, and cannot replace a personal knowledge of the actual data sets.

A number of investigators have pointed out the problems of varying source profiles, and changing particle composition after emission caused by condensation and transformation or agglomeration and settling. These cautions have been repeated in this conference. The Portland Aerosol Characterization Study (PACS) demonstrated the effort required to define source variations caused by fuel or operational variables, and to account for both source and receptor sample errors in applications of a conceptual model. Improved and more representative source profiles are needed, and the work in interpreting error contributions must be continued and expanded. Work should be continued to determine the usefulness of source profile catalogs as opposed to a new source sampling and analysis program for every new study.

Current efforts to obtain and analyze samples from plumes or from stacks with the use of dilution samplers should improve the available source composition data used in the chemical element or chemical mass balance (CEB or CMB) methods. The compositions from such samples represent diluted and cooled materials that contain the condensable inorganic and organic materials. Depending on the age of the plume or the dilution conditions, the agglomeration and settling effects may produce particles closer in physical properties and chemical composition to those of the particles in ambient air samples. Transformations that require more than a few minutes are not accounted for in dilution samplers with short residence times or in near stack plume samples.

A number of new approaches have been noted in the development of hybrid models and in the evaluation of multivariate statistical techniques. The hybrid model systems are particularly appealing, but more attention must be given to the assumptions involved and data sets required for these more complex models.

Several papers reported on efforts to extend our understanding of the applied mathematical techniques in relation to air quality measurements. The often proposed ridge regression method has been tested and found to offer no improvement over other methods. Two modes of fitting for factor analysis, Q and R, have been tested and found to give different results - a difference in the error estimates used in normalizing the data may be involved. The first report of a quantitative approach to analytical error estimates in CEB-CMB applications appears extremely useful. This latter effort will permit the determination, at a preset error level, of the number of sources that can be quantitatively identified.

Applications of optical and electron microscopy have been discussed with further emphasis on the critical sampling and analytical parameters. These methods, and other specialized techniques of sampling or source sample generation, are proving extremely useful in solving critical problems not generally resolvable by CEB, CMB or other approaches limited to compositional data.

Receptor modeling methods have been tested in a number of studies concerned with apportionment of the mass of suspended particulate matter. Results have been reported for total, respirable, fine or coarse particles. We have recently seen the initial evaluations of the techniques on other pollutants such as organic substances or sulfate ion. In princple, conceptual models could be developed for evaluating sources of acid aerosols, acid rain, visibility degradation or other effects. Applications for these effects may require reconsideration of the assumptions of the conceptual models, e.g., the almost universal assumption of linear additivity of sources.

It is clear from the discussions at this conference that we have evolved into a special subgroup of scientists with our own accumulated knowledge and specialized language. We need to improve our understanding of the objectives of the potential users and, most importantly, we must generate the means to educate the users so that they may contribute to the development of the methods and to the solutions for their problems. We must develop an awareness on the part of users as to the appropriate tools to use in relation to the problem at hand and funding available.

There is a final need that we all perceive and that must be delineated. In defining objectives, models, assumptions and data gathering programs, we must clearly differentiate research needs from applications, and, in particular, regulatory activities. In doing so we should find that careful definitions of the objectives of applied programs will aid us in understanding the deviations from the assumptions of conceptual models that are imbedded in such applications. These are precisely the actions called for in the keynote speech of the conference. This knowledge will be invaluable in defining future research objectives for improving both the accuracy and precision of source apportionment methods.

The need for much more theoretical work has been made clear in this conference. Definition of the quantitative limits of the multivariate statistical methods is clearly a pressing, but poorly funded area. Efforts to train a larger group of users, to define objectives, and to understand the differences between conceptual models and actual applications will inevitably contribute to a better understanding of the limitations of source apportionment techniques. This understanding is a major factor in focusing our efforts on the most critical research areas. We have demonstrated the power of the receptor modeling approach and must now contribute to the definition of its accuracy and limitations.

Specialty Conference On:

RECEPTOR MODELS APPLIED TO
CONTEMPORARY POLLUTION PROBLEMS

October 17-20, 1982

Danvers, Massachusetts

REGISTRANTS

ALONSO, Claudio, Humann Gr. St., Brazil
ANDERSON, Michael, East Hartford, CT

BALTENSPERGER, Urs, Institute of Organic Chemistry, Zurich, Switzerland
BATTERMAN, Stuart, MIT, Somerville, MA
BELOIN, Norman, U. S. EPA, Boston, MA
BIBBO, Robert, ERT, Concrod, MA
BOISSELLE, Robert, MA DEQE Air Quality Control, Boston, MA
BRIDGE, Jennie, Boston, MA
BRIGGS, S.L.K., Harvard University, Cambridge, MA
BUCHANAN, James, Boone, NC

CASUCCIO, Gary, Murrysville, PA
CHENG, Meng, B. B. N., Cambridge, MA
COOPER, Douglas, Harvard School of Public Health, Boston, MA
COOPER, John, NEA, Inc., Beaverton, OR
CORE, John, NEA, Inc., Beaverton, OR
CRAWFORD, John, New England Electric System, Westboro, MA
CRUTCHER, Ernest, MicroLab N. W., Redmond, WA
CUNNINGHAM, James, Texas Air Control Board, Austin, TX
CURRIE, Lloyd, National Bureau of Standards, Gaitersville, MD

DAISEY, Joan, N. Y. University Medical Center, Tuxedo, NY
DATTNER, Stuart, Texas Air Control Board, Austin, TX
DAVIS, Brian, SD School of Mines & Technology, Rapid City, SD
DELACY, Kathy
DENNIS, Stephen, M.D.E.Q.E., Boston, MA
DESCAMPS, Valentine, U. S. EPA, Wellesley Hills, MA
DIXON, David, Maine Board of Air Quality, Augusta, ME
DRAFTZ, Ronald, IIT Research Institute, Chicago, IL
DRIVAS, Peter, ERCO, LaJolla, CA
DUMYAHN, Thomas, Harvard School of Public Health, Arlington, MA
DUNCAN, Dan
DZUBAY, Thomas, U. S. EPA, Research Triangle Park, NC

ERNST, David, Bolt Beranek & Newman, Inc., Cambridge, MA
EVANS, Bernie, Radian Corp., Austin, TX
EVANS, Gary, U. S. EPA, Raleigh, NC

FAY, James, MIT, Cambridge, MA
FILHO, Benedito Conceicao
FOLEY, Mary, National Park Science, Boston, MA
FREDERICK, Edward, Air Pollution Control Assoc., Pittsburgh, PA
FRITTS, Thomas, Charles T. Main, Inc., Boston, MA

GENDRON, Leo, ERT, Concord, MA
GOLUMB, Daniel, MIT, Cambridge, MA
GORDON, Norma, State of Maine Dept. of Environ. Prot., Augusta, ME

GRAF, Jean, IIT Research Institute, Chicago, IL
GRASSICK, Daryl, MA Dept. of Environmental Quality Eng., Boston, MA
GRUHL, James, MIT, Cambridge, MA
GRUNINGER, John, Aerodyne Research Billerica, MA
GUERTIN, Jacques, EPRI, Palo Alto, CA
GULDBERG, Peter, Lincoln, MA

HALL, Robert, GCA/Technology Division, Bedford, MA
HATCHER, Robert, ERT, Concord, MA
HAUPT, Sue, New England Electric System, Westborough, MA
HEINOLD, David, ERT, Concord, MA
HEISLER, Steven, ERT, Concord, MA
HENRY, Ronald, ERT, West Lake Village, CA
HERMAN, Mark, Exxon Research & Engineering Co., Florham Park, NJ
HOPKE, Philip, University of Illinois, Urbana, IL
HWANG, Chang-Sih, University of IL, Urbana, IL

JANOCKO, Philip, Energy Technology Consultants, Murrysville, PA
JAVITZ, Harold, FRI Int., Menlow Park, CA
JOHNSON, David, SUNY-CESF, Syracuse, NY
JOHNSON, RONALD, S. D. School of Mines & Technology, Rapid City, SD

KEENER, Ronald, Duke Power Co., Hunterville, NC
KELLY, Robert, U. S. EPA, New York, NY
KINNY, Patrick, Harvard, Arlington, MA
KIRKPATRICK, Lane, GCA/Techn. Div., Bedford, MA
KLEINAU, J. H., Dorr-Oliver, Inc., Stanford, CT
KNEIP, Theo, N. Y. U. Medical Center, Tuxedo, NY
KOCZKUR, E., SENES Consultants Limited, Toronto, Ontario, Canada

LeBOURVEAU, J. W., New England Electric Systems, Westboro, MA
LELLENSON, William, N. Service Inc., Research Triangle Park, NC
LETENDRE, Nicole, N.E. Power Co., Somerset, MA
LILJESTRAND, Howard, University of TX, Austin, TX
LIN, Josephine, U. S. Steel Corp., Monroeville, PA
LOWENTHAL, Douglas, University of R. I., Naragansett, RI
LURIA, C., I. B. M. Corp., Dayton, NJ

MAGIL, Gary, Arthur D. Little, Cambridge, MA
MARSH, Stanley, Southern Cal. Edison, Rosemead, CA
MARTIN, Dennis, TRC, E. Hartford, CT
MARTIN, John, M. E. S., Inc., Amityville, NY
MATHUR, V. K.
MATTESON, John, DEQE-PVAPCD, Springfield, MA
McNAUGHTON, Dan, Salt Lake City, UT
MENTOS, Charles, Mass. DEQE, Boston, MA
MENZIE. John, New England Power Co., Westboro, MA
MEYER, Rob, Porter Rith Morris, Columbus, OH
MONTZ, Andres, Texas Air Control Board, Austin, TX
MORANDI, Maria, New York University, Berjenfield, NJ
MORGENSTERN, Paul, Atlantic Environmental Assoc., Waltham, MA
MORRIS, Charles, Environmental Res. & Tech., Concord, MA
MUELLER, Peter, Environmental Assessment Dept., Elec. Power Res. Inst., Palo Alto, CA

NASH, CARLTON, U. S. EPA, Chicago, IL
NASTA, Margaret, Mellon Institute, Pittsburgh, PA
NILL, Richard, Mabbett, Capaccio & Assoc., Cambridge, MA

OSA, Richard, Wisconsin Electric Power Co., Milwaukee, WI

PACE, Tom, U. S. EPA/OAQPS, Research Triangle Park, NC
PEACE, Earl, College of the Holy Cross, Worcester, MA
PERKINS, Stephen, U. S. EPA, Boston, MA
PETERSON, Terry
PIETARINEN, Charles, N. J. Dept. En. Pro., Trenton, NJ
POIROT, Rich, Vermont Agency of Ec., Montpeller, VT
POLLACK, Jon, U. S. EPA, Boston, MA

QUINK, Michael

RAHN, Kenneth, Grad School of Oceanography, Naragansett, RI
RHODES, Raymond, EPA, Research Triangle Park, NC
ROSCOE, Gary
ROSENSTEIN. Marvin, U. S. EPA, Boston, MA
ROTH, Philip, Systems Application, Inc., San Rafael, CA

SAMSON, Perry, University of Michigan, Ann Arbor, MI
SCHATZ, Arthur, Harvard School of Public Health, Boston, MA
SCHAUG, Jan, Norwegian Institute for Air Res., Norway
SIMON, Sarah
SMITH, Douglas, ERT, Concord, MA
SOUCY, Rick, Charles T. Main, Inc. Boston, MA
SOUTHERLAND, James, U. S. EPA, Research Triangle Park, NC
SPENGLER, John, Harvard University, Boston, MA
SPOONER, Bradley, New England Electric System, Westboro, MA
STACHOVIAK, Bradley, Wisconsin Dept. of Natural Res., Madison, WI
STAFFORD, Michael, Texas Air Control Board, Austin, TX
STEVENS, Robert, U. S. EPA, Research Triangle Park, NC

TATE, Sandra, Main D. E. P., Portland, ME
TESCHE, T. W., Systems Applications, Inc. San Rafael, CA
THOMPSON, Albert, Lawrence Berkeley Lab., Berkeley, CA
THURSTON, George, Harvard Kennedy School of Government, Cambridge, MA
TOMBACH, Ivar, AeroEnvironment, Pasadena, CA

VAN ARSDALE, Alan, Dept. of Environment Quality Eng., Boston, MA
VIAL, John, Michigan Dept. of Natural Resources, Lansing, MI
VIDULICH, George, College of the Holy Cross, Worcester, MA

WATSON, John, Desert Res. Inst., Reno, NV
WEISS, Adolph, APC Div. Inc., Indianapolis, IN
WISHINSKI, Paul, Vermont Air Program, Montpeller, VT
WOLFINGER, Thomas, MITRE Corp., McLean, VA
WOLJAN, Dennis, U. S. EPA, Denver, CO
WOODRUFF, Roger, Bottelle, Richard, WA

YAMARTINO, Robert, ERT, Inc., Concord, MA
YOCOM, John, TRC Env. Consultants, East Hartford, CT

EXHIBITORS

BGI, Inc./ETA Associates, Waltham, WA - Norman Levin

Environmental Research & Technology, Inc. - Jitendra Shah & Richard Knapp

LeMont Scientific, Inc. State College, PA - Paul Anderson

C. T. Main Crop., Boston, MA - John Takacs & Charles Kitson

Tekolum Associates, Inc., Lexington, MA - Daniel Mager

Tracor Northern, Inc., Middleton, WI - Daniel Dunkin

INDEX

Acid Deposition, 6
Aerosol, 46, 60, 72, 84, 163, 175, 224, 238
  Ambient, 84, 163, 238
  Collection, 163, 238
  Houston, 46, 60, 72, 224
Air Quality, 29, 249
  El Paso, 249
  Utility Contribution, 29
Analysis, 34, 60, 72, 84, 110, 127, 163, 200, 224, 249, 266, 296, 317, 348
  Atomic Absorption, 249
  Factor 34, 110, 200, 224
  Particulate, 72, 224, 266
  Microscopic, 249, 266
  Morphological, 266
  Reproductivity, 238
  Spectroscopic, 249, 317
  Street, 84
  Target Transformation Factor (TTFA), 60, 127, 200
Anisotropy, 266
APCA, 1
  Specialty Meeting, 1
  Committee (TP-5), 1
Apportionment, 110, 224
  Q-Model, 224
  Mass, 110
Atmospheric Deposition, 6

Balance, 18, 110, 141, 212, 238
  Chemical (CBM), 18, 110, 141, 212, 338
  Partical Class (PCB), 238
  Rules, 141
Benzo(a)Pyrene, 348
  Model, 348
Berlin, 285
  Sulfur Dioxide, 285

Cases, 141, 158, 249, 296, 317
Chemical, 34, 60, 96, 110, 127, 141, 158
  Balance, 34, 60, 96, 110, 127, 141, 158, 212, 358
  Profiles, 96
  Uncertainties, 141, 158
Chemistry, 163, 220
Coal Combustion, 29, 175
Control, 18
Chrysene, 348
  Model, 348
  Particulate, 18
Composition, 60
  Strategies, 18
Conference, 1, 46
  Quail Roost, 1, 46
  Purpose, 1
  Summary, 1

Data, 18, 46, 60, 110, 127, 163, 200, 249, 266, 285, 296, 317
  Chemical, 18, 163
  Correlations, 317
  National Bureau of Standards, 110,
  Precipitation, 200
  Rises, 60
  Simulated, 46, 96
  Sulfur Dioxide (SO_2), 285
Deposition, 29, 189
  Dry, 29
  Wet, 189
Diffractometry, 84
  X-Ray, 84

Electricity Generation, 29
Emissions, 96, 110, 238, 338
  Characteristics 238
  Particulate, 110
  Rates, 96
  Refinery, 338
  VOC, 338
Environmental Protection Agency (EPA), 46
Equations, 6, 141, 158, 296, 317, 348
Errors, 72
  Sources, 72
Examples, 141, 158, 249, 296, 317

Factor, 18, 34, 96, 110
  Analysis, 18, 34, 96
  Transformation, 110
Filters, 84
  Dichotomous Sampler, 84
Forces, 212
  Diffuscophoretic, 212
  Thermophoretic, 212
Fugitive Dust Impact, 34

Gas, 189, 212
  Scavenging, 189, 212
Geographic Areas, 29
  Canada, 163
  Eastern Seaboard, 29
  East Helena Montana, 34
  Harbour Island, Washington, 34
  Houston, Texas, 46, 60
  Kellogg, Idaho, 34

Midwest, 29
Ohio Valley, 29
White Face Mountain, 189

Hazardous Substances, 6
Houston, 46, 60
Study, 46, 60
Hydrocarbons, 338
Refinery, 338

Instruments, 72
Calibration, 72
X-Ray Diffraction (XRD), 72
Intercomparison, 46, 60

Light Microscopy, 266
Concept, 266

Manganese, 175
Mass Balance, 6, 18, 212
Meetings, 1
APCA, 1
ACS, 1
New York Academy of Sciences, 1
Methods, 6, 18, 46, 96, 249
Deconvolution, 96
Factor Analysis, 18
Mass Balance, 6, 18
Microscopy, 18, 249
Model, 46
Spectroscopy, 18, 249
Status, 46
Microscopy, 60, 224, 238, 266
Analytical Light (ALM), 224
Light, 60, 266
Scanning Electron (SEM), 224, 238
Model, 1, 6, 18, 46, 96, 127, 175, 212, 249, 285, 296, 338, 348, 358
Application, 249, 285
Benzo(a)Pyrene, 348
Chrysene, 348
Comparison, 46, 60
Concept, 6
Dispersion, 18, 96, 175, 296, 338
Equations, 285
Equilibrium, 212
Evaluation, 127
Measurement, 6
Polycyclic Aromatic Hydrocarbon (PAH), 348
Receptor, 1, 6, 18, 29, 34, 46, 60, 72, 84, 96, 110, 127, 141, 158, 163, 175, 189, 200, 212, 224, 238, 249, 266, 285, 296, 317, 338, 348, 358
Results, 285
Scavenging, 212
Texas Air Control Board (TACB), 249
Theoretical, 212
Time Series, 317
Morphology, 266
Artifacts, 266
Multilinear Regression (MLR), 60

Nitrogen, 338
Nitrogen Oxides (NO_X), 212
Scanenging, 212

Oil Combustion, 29
Oil Refinery, 338
Emissions, 338
Optical, 266
Properties, 266
Overview, 1, 358
Conference, 1
Panel, 358

Particulate Matter, 18, 72, 84, 110, 200, 212, 224, 238, 249, 266, 317, 338 348
Analysis, 72, 224, 249
Characterization, 238, 249, 266
Classification, 238, 249, 266
Composition, 200, 249, 338
Control, 18
Crustal, 84
Data, 200
Precipitation, 189, 212
Scavenging, 189, 212
Similarity, 18
Size, 18
Sources, 200, 238, 249
X-Ray Diffraction, 72
Pollutants, 175
Regional Transport, 175
Secondary, 175
Portland Aerosol Characterization Study, 127
Power Plants, 29
Precipitation, 29
Utility Contribution, 29

Quail Roost Conferences, 1, 46, 60, 96

Rainwater, 200, 212
Receptor, 18, 212, 296
Methods, 18
Receptor Models, 1, 6, 18, 29, 34, 46, 60, 72, 84, 96, 110, 127, 141, 158, 163, 175, 189, 200, 212,

224. 238, 249, 266, 285, 296, 317, 338, 348, 358
  Advantages, 6
  Applications, 6
  Assumptions, 6
  El Paso, 249
  Evolution, 6
  Intercomparison, 96
  Limitations, 6
  Methods, 18, 29
  Results, 60
  State-of-the-Art, 46
  Status, 46
  Theory, 296
  Usefulness, 1, 6, 18,
  Validity, 1, 96
Regression, 141, 158
Reproducibility, 127
Research, 29, 46

Sampling, 60, 72, 163, 175, 317, 348
Scavenging, 189, 212
  Sulfurdioxide (SO_2), 189, 212
Selenium, 175
Singular Value Decomposition (SUD, 141, 158
Smelters, 34, 163
  Canadian, 163
  Lead, 34
Source(s), 1, 6, 18, 34, 46, 60, 84 96, 110, 127, 141, 158, 175, 189, 238, 249, 266, 338
  Apportionment, 18, 34, 46, 60, 84, 127, 175
  Characterization, 1, 84, 127, 249, 266, 338
  Combination, 141, 158
  Contributing, 110, 189, 249, 266
  Geographic Placement, 96
  Hydrocarbons, 338
  Sectors, 175
  Types, 96
Source Receptor, 1, 6, 18, 29, 34, 46, 60, 72, 84, 96, 110, 127, 141, 158, 163, 175, 189, 200, 212, 224, 238, 249, 266, 285, 296, 317, 338, 348, 358
Spectroscopy, 72, 84, 224, 238, 249
Standards, 6
  Attainment, 6
Steel Industry, 175
Study, 46
  Houston, 46

Sulfate, 29, 84, 175, 338
  Lead, 84
Sulfur, 175
Sulfur Dioxide, 189, 212, 285, 317
  Scavenging, 189, 212, 285, 317
Summary, 1, 358
  Conference, 1
  Panel, 358

Theory, 296
Tracers, 163, 175, 212
  Elemental, 163,
  Indium, 163
  Manganese, 175
  Selenium, 175
  Vanadium, 175
Trajectory, 189
Transformation Factor, 110
  Analysis, 110
  Target, 110
Transport, 175, 189
  Regional Pollution, 175

Utility Industry, 29

Vanadium, 175
Visibility, 6, 29
Volatile Compounds (VDC), 60, 338
  Chemistry, 338
  Emissions, 338

Wet Scavenging, 189
White Face Mountain, 189
Wind Trajectory Analysis, 175
Workshop, 46, 60
  EPA, 1
  Quail Roost, 1, 96

X-Ray Diffraction, 72